ATTITUDES, APTITUDES, AND ASPIRATIONS OF AMERICAN YOUTH

Implications for Military Recruitment

Committee on the Youth Population and Military Recruitment

Paul Sackett and Anne Mavor, *Editors*

Board on Behavioral, Cognitive, and Sensory Sciences

Division of Behavioral and Social Sciences and Education

NATIONAL RESEARCH COUNCIL
OF THE NATIONAL ACADEMIES

THE NATIONAL ACADEMIES PRESS
Washington, D.C.
www.nap.edu

THE NATIONAL ACADEMIES PRESS • 500 Fifth St., N.W. • Washington, DC 20001

NOTICE: The project that is the subject of this report was approved by the Governing Board of the National Research Council, whose members are drawn from the councils of the National Academy of Sciences, the National Academy of Engineering, and the Institute of Medicine. The members of the committee responsible for the report were chosen for their special competences and with regard for appropriate balance.

The study was supported by Contract No. M67004-00-C-0030 between the National Academy of Sciences and the U.S. Marine Corps. The views, opinions and/or findings contained in this report (paper) are those of the author(s) and should not be construed as an official OASD position, policy or decision unless so designated by other official documentation.

Library of Congress Cataloging-in-Publication Data

National Research Council. Committee on the Youth Population and
Military Recruitment.
Attitudes, aptitudes, and aspirations of American youth : implications
for military recruiting / Committee on the Youth Population and Military
Recruitment ; Paul Sackett and Anne Mavor, editors
p. cm.
Includes bibliographical references and index.
ISBN 0-309-08531-4 (hardcover)
1. United States—Armed Forces—Recruiting, enlistment, etc.
2. Manpower—United States. 3. Youth—United States. I. Sackett, Paul.
II. Mavor, Anne, 1952- III. Title.
UB323 .N38 203
355.2'23'0973—dc21
2002015091

Additional copies of this report are available from the National Academies Press, 500 Fifth St., N.W., Box 285, Washington, DC 20055

Call (800) 624-6242 or (202) 334-3313 (in the Washington metropolitan area)

This report is also available online at **http://www.nap.edu**

Printed in the United States of America

Suggested citation: National Research Council (2003) *Attitudes, Aptitudes, and Aspirations of American Youth: Implications for Military Recruitment.* Committee on the Youth Population and Military Recruitment. Paul Sackett and Anne Mavor, editors. Division of Behavioral and Social Sciences and Education. Washington, DC: The National Academies Press.

THE NATIONAL ACADEMIES

Advisers to the Nation on Science, Engineering, and Medicine

The **National Academy of Sciences** is a private, nonprofit, self-perpetuating society of distinguished scholars engaged in scientific and engineering research, dedicated to the furtherance of science and technology and to their use for the general welfare. Upon the authority of the charter granted to it by the Congress in 1863, the Academy has a mandate that requires it to advise the federal government on scientific and technical matters. Dr. Bruce M. Alberts is president of the National Academy of Sciences.

The **National Academy of Engineering** was established in 1964, under the charter of the National Academy of Sciences, as a parallel organization of outstanding engineers. It is autonomous in its administration and in the selection of its members, sharing with the National Academy of Sciences the responsibility for advising the federal government. The National Academy of Engineering also sponsors engineering programs aimed at meeting national needs, encourages education and research, and recognizes the superior achievements of engineers. Dr. Wm. A. Wulf is president of the National Academy of Engineering.

The **Institute of Medicine** was established in 1970 by the National Academy of Sciences to secure the services of eminent members of appropriate professions in the examination of policy matters pertaining to the health of the public. The Institute acts under the responsibility given to the National Academy of Sciences by its congressional charter to be an adviser to the federal government and, upon its own initiative, to identify issues of medical care, research, and education. Dr. Harvey V. Fineberg is president of the Institute of Medicine.

The **National Research Council** was organized by the National Academy of Sciences in 1916 to associate the broad community of science and technology with the Academy's purposes of furthering knowledge and advising the federal government. Functioning in accordance with general policies determined by the Academy, the Council has become the principal operating agency of both the National Academy of Sciences and the National Academy of Engineering in providing services to the government, the public, and the scientific and engineering communities. The Council is administered jointly by both Academies and the Institute of Medicine. Dr. Bruce M. Alberts and Dr. Wm. A. Wulf are chair and vice chair, respectively, of the National Research Council.

www.national-academies.org

COMMITTEE ON THE YOUTH POPULATION AND MILITARY RECRUITMENT

PAUL R. SACKETT (*Chair*), Department of Psychology, University of Minnesota, Twin Cities
DAVID J. ARMOR, School of Public Policy, George Mason University, Fairfax, VA
JERALD G. BACHMAN, Institute for Social Research, University of Michigan, Ann Arbor
JOHN SIBLEY BUTLER, Department of Sociology and Department of Management, University of Texas at Austin
JOHN EIGHMEY, Greenlee School of Journalism, Iowa State University, Ames
MARTIN FISHBEIN, Annenberg School for Communication, University of Pennsylvania
CAROLYN SUE HOFSTRAND, Taylor High School, Volusia County, FL
PAUL F. HOGAN, Lewin-VHI Inc., Fairfax, VA
CAROLYN MADDY-BERNSTEIN, Education Consultant, Oro Valley, AZ
ROBERT D. MARE, Department of Sociology, University of California at Los Angeles
JEYLAN T. MORTIMER, Life Course Center, University of Minnesota, Twin Cities
CAROL A. MUTTER, LTG, retired, U.S. Marine Corps
LUTHER B. OTTO, emeritus, Department of Sociology, North Carolina State University, Raleigh
WILLIAM J. STRICKLAND, HumRRO, Alexandria, VA
NANCY T. TIPPINS, Personnel Research Associates, Dallas, TX

ANNE S. MAVOR, *Study Director*
MARILYN DABADY, *Senior Research Associate*
WENDY E. KEENAN, *Senior Project Assistant*
CECILIA ROSSITER (until January 2002), *Senior Project Assistant*
SUSAN R. McCUTCHEN (until January 2001), *Senior Project Assistant*

Preface

This report is the work of the Committee on the Youth Population and Military Recruitment. The committee was established by the National Research Council (NRC) in 1999 in response to a request from the Department of Defense. The impetus for the study was the recruiting problems encountered by the Services in the late 1990s. The central question is how to attract qualified youth to serve their country and, if necessary, be willing to put themselves in harm's way. Although military missions have diversified since the end of the cold war, the primary function of the Services remains the provision of the nation's warriors and protectors.

The charge to the committee was to provide information about the demographic characteristics, skill levels, attitudes, and values of the youth population, to examine options available to youth following high school graduation, and to recommend various recruiting and advertising strategies and incentive programs based on sound scientific data with the goal of increasing propensity and facilitating enlistment. The focus was limited to policy options that could be implemented within the current institutional structure of the military services. This report is intended not only for defense policy makers and recruiting personnel but also for the research community.

Committee members represent expertise in demography, military manpower, military sociology, psychology, adolescent development, economics, advertising and communication, and private-sector management. In developing our conclusions and recommendations we synthesized data and research from a variety of sources. First, we reviewed documents from the Department of Defense specifying current and projected defense goals, strategies, and manpower requirements as well as historical trends in force size, structure and quality, and turnover. This review provided a baseline description of current and projected recruiting demand: that is, will the United States need a larger force in the future and will a different level of personnel quality or skill mix be required?

Second, the committee analyzed data from large government databases on demography (the Current Population Survey, National Youth Longitudinal Survey), education (National Center for Education Statistics), and the quality characteristics of military applicants and enlistees (Defense Manpower Data Center). In some cases the existing data compilations and interpretations were used, and in other cases data were reanalyzed to answer the specific questions posed in this report. Projections based on demographic trends and trends in basic skill levels and moral characteristics of the youth population were used to determine the adequacy of the supply of qualified youth for military service. Data on civilian work and education options were used to assess the degree to which these opportunities compete with military service for qualified youth.

Third, the committee examined large-scale national surveys on youth attitudes and values (Monitoring the Future from the University of Michigan, the Youth Attitude Tracking Survey from the Defense Manpower Data Center, and the Sloan Study from the National Opinion Research Center) as well as locally based cross-sectional studies. These data were used to describe trends, over the last 26 years, in youth attitudes toward citizenship, work, education, and military services. In support of these trends, we reviewed the professional literature on socialization, attitude formation, and youth development. This literature also served to assess the degree of influence that parents, counselors, and peers have on youth attitudes, values, and career choices.

Fourth, the committee reviewed and analyzed current military advertising and recruiting in light of current trends in youth propensity to enlist in the Service. These analyses were informed by theory and research on the relationship between intentions and behavior, by principles of advertising, and by research relevant to the selection, training, and motivation of productive recruiters.

Several individuals representing recruiting activities in the Services provided important context for the committee's work. First, we would like to thank Col. Greg Parlier, Lt. Col. Mark Young, and the staff of the U.S. Army Recruiting Command at Fort Knox, Kentucky, for hosting a site visit for a subgroup of the committee and later presenting material to the full committee. Second, we would like to thank the members of the recruiting panel who discussed current practices and issues. They are Col. G. K. Cunningham, U.S. Marine Corps Recruiting Command; Ed Kearl, U.S. Navy Recruiting Command; Kevin Lyman, U.S. Army Recruiting Inspection Branch; and Lt. Col. Martin Pullum, U.S. Air Force Recruiting Command.

Many individuals provided the committee with useful information through presentations and written materials. We would like to particularly thank Monica Gribben, HumRRO, and Barbara Snyder, University of Chicago, for preparing papers and presentations. We also extend our gratitude to Beth Asch, James Dertouzos, Susan Everingham, and Bruce Orvis of the RAND Corporation; Robert Clark and Lt. Col. Jeff Perry, Office of the Assistant Secretary of Defense, Force Management Policy; Karen Horton, GTE Service Corporation; James Hoskins, Wirthlin Worldwide; Deirdre Knapp, HumRRO; Anita Lancaster and Jerome Lehnus of the Defense Manpower Data Center; Reed Larsen, University of Illinois; Janet Stanton, Bates USA; David Treteler, National War College; John Warner, Clemson University; and James Youniss, Catholic University.

To our sponsor, the Office of the Assistant Secretary of Defense for Force Management Policy, we are most grateful for their interest in the topic of this report and for their many useful contributions to the committee's work. We are particularly indebted to W.S. Sellman for his insight, his encouragement, and his dedication to science and to Jane Arabian for her guidance and support. Finally we wish to extend a special thanks to Vice Admiral Patricia Tracey for her vision and to Lt. General John Van Alstyne for his continuing interest.

In the course of preparing this report, each member of the committee took an active role in drafting chapters, leading discussions, and reading and commenting on successive drafts. We are deeply indebted to all for their broad scholarship, their critical examination of the evidence, and their cooperative spirit.

This report has been reviewed in draft form by individuals chosen for their diverse perspectives and technical expertise, in accordance with procedures approved by the NRC's Report Review Committee. The purpose of this independent review is to provide candid and critical comments that will assist the institution in making its published report as sound as possible and to ensure that the report meets institutional standards for objectivity, evidence, and responsiveness to the study charge. The review comments and draft manuscript remain confidential to protect the integrity of the deliberative process. We wish to thank the following individuals for their review of this report: John P. Campbell, University of Minnesota; William C. Howell, Arizona State University; Gerald B. Kauvar, The George Washington University; Catherine Milton, Friends of the Children, New York, NY; Kevin R. Murphy, Pennsylvania State University;

Michael Pergamit, National Opinion Research Center, Washington, DC; Kim B. Rotzoll, University of Illinois; David R. Segal, University of Maryland; and Sheila E. Widnall, Massachusetts Institute of Technology.

Although the reviewers listed above have provided many constructive comments and suggestions, they were not asked to endorse the conclusions or recommendations, nor did they see the final draft of the report before its release. The review of this report was overseen by Marshall S. Smith, of the William and Flora Hewlett Foundation, and Alexander H. Flax, consultant, Potomac, MD. Appointed by the NRC, they were responsible for making certain that an independent examination of this report was carried out in accordance with institutional procedures and that all review comments were carefully considered. Responsibility for the final content of this report rests entirely with the authoring committee and the institution.

Staff of the NRC made important contributions to our work in many ways. We extend particular thanks to Marilyn Dabady for her outstanding efforts as a senior research associate. We are also grateful to Wendy Keenan and Susan McCutchen, the committee's senior project assistants, who were indispensable in organizing meetings, arranging travel, compiling agenda materials, and in managing the preparation of this report. Finally we wish to thank Alexandra Wigdor for initiating the project and Chris McShane for her thoughtful editing of this report.

Paul Sackett, *Chair*
Anne Mavor, *Study Director*
Committee on the Youth Population and Military Recruitment

Contents

ATTITUDES, APTITUDES, AND ASPIRATIONS OF AMERICAN YOUTH

Executive Summary

The Department of Defense (DoD) is the largest employer in the United States. The current size of the enlisted military force is 1.2 million and approximately 200,000 new recruits are needed each year to maintain this level. Since the end of the draft in 1973, the supply and demand for recruits has fluctuated based on mission requirements, budgetary constraints, and economic conditions. In recent times, recruiting has been made more difficult by a healthy economy, the growing number of youth who aspire to a college education, and a steadily downward trend in interest in military service. Even with the instability in the economy and the loss of civilian jobs in many sectors in 2000–2001, interest in the military has not increased.

THE PROBLEM

In the late 1990s, the Services struggled to meet their recruiting goals and in some cases fell short. This led to the question of how the recruiting process and the recruiters' job could be better supported in order to ensure that force strength, force quality, and the required skill mix of personnel will be available to meet the ever-changing security and defense challenges. Military officials recognized that a fundamental understanding of the youth population and of the effectiveness of various advertising and recruiting strategies used to attract them would be extremely valuable in addressing these questions. As a result, in 1999, the DoD asked the National Academy of Sciences, through its National Research Council, to establish the Committee on the Youth Population and Military Recruitment to provide information about the demographic characteristics, skill levels, attitudes, and values of the youth population, to examine options available to youth following high school graduation, and to recommend various recruiting and advertising strategies and incentive programs based on sound scientific data with the goal of increasing propensity and facilitating enlistment. It was decided early in the process that the focus would be limited to policy options that could be implemented within the current institutional structure. Thus, such topics as changing the length of the enlistment term or the coordination among various government agencies were not addressed.

The committee is composed of 15 experts in the areas of demography, military manpower, military sociology, psychology, adolescent development, economics, advertising and communication, and private sector management. The sections that follow provide an overview of our major findings and recommendations.

MILITARY MANPOWER REQUIREMENTS

In spite of the terrorist attacks on the World Trade Center and the Pentagon on September 11, 2001, and in spite of changing missions and requirements for the U.S. military, we see no clear indication that the size of the military will change significantly over the next 20 years. It is expected that as priorities change the force mix among the active duty, Reserve, and National Guard structures will change, with increasing emphasis on the Reserves. All of the subsequent discussion is based on this premise that overall force size will not change. The challenge for recruiting and retention will be manning selected career areas to ensure that shortfalls in critical areas do not compromise unit readiness.

There have been few major changes in the occupational distribution of first-term personnel in the past 10 years, but future military missions coupled with advances in technology are expected to require military personnel to make greater use of technology. Technological changes will make some jobs in the future easier and others more difficult, but overall minimum aptitude requirements are unlikely to change much over the next 20 years. However, the mental, physical, and moral requirements for military service are unlikely to change over this time period. The Services are currently accessing recruits who have sufficient aptitude and can be trained to perform military tasks adequately even if occupational requirements do increase somewhat. Recruits satisfying current quality levels can be trained to meet future demands.

Since retention is critical to maintaining required force size, experience, and skill mix, it is important to better understand first-term attrition and voluntary separation at the end of a term of duty. More valid and reliable data from exit surveys are needed to achieve this understanding than are available at present.

THE YOUTH POPULATION

Demographic Characteristics

The demographic context for armed forces recruitment is determined by the size and composition of the youth population. These are fundamental constraints on future recruitment efforts. Cohorts of persons reaching 18 years of age are expected to grow significantly over the next 10 years and then remain approximately at a plateau during the following decade. Approximately 3.9 million youth reached age 18 in 1999, a number that will increase to approximately 4.4 million by 2009 and trail off only slightly during the subsequent decade.

The ethnic composition of the youth population will change significantly over the next 15 to 20 years. Even in the absence of changes in immigration patterns, the ethnic makeup of the youth population will change because of recent changes in the ethnic makeup of women of childbearing age and ethnic differences in fertility rates. Based on recent fertility patterns, the percentage of young adults who are Hispanic, of whom the largest subgroup is of Mexican origin, will increase substantially. A growing percentage of youth will be raised by parents who are immigrants to the United States, a result of high rates of recent immigration and relatively high fertility levels of foreign-born women.

The socioeconomic characteristics of parents, such as their levels of educational attainment, have a large effect on the aspirations and decisions of youths, especially concerning higher education. Average levels of maternal education for teenagers have increased markedly and will continue to do so over the next two decades, a result of secular increases (i.e., trends over time) in educational attainment in the population. Within the next two decades, the majority of youths will be raised by mothers who have completed at least some college.

Trends in numbers of births and in the composition of the child population have offsetting effects on potential enlistment trends. Although the annual number of births has increased in recent years, children are increasingly raised by highly educated parents and by parents who have no direct experience with the armed forces. The net impact of these offsetting trends is a small increase in expected numbers of enlistees in the next decade. Thus, demographic trends do not emerge as factors that will contribute to increasing difficulty in meeting enlistment goals. Other factors discussed in this report, including advertising and recruitment practices, will determine whether potential enlistees actually enlist at a rate necessary to meet the Services' goals.

Qualifications

Current statutory enlistment standards for education and aptitude levels are somewhat higher than the distribution of education and aptitudes in the general population. However, it is expected that potential supply of highly qualified youth, those who meet or exceed enlistment standards, in terms of education and aptitude, will remain fairly stable over the next 10 years, and there is no reason to expect any decline over the next 20 years. If anything, the proportion of highly qualified youth may increase slightly, particularly if the high school graduation rate remains high.

An analysis of current enlistment force composition shows that the enlistment rates of highly qualified males and females are virtually the same. Regarding ethnicity, black and Hispanic groups have lower enlistment rates of highly qualified youth than whites, with Hispanics showing higher rates than blacks. Although the percentage of Hispanic youth in the enlisted force is increasing, they are still underrepresented compared with the general population of the same age. At the same time, blacks are somewhat overrepresented. In 2000, the force was 18 percent female compared with 51 percent female in the general civilian population of the same age. The ethnic distribution of the force was 62 percent white, 20 percent black, and 11 percent Hispanic compared with 65 percent, 14 percent, and 15 percent, respectively, in the civilian population ages 18–24.

With regard to physical problems, current population trends suggest increases in certain health conditions, such as obesity and asthma, both of which make the individual ineligible for military service. Concerning moral issues, current trends in illicit drug use and criminal activity appear to have little effect on recruiting issues.

Opportunities and Options

The primary alternatives to military service are civilian employment and postsecondary education. Currently, the greatest challenge to military recruiting is attracting college-bound youth. The armed forces compete directly for the same portion of the youth market that colleges attempt to attract—high school graduates who score in the upper half of the distribution on the Armed Forces Qualification Test (AFQT). During the 1990s, rates of college enrollment and levels of education completed increased dramatically as a result of three broad trends: (1) changes over time in parental characteristics, especially parents' educational attainment, which increased youths' resources and aspirations for education; (2) the greater inclusion in higher education of women and some ethnic minorities; and (3) increased economic incentives to attend and complete college, a result of changes in the labor market for college- and noncollege-educated workers. In 1999, 63 percent of high school graduates enrolled in college in the same year they graduated from high school.

Opportunities for employment in the civilian sector for high school graduates are highly variable and do not have a clear economic advantage over military service. Although compensation strategies can respond to market demands in the civilian sector, they are not necessarily superior to the compensation offered in the military. There are two areas, however, in which substantial differences between the military and civilian employment exist: working conditions and the opportunity to serve a higher purpose. An individual might prefer working conditions in the civilian sector to those in the military, where conditions may be onerous or life-threatening. However, that individual is more likely to find a transcendent purpose (e.g., duty to country) by serving in the military rather than being employed in the civilian sector.

Attitudes and Values

Youth attitudes toward the importance of various goals in life, preferred job characteristics, and work setting have changed very little over the past 25 years. The value that has been changing most significantly is college aspirations and college attendance. One useful finding regarding education and military service is that, in recent years, the majority of high school seniors (both male and female) who reported the highest propensity to join the military also expected to complete a four-year college program.

One critical finding regarding youth attitudes is that the propensity to enlist in the military among high school males has been declining since the mid-1980s, while prior to that time, propensity had been increasing. In this key group for recruiting, the proportion indicating that they "definitely will" join a military Service has declined from 12 to 8 percent during that time period. There has also been a shift in interest reflected by a decline in those indicating they "probably won't" and an increase in those saying "definitely won't" join. The proportion least interested in military service has increased in the past two decades from about 40 to about 60 percent. The percentage of females who say they "definitely will" serve has remained at 5 percent over the past several years; however, since 1980, the percentage who say they "definitely won't" serve has increased from 75 to approximately 82 percent.

Several other aspects of youth attitudes and behavior provide potential guidance for the design of military recruiting and advertising messages: (1) the time in which youth make decisions about education and careers has extended well into their 20s; (2) there has been little or no change in youths' views about the military service as a workplace or the value and appropriateness of military missions; (3) there has been some increase in the desire of youth to have two or more weeks vacation—a benefit of military service over the civilian sector; (4) there is a possible link between youth attitudes toward civic duty and volunteerism and military service (the potential of this link requires further study); and (5) parents, particularly mothers, and counselors have a strong influence on youth decision making with regard to career and educational choices.

ADVERTISING AND RECRUITING

Most attempts to predict propensity have occurred at the aggregate level, examining the relation between the proportion of people with a propensity to enlist (or who have actually enlisted) and a large array of demographic, economic, and psychosocial variables (e.g., percent unemployed, civilian/military pay differentials, educational benefits offered, percentage of the population holding a given belief, attitude, or value) over time. Such aggregate-level analysis can disguise important effects at the individual level. For purposes of designing interventions to increase the proportion of the population with a propensity to enlist at any given point in time or to increase the likelihood that those with a propensity will, in fact, enlist, individual-level analyses that identify the critical determinants of propensity are critical. These types of analyses have not been done. Thus, the most relevant data for guiding the development of effective messages to increase propensity are currently not available.

As a matter of military readiness, DoD may wish to consider maintenance of the level and direction of the propensity to enlist to be a primary responsibility. Societal factors are also important, but some organization or voice in the nation must take the lead in maintaining propensity at a level needed to effectively sustain the military services. Advertising can be used to support the propensity to enlist in the youth population of interest for military recruiting. Rather than allocate advertising expenditures on the basis of immediate recruiting goals, advertising can be more usefully deployed as a means of supporting and maintaining the level of propensity to enlist. That is, military readiness may be best served when the first role of military advertising is to support the overall propensity to enlist in the youth population and to maintain a propensity level that will enable productivity in military recruiting.

Advertising planners should consider the trade-offs between primary and selective demand. An assessment should be made regarding the portion of military advertising to be devoted to supporting the overall propensity to enlist in the military (primary demand) and the proportion to be devoted to attracting needed recruits to the individual military Services (selective demand).

OVERARCHING RECOMMENDATIONS

Two classes of factors appear to be linked to recruiting outcomes. The first class involves "doing more," meaning investing more resources in traditional recruiting activities. The second class involves "doing differently," meaning engaging in new recruiting activities or modifying the way traditional activities are carried out.

In terms of doing more, research indicates that recruiting success is responsive to additional expenditures in the number of recruiters, dollars spent on advertising, size of enlistment bonuses, dollars spent on funding subsequent education, and pay. The marginal cost of increasing recruiting effectiveness via pay is markedly higher than that for the other options. For this reason, we recommend that the Services and DoD periodically evaluate the effects of increased investment in recruiters, educational benefits, enlistment bonuses, and advertising as well as the most efficient mix of these resources.

In terms of doing differently, we offer a variety of conclusions and recommendations. First, in the important domain of education, we recommend increasing mechanisms for permitting military service and pursuit of a college degree to occur simultaneously. In light of the higher education aspirations of youth and their parents, such mechanisms are central to recruiting success. Perhaps the most dramatic attitudinal and behavioral change over the past several decades is the substantial increase in educational aspirations and college attendance.

We also note that the numbers of college dropouts and stopouts (those who leave and return later) and the numbers of youth delaying the traditional activities that signal a transition from adolescence to adulthood (e.g., career choice, mate choice) suggest increasing attention to a broader market than the traditional focus on those just completing high school. For this reason, we recommend that DoD investigate mechanisms for cost-effective recruiting of the college stopout/dropout market and that DoD continue to link Service programs with existing postsecondary institutions offering distance degree programs.

Second, in the domain of advertising, we recommend attention to three key issues. One is the balance between a focus on military service as a whole and Service-specific advertising. Advertising theory and research suggest the value of supporting overall propensity for military service in addition to Service-specific advertising. Another is a balance between a focus on the extrinsic rewards of military service (e.g., funds for college) and intrinsic rewards, including duty to country and achieving purpose and meaning in a career. While many youth are responsive to an extrinsic focus, an additional segment of the youth population sees intrinsic factors as the primary appeal of military service. A final issue is the role of parents in the enlistment decisions of their sons and daughters. Their key role suggests that attention be paid to the effects of advertising on parental perceptions of military service. We recommend that a key objective of the Office of the Secretary of Defense should be to increase the propensity to enlist in the youth population. We further recommend that advertising strategies increase the weight given to the intrinsic benefits of military service.

Third, in the domain of recruiting practices, we recommend that attention be paid to the selection and training of recruiters. We note that there are substantial differences in recruiter performance, yet the process of staffing the recruiting services does not focus centrally on selecting individuals on the basis of expected productivity. We also note the importance of rewarding and providing incentives to recruiter performance. Specifically, we recommend that the Services develop and implement recruiter selection systems that are based on maximizing mission effectiveness; that they develop and implement training systems that make maximum use of realistic practice and feedback; and that they explore innovative incentives to reward effective recruiting performance.

Recruiting is a complex process, and there is no single route to achieving success in achieving recruiting goals. Nonetheless, we believe that progress has been made toward a better understanding of the process, and that useful avenues for exploration have been identified.

1

Introduction

The Department of Defense (DoD) faces short- and long-term challenges in selecting and recruiting an enlisted force to meet the personnel requirements associated with diverse and ever-changing missions at home and abroad. The country's awareness of the need for a high-quality and well-trained military was significantly heightened by the September 11, 2001, terrorist attacks on the World Trade Center and the Pentagon. Recruiting an all-volunteer force is a difficult task: in order to get one eligible recruit, an Army recruiter must contact approximately 120 young people. In 1997 and 1998, some of the Services fell short of their recruiting goals. During this time, the economy was strong, jobs in the private sector were plentiful, over 60 percent of high school graduates were attending their freshman year of college, and the interest or propensity of youth to enlist in one of the military Services was decreasing. Since then, the economy has suffered some setbacks, college aspirations have remained constant or increased, and, importantly, propensity has continued to decrease, particularly among those youth who meet the military standards of graduating from high school with a diploma and scoring in the top half of the Armed Forces Qualification Test (AFQT).

THE CHALLENGE AND THE CHARGE

In light of these concerns, the DoD asked the National Academy of Sciences, through its National Research Council, to establish the Committee on the Youth Population and Military Recruitment to provide information about the demographic characteristics, skill levels, attitudes, and values of the youth population, to examine options available to youth following high school graduation, and to recommend various recruiting and advertising strategies and incentive programs with the goal of increasing propensity and facilitating enlistment. Specifically the committee was asked to undertake a series of interrelated tasks designed to provide a sound scientific basis for both short- and long-range planning.[1] These tasks include the following:

1. Develop a complete profile of American youth today and in the future. Create a multidimensional characterization of youth using scientific literature that offers insight into their motivations, interests, behavior, values, and attitudes. Evaluate demographic trends in light of existing and potential recruiting strategies, training, and retention.

2. Examine survey data on perceptions of military service, primary influencers of these perceptions, and the desirability of enlisting, with specific attention to subgroup differences. Identify further analyses and interpretations of survey items and data and consider valid ways of getting the data into practice quickly.

3. Examine the changing nature of work generally and the new demands placed on the military in the post-Cold War era. Also, consider alternative choices for youth—the civilian workforce and postsecondary education—and explore the implications of current and projected trends in work and education as they influence approaches to selecting and attracting youth with the needed skills, abilities, and attitudes.

4. Review and evaluate advertising programs directed at youth (e.g., anti-drug and anti-smoking campaigns, education, safe sex) to determine methods for creating an effective recruiting message. Examine the literature on advertising, communication, and attitude theory and measurement to determine how the characteristics and motivations of current and future generations can be linked to the design of military recruiting efforts.

5. Develop policy options. Consider a full range of personnel options for expanding the pool of recruits, including greater coordination of procedures among the military Services; coordination with alternative government agency programs, such as the Job Corps and the Corporation for National Service; material and psychological incentives that may enhance recruiting success by subgroup; and changes in pay scales and the duration of enlistment.

---

[1]These tasks comprise substantially the effort referred to as phase 1 in the original version (or earlier versions) of the task statement.

The committee assembled to accomplish the charge is composed of experts from the following disciplines: demography, military sociology, adolescent development, economics, advertising and communication, and private sector management. The committee decided early in its deliberations to work on ways to improve recruiting within the current military structure. Such issues as length of term and coordination with agencies outside the military establishment would require additional studies and were beyond the scope of the current committee. The policy options examined in the report include maintaining competitive pay and benefits, employing structural changes in compensation, pursuing ways for enlistees to gain a college degree while serving in the military, developing consistent advertising themes that build on strengths offered by military service, and employing better methods to select, train, and motivate recruiters.

CONTEXT

Current and projected military manpower requirements and historical trends in force characteristics provide a picture of the demand side for military recruiters and one element of the context for the committee's work. Topics of particular importance with regard to manpower requirements include the kinds of missions the Services expect to perform, the characteristics of military jobs and the skills that will be needed to perform them, and the anticipated size and structure of the force. Historical trends and the current status of military enlisted personnel provide important information about changes in force size, distribution of occupational specialties and skills, levels of first-term attrition, and reenlistment percentages.

Since the end of the Cold War, the active military force has been reduced by more than one-third and at the same time has been asked to respond to a large number of diverse missions. As a result of the attacks of September 11, 2001, the United States has focused its commitments on defending against terrorism at home and fighting it around the world. However, fighting the global war on terrorism does not diminish other commitments for peacekeeping, providing humanitarian aid, and war fighting. According to the latest Quadrennial Defense Review (U.S. Department of Defense, 2001), DoD is now required to provide a large rotational base of military personnel to support long-standing commitments as well as to respond to small-scale contingency operations. The Services have calculated deployment numbers in multiples of three for each mission—one Service member engaged in a mission, one returning, and one in training. In recent testimony before the Senate Armed Services Committee, each Service chief requested personnel increases for the upcoming year (March 7, 2002).

The Services are currently in the process of transformation. This involves, among other things, the development of smaller, more flexible units and the application of technology advances to equipment design. The Army is developing a family of ground combat systems that provide digitized information exchange; the Navy is automating a wide range of functions in smart ships such as the DD-X, and the Air Force and the Navy are exploring the development of remotely piloted, armed aircraft. Both the advances in technology and anticipated missions provide critical requirements for determining the size and shape of the force and specifying both recruiting and retention goals.

Recruiting goals and the resulting challenge for recruiters are determined by anticipated shortfalls in the specified force size (end strength) and skill distribution. These shortfalls can be a result of attrition during the first term of service, voluntary losses based on those who do not reenlist when their term of service expires, or an overall increase in the number of personnel required. First-term attrition is primarily due to behavioral and medical problems for young men and behavioral problems and pregnancy for young women. Voluntary separations for those personnel whom the Services would like to retain are generally attributed to dissatisfaction by either the enlisted personnel or their spouses with compensation or quality of life. The Services have developed several programs to reduce unwanted losses; however, the data show that first-term attrition remains high—at approximately 30 percent—and many personnel whom the Services would like to retain are leaving for civilian jobs or to pursue additional education.

The second element of context for the committee's work is the source of supply for military jobs—the youth population. The characteristics of youth and the options available to them narrow the pool of individuals who are interested in serving and those who ultimately are selected. The important factors to be considered about the youth population include (1) the projected size and demographic composition; (2) the trends in basic knowledge, skill, physical, and moral characteristics; (3) opportunities and aspirations regarding employment and education; and (4) values and attitudes regarding service to country and the features of military service, including quality of life.

APPROACH AND REPORT ORGANIZATION

There are a large number of hypothesized causes for the military's recruiting difficulties over the past several years, as well as a similar set of hypotheses about potential effects on recruiting effectiveness in the future. The committee's goal in this report is to identify and examine a wide variety of such causal factors. Some of these are factors within the control of decision makers (e.g., pay and benefits, recruiting practices, advertising messages), while others are not (e.g., changes in the size of the cohort eligible for military service, changes in the skill levels of American youth). In examining each of these factors, the committee's perspective is forward-looking, that is, the focus is not on producing a definitive answer to the question of what caused the military recruiting shortfalls of recent years, but rather on identifying factors likely to influence future recruiting effectiveness. Given the broad set of individual, situational, organizational, and societal influences on decisions about military service, there was no expectation that a single factor would account for recruiting effectiveness. We hoped, however, to identify a small set of important variables from among a broader array of possible factors affecting recruiting effectiveness.

The report is structured around a set of potential contributing factors. Following this introduction, the report begins, in Chapter 2, with a discussion of demand factors, inquiring about possible changes in overall force size and structure, in the aptitude levels needed for effective performance, in the physical demands of military work, in the moral and character requirements of military work, and in the levels of attrition and retention.

Chapter 3 examines the demographic context for armed forces recruitment. The size and composition of the youth population are fundamental constraints on future recruitment efforts. Forecasts are provided for the size and some aspects of the composition of the youth population for the next 15 to 20 years. The accuracy of these predictions is high because they are based on persons who are already born.

Chapter 4 reviews the four major domains in which military applicants are screened: aptitudes (indexed by the AFQT), educational attainment (possession of a high school diploma), physical and medical qualification, and moral character (e.g., lack of a criminal record). In each of these domains, the Services' current enlistment requirements are examined and evaluated for possible changes in the future. The chapter then turns to a review of the evidence regarding the current supply of youth possessing the needed characteristics and the likelihood that the proportions will change over time.

Chapter 5 includes an examination of the three major options available to the youth who make up the prime military recruiting market: joining the military, entering the civilian labor market, or pursuing higher education. Particular emphasis is placed on the changing landscape regarding (1) participation in postsecondary education and opportunities available to youth in the civilian labor market and (2) aspects of these alternatives that compete with the service options or may be fruitfully combined with them.

Chapter 6 presents data on the changes in youth attitudes, values, perceptions, and influencers over the past two decades, including major findings from extensive long-term longitudinal and cross-sectional research on youth attitudes, as well as on the relationship between youth attitudes and the propensity to enlist.

Chapter 7 offers an integrated theory of behavioral choice that can productively guide future research on the determinants of propensity and of actual enlistment.

Chapter 8 examines a range of issues involving military advertising and recruiting, including goals, strategies, and messages. Advertising is a part of the broad recruitment process, and we examine this process more generally, including a comparison with recruiting practices in the civilian labor market.

Chapter 9 presents the committee's conclusions and recommendations based on the analyses provided in each of the chapters. The final section of this chapter draws the implications of the committee's overarching conclusions and recommendations.

In the course of the study, the committee prepared two letter reports, one on evaluation of the Youth Attitudes Tracking Study (Appendix A) and the other on the scientific underpinnings of the popular literature on generations (Appendix B). Biographical sketches of committee members also are included (Appendix C).

2

Military Manpower Requirements

This chapter outlines the "demand" for military manpower and the processes that are used to determine force size and structure, recruiting challenges, and retention needs. The two central questions are: (1) Will the United States need a larger force in the future? (2) Will a different level of personnel aptitude or skill mix be required? Answers to these questions depend on the missions to be performed. As missions and priorities change, so do the recruiting requirements. Thus, future recruiting need is a function of any mission changes that would dictate a change in force size as well as any changes in retention that would dictate a change in the number of positions to be filled by new recruits.

Our review of these factors is divided into three major sections. The first section discusses future defense strategies and force structure policy decisions presented by the President of the United States, the National Security Council, and the Department of Defense (DoD). The second section provides a discussion of force size and level of recruit qualifications from an historical perspective. The third section examines the implications of past and present trends in attrition and retention for recruiting. Each offers guidance concerning both current and future military recruitment needs. Figure 2-1 provides an overview of the major factors involved in the process of establishing recruiting requirements.

The size and shape of the military services are determined by the National Security Strategy (NSS) and the National Military Strategy (NMS). The NSS is developed by the National Command Authority (the President and the National Security Council) and is usually prepared by each new administration every four years. Although national security is

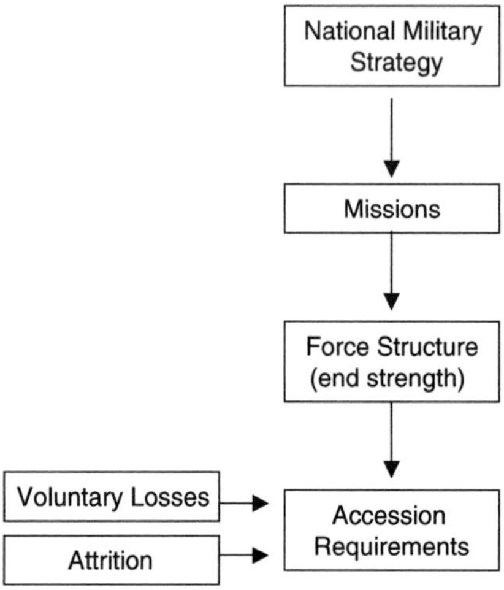

FIGURE 2-1 Manpower planning.

certainly a responsibility of the military, it also has diplomatic, economic, and informational dimensions as well. The NSS provides the overall guidance to all agencies of the federal government with roles in national security. Its function is to:

- ensure U.S. security and freedom of action,
- honor international commitments, and
- contribute to economic well-being.

The NMS is derived from the NSS by the secretary of defense and the chairman of the Joint Chiefs of Staff. The NMS describes the use of military power in peace and war to help meet national security objectives. The commanders in chief of the regional combatant commands develop their theater strategies from the NMS. The actual military force structure (size—i.e., end strength) and shape (skills, organizational structure, etc.) are based on the specified and implied missions in the NMS. For example, if the U.S. Navy is required to maintain a continuous forward presence in the Atlantic, the Pacific, the Mediterranean, and the Persian Gulf, then a minimum number of ships is required, which in turn requires a certain number and skill mix of people to man those ships.

The Services, through the DoD budget process, request authorization for their required personnel end strength (size) and for the necessary appropriations (funds to pay for that end strength) each year. The approved DoD budget is authorized and appropriated by Congress and signed by the President. In some cases, the funds appropriated do not match the numbers authorized—in these cases each Service has the option, up to a point, of taking some funds appropriated for other programs to pay for the additional personnel authorized.

Recruiting requirements are determined by anticipated shortfalls in authorized end strength.[1] This involves comparing the current number on active duty and the end strength authorized for the coming fiscal year and projecting personnel losses for that year. Personnel losses are calculated by adding estimates of voluntary losses and estimated losses based on those who do not reenlist when their terms of service expire and to estimated losses from attrition prior to the end of an enlistment term (for medical, disciplinary, or other reasons). End strength requirements, that is the raw numbers needed, are necessary but not sufficient to the process. Recruiting goals also take into account specific skills required, some of which involve higher initial qualifications than others. Qualified recruits are channeled into schools to train them for specific jobs.

The Services can reduce the number required to recruit by reducing the number of voluntary losses and attrition, or increasing retention, or both. However, there is a limit to this strategy. The military requires a constant influx of new young people, and therefore it is as important to bring in new young men and women as it is to keep an optimal number of trained and experienced military members. The desired retention rate differs for each Service on the basis of its mission (i.e., the Army and the Marine Corps require more young people to perform very physically demanding duties; Air Force and Navy jobs are generally more technically oriented, require more experienced individuals, and thus have higher targets for first-term retention rates).

SHAPING THE FORCE

Since the end of the Cold War, the U.S. military force has become significantly smaller and has been asked to respond to numerous diverse situations around the world. Some of these situations have required war fighting, others have involved peacekeeping, and still others have focused on humanitarian aid. In many situations, there has been uncertainty concerning whether, how, or at what level U.S. troops should be engaged.

The terrorist attacks of September 11, 2001, on the World Trade Cen-

---

[1]This is certainly a "crystal ball" exercise, but it is based on historical data that give good indications for future activity. It can, however, be changed significantly by uncontrollable events during the year (e.g., the unemployment rate, quality-of-life issues).

ter and the Pentagon have put into motion a series of political, economic, and military actions to secure the homeland and to combat terrorism around the world. In response to the terrorist attacks, President Bush and his administration have established a new Office of Homeland Security, obtained international agreements to freeze the assets of terrorist organizations, obtained support from U.S. allies to grant access to airfields and seaports, deployed troops, and ousted the Taliban regime from Afghanistan. In June 2002, the President asked that the Congress join him in creating a single, permanent department (the Department of Homeland Security) with an overriding and urgent mission: securing the homeland of America and protecting the American people. Indications are that this new department will have little impact on the size or shape of the DoD or the military. The war against terrorism continues on many fronts and will be a long one, requiring all national security tools, including the military.

It is difficult to predict what the demands for military forces will be at any given time. The discussion below illustrates current thinking regarding strategies and personnel deployment and the kinds of decisions that must be made. We are aware that the details presented will change as a function of world events.

The most reliable and complete source for current military thinking is the Quadrennial Defense Review or QDR (U.S. Department of Defense, 2001), which addresses certain elements of the NSS and covers virtually all elements of a NMS; we draw primarily on the QDR in this discussion. Much has been said about potentially major changes for the military that could have an impact on the numbers and types of young men and women needed in the "transformed" military. Secretary of Defense Donald Rumsfeld states specifically that a central objective is to shift the basis of defense planning from a "threat-based" model, which has dominated military thinking in the past, to a "capabilities-based" model for the future.

The events of September 11 signaled that the security of the homeland must be the first priority. Government officials now realize that they cannot predict where or when or how America, Americans, or American interests will be attacked. They can no longer focus on the Cold War notion of two nearly simultaneous major theater wars as the sole model for building, equiping, and training U.S. military forces. New strategic planning principles include emphasis on homeland defense, on surprise, on preparing for asymmetric threats, on the need to develop new concepts of deterrence, on the need for a capabilities-based strategy, and on the need to balance deliberately the different dimensions of risk.

The strategy for America's defense seeks to assure allies and friends that the United States can fulfill its security commitments, to dissuade adversaries from undertaking operations that could threaten U.S. interests, to deter aggression by swiftly defeating attacks and imposing severe penalties for aggression, and to decisively defeat any adversary if deterrence fails.

DoD has also announced the establishment of a new unified combatant commander for homeland security to help coordinate military efforts. The intent is to accomplish this with the current end strength. However, "Preparing forces for homeland security may require changes in force structure and organization. . . . U.S. forces . . . require more effective means, methods, and organizations to perform these missions" (U.S. Department of Defense, 2001:19).

DoD is now specifically required to plan for a larger base of forces from which to provide forward-deployed forces, in order to support long-standing contingency commitments in the critical areas of interest. DoD must ensure that it has sufficient numbers of specialized forces and capabilities so that it does not overstress elements of the force when it is involved in smaller-scale contingency operations (U.S. Department of Defense, 2001:21).

This larger base of forces is to be provided on a rotational basis. As mentioned earlier, the Services calculate deployment numbers in multiples of three: for every soldier on a mission, a second must be in training for that role, and a third will have just returned from the mission assignment and is scheduled for retraining. A recent newspaper article claimed that for every deployed soldier, there are seven in support, which presumably include trainers and base support personnel, in addition to those in the rotation described above.

DoD is developing more effective ways to compute the required rotational base across various types of forces to support forward posture. Changes potentially affecting force structure include: (1) a more streamlined organizational structure with a reduction in headquarters staffs, a consolidation of overlapping functions of Office of the Secretary of Defense, the Services, and the Joint Chiefs of Staff, and changes in military departments; (2) the possibility of a joint Services training capability, the establishment of a Joint Opposing Force, and the creation of a Standing Joint Task Force; and (3) Service-specific decisions to increase presence in various parts of the world.

The current focus on combating terrorism does not change the need to respond to other events around the world as they call for attention. Many of the planning documents in place prior to September 11 still apply. For example, *Joint Vision 2020* (Joint Chiefs of Staff, 2000) suggests that U.S. international interests in security, politics, and economics will further expand as a result of the wide availability of new information, communication, and transportation technologies. Many countries will have access to the same technology, so it will not be possible for the United States and its allies to maintain a total technological advantage over their adversaries. This document also anticipated the increasing use of asymmetric methods, such as the terrorist attacks on September 11.

As plans for force structure are being formulated, advances in technology are making it possible to automate many functions currently performed by personnel. Among these are smart ships, such as the DDX class warships, and remotely piloted, armed aircraft—under development for both the Air Force and the Navy (Hey, 2001). It is anticipated that remotely piloted aircraft being built by Boeing as a Defense Advanced Research Projects Agency effort will cost less to operate and support than manned vehicles; the goal is to have 30 aircraft in operation by 2010 (Hebert, 2001). However, according to Blake Crane, a Heritage Foundation defense analyst, remotely piloted planes are seen as a supplement to manned military craft, not as a replacement (Hey, 2001).

Another initiative is the establishment of two new Air Force squadrons devoted to preparing U.S. defenses against attacks in space on commercial satellites and other spacecraft. One squadron will explore future space technologies, and the other will play the enemy role in war game exercises. Although planning documents discuss many initiatives that could require more people, they also suggest reallocating personnel resources rather than increasing overall end strength.

FORCE CHARACTERISTICS

Force Size

As described above, the size of the U.S. military force is driven by many factors. In response to those factors, the Services are now dramatically smaller than they were a few years ago. After a period of relative stability in force size during the 1970s and early 1980s, the Services have seen a consistent drop since 1987. Between 1987 and 2000, the active component of the enlisted force decreased in size by 38 percent, from 1.85 to 1.15 million enlisted members. During those same years, the reserve component of enlisted strength decreased 26 percent, from 989,000 to 733,000 (U.S. Department of Defense, 2002). (The role of the National Guard has taken on increased importance with the new emphasis on homeland security; however, in this study we did not review data on the National Guard.) The reductions in force size observed during the 1990s appear at this point to have leveled off; in fact, the chiefs of staff of each of the Services suggested in testimony before the Senate Armed Services Committee (March 7, 2002) that force size needs to grow somewhat over the next few years. Table 2-1 displays the active component of enlisted strength by Service from 1980 through 2000. Table 2-2 displays the reserve component of enlisted strength by Service for the same time period.

Given recent terrorist threats to the United States, it seems unlikely that force sizes will be significantly reduced in the near term from their current levels. In fact, during the development of the QDR, DoD specifically "assessed the current force structure across several combinations of scenarios on the basis of the new defense strategy and force sizing construct." Some scenarios resulted in moderate operational risk in mission accomplishment; some resulted in "high risk" (U.S. Department of Defense, 2001:22). One could argue that this makes the case for increased force structure—certainly to maintain the current force structure as a minimum. In fact, Congress authorized, in the fiscal year 2002 (FY 2002) budget, each of the Services to have on hand personnel up to 2 percent above their authorized strength. But the money to pay for those additional people was not appropriated; the Services had to take money from other

TABLE 2-1 Active Component Enlisted Strength (in thousands)

Fiscal Year	Army	Navy	Marine Corps	Air Force	Total DoD
1964	860.5	585.4	172.9	720.6	2329.4
1980	673.9	459.6	170.3	455.9	1759.7
1981	675.1	470.2	172.3	466.5	1784.0
1982	672.7	481.2	173.4	476.5	1803.8
1983	669.4	484.6	174.1	483.0	1811.1
1984	667.7	491.3	175.9	486.4	1821.3
1985	666.6	495.4	177.9	488.6	1828.5
1986	666.7	504.4	178.6	494.7	1844.3
1987	668.4	510.2	177.0	495.2	1853.3
1988	660.4	515.6	177.3	466.9	1820.1
1989	658.3	515.9	176.9	462.8	1813.9
1990	623.5	501.5	176.5	430.8	1732.4
1991	602.6	494.5	174.1	409.4	1680.5
1992	511.3	467.5	165.2	375.7	1519.8
1993	480.3	438.9	160.1	356.1	1435.4
1994	451.4	401.7	156.3	341.3	1350.7
1995	421.5	370.9	156.8	317.9	1267.2
1996	405.1	354.1	157.0	308.6	1224.9
1997	408.1	334.2	156.2	299.4	1197.9
1998	402.0	322.1	155.3	291.6	1170.9
1999	396.2	314.3	154.8	286.2	1151.4
2000	402.2	314.1	155.0	282.3	1153.6

SOURCE: Office of the Assistant Secretary of Defense (Force Management Policy) (2000).

TABLE 2-2 Reserve Component Enlisted Strength

Fiscal Year	Army National Guard	U.S. Army Reserve	U.S. Navy Reserve	U.S. Marine Corps Reserve	Air National Guard	U.S. Air Force Reserve	Total DoD
1980	329,298	169,165	70,010	33,002	84,382	45,954	731,811
1981	350,645	188,103	72,608	34,559	85,915	52,686	784,516
1982	367,214	208,617	75,674	37,104	88,140	50,553	827,302
1983	375,500	216,218	88,474	39,005	89,500	52,810	861,507
1984	392,412	222,188	98,187	37,444	92,178	55,340	897,749
1985	397,612	238,220	106,529	38,204	96,361	59,599	936,525
1986	402,628	253,070	116,640	38,123	99,231	62,505	972,197
1987	406,487	255,291	121,938	38,721	100,827	63,855	987,119
1988	406,966	253,467	121,653	39,930	101,261	65,567	988,844
1989	406,848	256,872	122,537	39,948	101,980	66,126	994,311
1990	394,060	248,326	123,117	40,903	103,637	66,566	976,609
1991	395,988	249,626	123,727	41,472	103,670	67,603	982,086
1992	378,904	245,135	115,341	38,748	104,758	65,806	948,692
1993	363,263	219,610	105,254	38,092	102,920	64,720	893,859
1994	351,390	206,849	86,300	36,860	99,711	63,411	844,521
1995	331,559	191,558	79,827	36,292	96,305	62,144	797,685
1996	328,141	179,967	77,376	37,256	97,153	57,615	777,508
1997	329,288	168,596	75,373	37,254	96,713	56,068	763,295
1998	323,150	161,286	73,490	36,620	94,861	56,032	745,439
1999	319,161	161,930	69,999	35,947	92,424	55,557	735,018
2000	315,645	165,053	67,999	35,699	93,019	55,676	733,091

SOURCE: Office of the Assistant Secretary of Defense (Force Management Policy) (2000).

portions of their budget if they wished to increase their end strength. In addition, the extra authorization was not included for FY 2003—adding people in one year and having to go back to original end strength the next year is not easily done, even for a few thousand people.

After six months of the war on terrorism, along with all the other military commitments, the *Washington Times* on April 10, 2002, reported the Services as needing as many as 51,400 more troops (an increase of about 5 percent to current end strength). But the cost of personnel is very high and already a significant proportion of the DoD budget. The article states, "it would cost $40,000 [per year] to add each enlisted person, in addition to the $10,000 to recruit that person."

Note that the QDR looks for ways other than just increased force structure to mitigate the operational risk, to include possible "changes in capabilities, concepts of operations, and organizational designs" (p. 61). Secretary Rumsfeld is on record as saying that he is not yet ready to increase end strength—he wants to look for other ways to address these issues. One way is through reductions in current commitments; another is to reduce the numbers of personnel required to accomplish certain functions. A good example is the Navy's DD-X warship, being built from the ground up to require a crew of only 95, compared with the 300 needed today.

At the same time that force levels were dropping, the average age of Service members was increasing. For example, between 1980 and 1997, the average age of members of the enlisted force rose from 25 to 27 (U.S. Department of Defense, 2002). The overall DoD force profile reveals that about half the enlisted force has less than 6 years of service, 45 percent has between 6 and 19 years of service, and the remaining 4 percent has 20 or more years of service. According to the Office of the Assistant Secretary of Defense (2001), in 2000, the force was 18 percent female as contrasted with 51 percent female in the general civilian population ages 18–24. The ethnic distribution of the force was 62 percent white, 20 percent black, and 11 percent Hispanic as contrasted with 65 percent, 14 percent, and 15 percent, respectively, in the general civilian population ages 18–24.

Even given the uncertainty of future demand and concerns about the adequacy of current end strength numbers, there is no compelling evidence that the requirement for end strength will change radically in the future.

Recruiting Results

The Services have generally responded to decreasing force sizes by lowering the number of new recruits that they acquired in any given year. Some discharge policies were also liberalized during times of downsizing. While the Services required over 388,000 non-prior-service and prior-service recruits in 1980, that number had dropped to 175,000 recruits in 1995—a 55 percent reduction. As a result of these lower numbers of new recruits needed, Service-recruiting commands were generally very successful in meeting their goals during the late 1980s and early 1990s. As Table 2-3 shows, however, some of the Services failed to meet their goals once force sizes leveled off and the numbers of new recruits needed to meet end strength started increasing.

What the table does not show, however, is the relative difficulty of achieving these goals. For example, even if recruiting is very easy in any given year, the Services would not substantially exceed their goals because of congressional limits on their end strength. Thus, one cannot tell from the table alone whether any Service struggled to meet its goal on the last day of the year or was certain it would meet its goal early in the year.

TABLE 2-3 Active Enlisted Recruiting Goals and Success by Service

	Army		Navy		Marine Corps		Air Force		Total DoD		
Fiscal Year	Goal	Actual	Goal	Actual	Goal	Actual	Goal	Actual	Goal	Actual	%
1980	172,800	173,228	97,627	97,678	43,684	44,281	74,674	74,674	388,785	389,861	100
1981	136,800	137,916	101,904	104,312	42,584	43,010	81,044	81,044	362,332	366,282	101
1982	125,100	130,198	81,922	92,784	40,558	40,141	73,620	73,620	321,200	336,743	105
1983	144,500	145,287	82,790	82,790	37,690	39,057	63,591	63,591	328,571	330,725	101
1984	141,757	142,266	82,907	82,907	38,665	42,205	61,079	61,079	324,408	328,457	101
1985	125,300	125,443	87,592	87,592	36,536	36,620	67,021	67,021	316,449	316,676	100
1986	135,250	135,530	94,878	94,878	36,682	36,763	64,400	66,379	331,210	333,550	101
1987	132,000	133,016	92,909	92,909	34,713	34,872	55,000	56,029	314,622	316,826	101
1988	115,000	115,386	93,939	93,939	35,911	35,965	41,200	41,500	286,050	286,790	100
1989	119,875	120,535	94,286	95,186	34,130	34,424	43,730	43,751	292,021	293,896	101
1990	87,000	89,620	72,402	72,846	33,521	33,600	36,249	36,249	229,172	232,315	101
1991	78,241	78,241	68,311	68,311	30,015	30,059	30,006	30,006	206,573	206,617	100
1992	75,000	77,583	58,208	58,208	31,851	31,852	35,109	35,109	200,168	202,752	101
1993	76,900	77,563	63,073	63,073	34,802	34,776	31,515	31,515	206,290	206,927	100
1994	68,000	68,039	53,964	53,982	32,056	32,056	30,000	30,019	184,020	184,096	100
1995	62,929	62,929	48,637	48,637	32,346	33,217	30,894	31,000	174,806	175,783	101
1996	73,400	73,418	48,206	48,206	33,173	33,496	30,867	30,867	185,646	185,987	100
1997	82,000	82,088	50,135	50,135	34,512	34,548	30,310	30,310	196,957	197,081	100
1998	72,550	71,733	55,321	48,429	34,244	34,285	30,194	31,685	192,309	186,132	97
1999	74,500	68,209	52,524	52,595	33,668	33,703	34,400	32,673	195,092	187,180	96
2000	80,000	80,113	55,000	55,147	32,417	32,440	34,600	35,217	202,017	202,917	100
2001	75,800	75,855	53,520	53,690	31,404	31,429	34,600	35,381	195,324	196,355	101

SOURCE: Office of the Assistant Secretary of Defense (Force Management Policy) (2001).

What is shown by the data, however, is that fiscal years 1998 and 1999 were very difficult for many Services since they did not meet their goals.

Force Quality

The size of the force (end strength) is specified by Congress, in response to many factors, as described earlier. Service effectiveness, however, depends on more than just end strength; the education and aptitude of Service personnel has a direct relationship to mission performance (Armor and Roll, 1994). As noted in the QDR (U.S. Department of Defense, 2001:9):

> The Department of Defense must recruit, train, and retain people with the broad skills and good judgment needed to address the dynamic challenges of the 21st century. Having the right kinds of imaginative, highly motivated military and civilian personnel, at all levels, is the essential prerequisite for achieving success. Advanced technology and new operational concepts cannot be fully exploited unless the Department has highly qualified and motivated enlisted personnel and officers who not only can operate these highly technical systems, but also can lead effectively in the highly complex military environment of the future.[2]

Typically, two measures are used to assess the overall education and aptitude of the enlisted force: cognitive ability (as measured by scores on the Armed Forces Qualification Test [AFQT]) and educational attainment. DoD sets minimum levels required in both areas for new recruits; the Services can—and often do—set their own enlistment standards that can exceed the minimums established by DoD (See Chapter 4, this volume).

As measured by either AFQT or educational attainment levels, the Services today have a highly qualified enlisted force. Of the 1.15 million active component enlisted members in the Services at the end of FY 2000, 96 percent held at least a high school diploma, while an additional 3.3 percent held an alternative high school credential (e.g., a GED certificate). The combined 99.3 percent of enlisted members holding some type of high school credential compares very favorably with the civilian population rate (ages 18–44) of 88.6 percent (Office of the Assistant Secretary of Defense, 1999:Table 3-7).

Because the AFQT is used primarily as a qualifying standard for enlistment, DoD does not routinely report the AFQT levels of the entire enlisted force. DoD does report, however, the AFQT levels for each acces-

[2]The skills needed to operate this technology are anticipated by a number of studies, including Soldier 21 and Sailor 21.

sion cohort. Although a more complete discussion of military service enlistment standards—including AFQT—is presented in Chapter 4, Table 2-4 clearly shows that qualification levels for entering cohorts generally increased until 1992 then began a slight decline. Even considering the more recent decline, however, current education achievement and aptitude levels compare very favorably with those of the civilian population ages 18–23. This is by design, since the military requirement is to recruit 60 percent from the population scoring in the top 50 percent of the AFQT.

Although the force is highly qualified today, as noted above, there has been speculation that future military service roles and missions may require even higher qualifications. The heavy emphasis on technical skills in the military of the future adds to this speculation. Alternatively, recent RAND research concluded that future military jobs would require no *significant* changes in abilities from those observed in today's military;

TABLE 2-4 Active Enlisted Accessions by AFQT Category (percentage)

Year	I-IIIA	IIIB	IV	Unknown
1980	48.7	42.2	8.7	0.4
1981	47.3	31.1	21.3	0.3
1982	52.0	32.4	15.2	0.4
1983	57.5	31.4	10.7	0.4
1984	58.2	32.0	9.5	0.4
1985	60.2	31.9	7.5	0.3
1986	62.4	32.5	4.9	0.2
1987	67.0	28.2	4.7	0.2
1988	66.5	28.1	4.9	0.5
1989	64.4	28.4	6.4	0.8
1990	68.0	28.3	3.1	0.7
1991	72.1	26.5	0.5	0.9
1992	74.9	24.5	0.2	0.4
1993	71.1	27.7	0.8	0.4
1994	70.6	28.3	0.7	0.4
1995	70.1	28.8	0.7	0.5
1996	68.5	30.0	0.7	0.7
1997	68.3	30.3	1.0	0.5
1998	67.5	31.1	1.0	0.4
1999	65.1	33.0	1.4	0.6
2000	65.8	33.1	0.7	0.4

Civilian Population Ages 18–23

Year	I-IIIA	IIIB	IV	Other
1980	51.4	18.28	21.03	9.28

SOURCE: Office of the Assistant Secretary of Defense (Force Management Policy) (2000).

however, in a few specific military occupations, the abilities of the current workforce might not be sufficient for future jobs in those occupations (Levy et al., 2001). The answer may be in evolving changes to military training for basically qualified recruits.

Skills

One question that arises when attempting to determine the abilities required by future military jobs is whether (and how) the distribution of jobs has changed in the past. Gribben (2001) directly addressed this question with regard to military occupational categories. Table 2-5 displays the results of her analysis. Even given the dramatic drop in numbers of Service personnel over the last 25 years, there has been relatively little change in the distribution of military personnel in these occupational areas. Gribben notes that the "dramatic increases in technology-oriented positions in the civilian workplace during the last 25 years are not very closely tied to occupational area trends in the military during the same time period."

Interviews with senior personnel managers in DoD and the Services confirm that no special skills are anticipated to be a prerequisite for military service—given young men and women who meet the minimum qualifications, they can and will be trained to meet changing military requirements. This emphasizes the importance of the skills training mission by the Services and requires periodic evaluation to ensure that training fills the gaps between the abilities that recruits bring to the Services

TABLE 2-5 Enlisted Member Occupational Distribution by Year (percentage)

Code	Title	1976	1980	1985	1990	1995	2000
0	Infantry specialists	12.9	13.9	14.5	17.1	17.1	16.9
1	Electronic equipment repairers	9.2	8.8	9.5	9.9	9.8	9.7
2	Communications and intelligence specialists	8.2	8.3	9.5	9.8	8.7	8.8
3	Health care specialists	4.4	4.4	4.8	5.6	6.7	6.7
4	Other technicians	2.2	2.1	2.4	2.3	2.5	3.0
5	Support and administration	15.5	15.3	15.8	15.5	16.1	16.1
6	Electical/mechanical equipment repairers	19.0	19.8	20.1	20.1	19.5	19.7
7	Craftsworkers	4.2	4.0	4.2	4.0	4.0	3.5
8	Service and supply handlers	10.3	8.9	9.4	8.9	8.9	8.4
9	Nonoccupational	14.1	14.4	9.8	6.9	6.6	7.2

SOURCE: Gribben (2001).

and what is required. The military has no control over the subject matter taught in the high schools and thus there are no guarantees that high school graduates will have the necessary skills and motivation to perform military jobs.

At the Army 2010 Conference: Future Soldiers and the Quality Imperative, (May–June 1995), discussions included the projection that future soldiers "will require better problem solving and decision making skills that will enable them to go beyond the application of rote procedures to 'think outside the box'" (paper presented by Donald Smith, HumRRO). At the same conference, Michael G. Rumsey of the U.S. Army Research Institute reported on a project conducted in 1994. Rumsey and five other behavioral scientists prioritized 15 attributes based on the Army's Project A job analyses. They overwhelmingly viewed cognitive ability to be the most important attribute for future success (Rumsey, 1995).

A more recent and more detailed study on the characteristics of 21st century Army soldiers followed up Rumsey's work and confirmed the finding that cognitive aptitude is of the highest priority for the enlisted soldier of the future (Ford et al., 1999). Other high-priority knowledge, skill, and ability areas identified by Ford and her colleagues include conscientiousness/dependability, selfless service orientation, and a good working memory capacity. These represent a slight departure from skill, knowledge, and ability ratings for the force in the 1990s, for which conscientiousness/dependability was ranked first, followed by general cognitive aptitude and the need to achieve. The methodology for this study included evaluation and ranking of job characteristics and job skill, knowledge, and aptitude requirements for the next 10 and 25 years, respectively, by three panels—one composed of subject matter experts (active Army senior noncommissioned officers representing 21 military occupational specialties), one composed of psychologists who worked from the lists prepared by the subject matter experts, and one composed of project staff who worked from the outputs of the previous two panels. The baseline lists were drawn from earlier Army projects and visits of Army facilities.

The Navy Personnel Research and Development Center published *Sailor 21: A Research Vision to Attract, Retain, and Utilize the 21st Century Sailor* on December 14, 1998. A key framing concept for this study is that these new technologies may require fewer people, but they must be more capable, faster, and able to perform a much broader range of tasks (Navy Personnel Research and Development Center, 1998).

Regarding the Marine Corps, then Commandant General Charles C. Krulak wrote in the January 1999 edition of the *Marine Corps Gazette*, "In many cases, the individual Marine will be the most conspicuous symbol of American foreign policy and will potentially influence not only the immediate tactical situation, but the operational and strategic level as well." Hence the term, "the strategic corporal." In order to meet these needs, timely and responsive changes to training are necessary. New systems are especially problematic as schedules slip and funding is used for other priorities. To the extent possible, training changes should be anticipatory, especially for new systems.

Operations in Afghanistan in 2001–2002 have highlighted the challenges awaiting military men and women of the future. The environment was much different from what they had trained for. Soldiers adapted, adopted, and prevailed. For example, they traveled on horseback while using 21st century technology to designate targets.

There are two reasons why the emerging technology-oriented military is unlikely to require higher-level technical skills in the aggregate from its recruits at the point of entry into military service. The first is the observation that with increased technologies, the general level of computer and other forms of technical literacy in the general population has been rising concurrently. We live in a computer ambience that readily takes care of most basic skill requirements. This is particularly true of the nation's youth. Rather than a need, for example, for massive new school programs to teach computing to young people, youth have typically taught their elders, including parents and teachers. A second reason the military is unlikely to need to ratchet up its average level of technical skills requirements follows from what has been learned about civilian jobs that underwent technological change and what changes in the labor force were likely to be due to technology. Although there have been substantial changes in the education and skill requirements for particular occupations, industries, and sectors of the economy, skills upgrading in some areas has been accompanied by downgrading in others, leaving only a slow, upward drift in skill requirements in the aggregate economy over time (Cyert and Mowery, 1987; Handel, 2000). Overall, there was not much net change. The notion that high-tech jobs require new batteries of sophisticated skills is not supported by the evidence.

In sum, observational and research studies of civilian jobs and workplaces that have gone through technological evolutions suggest that skill levels increase in some jobs but decrease in others. On balance, the civilian labor force hasn't been affected very much by technological change, and it is unlikely that the military experience will be any different.

RETENTION AND RECRUITING

As mentioned earlier in this chapter, fully staffed military services anticipate personnel turnover, and unit readiness is predicated, in part, on maintaining acceptable rates of retention. Retention is negatively affected operationally by attrition. Attrition rates that exceed expectations translate into heightened recruitment needs and goals.

Retention relates to recruitment in five ways. First, retained personnel are the largest numerical component of a fully manned service. As attrition rates increase, the need for replacements increases, which puts pressure on recruitment capabilities. Second, to the extent that attrition rates are differentiated across occupational specialties within a Service, unplanned attrition further compromises mission readiness. Retention is more than maintaining total head counts within a Service. The mix of personnel is also critical to readiness and mission accomplishment. Retaining an appropriate force mix has become increasingly challenging.

Third, skill levels must be maintained at appropriate levels. The Services "grow their own," and newly minted recruits cannot immediately replace seasoned veterans. For example, press reports indicated that in the late stages of the most recent Bosnian conflict, advanced military helicopters were on site and combat ready, but the equipment remained grounded for lack of combat-experienced pilots. Recruitment is not a quick fix for retention shortfalls.

Fourth, the problems associated with within-Service supply-demand imbalances may be generalized across the Services. Each Service is part of a team with the other Services, and differential rates of attrition among the Services may also compromise readiness. Differential rates of attrition by military occupational specialties coupled with supply-demand imbalances across the Services dictate the qualifications that must be sought in recruits in a given period.

A similar problem exists in the relationship between the active-duty forces and the Reserves. The increasing operational tempo in recent years has necessitated the increased participation of the Reserves. From November 1994 through December 2000, the use of Reserve forces increased from 8 to 12 million man-days per year (U.S. Department of Defense, 2001:58). With the Reserves primarily responsible for a key component of the highest defense priority, homeland defense, increased attention must be paid to Reserve recruiting.

Fifth, the expressed quality of military experience contributes to the attractiveness of military service for potential recruits. Recruitment goals are affected by policy changes, such as the planned 1991–1997 downsizing of the military, and the mix of occupational specialties changes in response to, for example, technological innovations and applications. However, assuming reasonably stable personnel demands over time, the extent to which the Services can maintain acceptable retention rates by enhancing the perceived quality of the military experience may effectively lower numerical recruitment requirements and build recruitment capabilities.

There's an old saying that when you recruit the military member, you retain the family. The QDR specifically focuses on quality of life issues as "critical to retaining a Service member and his or her family. Recent surveys conducted by the Department indicate that the two primary reasons that Service members leave or consider leaving are basic pay and family separation" (U.S. Department of Defense, 2001:9). Also, when the U.S. economy is strong, military members (and potential military members) have more alternatives. When jobs in the economy are hard to find, more young men and women view the military as a viable alternative for employment (Warner et al., 2001).

In sum, recruitment needs are in part a function of service retention capabilities: each Service's ability to retain personnel within acceptable rates of attrition and voluntary losses at the level required in each military occupational specialty (MOS). Attrition rates that exceed expectations translate into additional, unexpected recruitment needs.

First-Term Attrition

Table 2-6 presents aggregate rates of attrition for the military at 6, 12, 24, and 36 months over the 15-year period, 1985–1999. The size of the entry cohorts ranged from a high of 315,000 in 1986 to about half that many, 168,000, in 1995. The size of the most recent cohorts was about 10 percent higher than the smallest cohorts since 1985, averaging about 185,000—well below the size of the entry cohorts at the beginning of the period. Following the 1991–1997 drawdown, recruitment goals have remained modest, which suggests that recruitment difficulties cannot be attributed to increasing recruitment goals.

From 1985 to 1999, aggregate attrition in the military rose systematically over each period for which losses were calculated—6, 12, 24, and 36 months. Attrition at 6 months rose from about 11 percent to about 15 percent; at 12 months from about 14 to about 19 percent; at 24 months from about 22 to about 26 percent; and at 36 months from about 28 to about 31 percent. Taking both cohort size and attrition rates into account indicates an inverse relationship between the two.

The incremental increases in the cumulative attrition percentages remained relatively constant over the 15-year period. Attrition at the end of 6 months ranged from 11 to 16 percent. Losses over the period 6 to 12 months contributed about another 4 percent. Losses over 12 to 24 months contributed about 8 percent more, and losses over 24 to 36 months contributed about another 6 percent. In sum, the largest aggregate military attrition occurred within the first 6 months. About half of the cumulative attrition occurred by 12 months and, beginning about 1990, the period of

TABLE 2-6 Aggregate Military Attrition at 6, 12, 24, and 36 Months, 1985–1999

Enlisted during FY	1985	1986	1987	1988	1989	1990	1991	1992	1993	1994	1995	1996	1997	1998	1999
Percentage lost at 6 mos.	11	12	11	12	11	11	12	12	15	15	15	14	14	16	15
Percentage lost at 12 mos.	14	15	15	15	16	15	16	16	18	19	19	19	18	20	19
Percentage lost at 24 mos.	22	23	22	22	23	23	24	23	26	26	27	25	26	26	—
Percentage lost at 36 mos.	28	29	28	28	30	29	30	29	32	33	33	31	31	—	—
Size of entry cohort	301K	315K	297K	273K	278K	224K	201K	201K	200K	176K	168K	180K	189K	180K	184K

SOURCE: Correspondence from Vice Admiral P.A. Tracey, U.S. Department of Defense, Washington, DC (2001).

the drawdown, the proportion of cumulative attrition occurring by 12 months edged even higher.

There is preliminary indication that a turnaround in attrition rates may be taking place. All Services met their fiscal year 2001 recruiting goals. Furthermore, all Service branches were meeting their cumulative goals for retaining first-term enlisted personnel. However, while the Marine Corps and the Army were exceeding goals for second-term and career enlisted personnel, the Navy and the Air Force fell short.

Attrition rates for the separate Services over the 15-year period follow the general pattern of the aggregate rates, as expected, although there are exceptions and countertrends (Table 2-7).

Thus, attrition rises over the measurement period for each Service. The incremental increases in cumulative attrition at 6, 12, 24, and 36 months for each Service remained relatively constant. The largest attrition rates for the Services occurred within the first 6 months. As noted below, there are some differences by Service in the proportion of total attrition that occurred by 12 months. Finally, beginning about 1990, the proportion of cumulative attrition occurring by 12 months increased across the Services except for the Marine Corps. (This could be partially explained by differing Service policies during the drawdown, some having very liberal discharge policies in order to reduce their numbers.) It can further be seen from Table 2-7 that there are some differences in attrition patterns among the Services.

The Services met their aggregate and separate service recruitment goals for the year 2001. Nonetheless, the DoD's April Readiness Report (Grossman, 2001:S1) indicated that the Navy and the Air Force experienced some difficulty in meeting recruitment targets for specific military occupation specialties. Thus, the immediate recruitment and retention challenge for the Services may no longer be meeting end strength goals but manning selected career areas to ensure that shortfalls in critical areas do not compromise unit readiness. The Readiness Report (Grossman, 2001:S1) noted several enlisted retention challenges by Service. For example, the Army needed soldiers with skills in specialized languages, signal communications, information technology, and weapon systems maintenance. The Navy needed some enlisted personnel with high-tech ratings. The Marines needed technical specialists, such as intelligence, data communications experts, and air command and control technicians. The Air Force needed air traffic controllers and communications/computer system controllers with 8 to 10 years of military service.

As the military's warrior mission evolves into an efficient, effective, technologically sophisticated, fast-strike force, its susceptibility to paralysis for lack of capacity to man critical occupational specialties becomes increasingly apparent. Power outages for human resource reasons are no

TABLE 2-7 Service Attrition at 6, 12, 24, and 36 Months, 1985–1999

Enlisted during FY	1985	1986	1987	1988	1989	1990	1991	1992	1993	1994	1995	1996	1997	1998	1999
Army															
Percentage lost at 6 months	10	10	9	10	10	11	13	13	15	16	15	15	12	18	16
Percentage lost at 12 months	14	14	13	14	14	14	17	17	19	20	19	20	16	22	20
Percentage lost at 24 months	23	23	22	22	23	24	26	25	28	28	28	26	25	30	—
Percentage lost at 36 months	29	29	29	29	31	32	33	32	35	36	33	33	32	—	—
Navy															
Percentage lost at 6 months	11	13	13	14	13	10	10	13	16	16	16	14	16	15	17
Percentage lost at 12 months	15	18	17	19	18	15	14	17	20	21	21	21	22	20	22
Percentage lost at 24 months	23	26	25	26	25	23	23	25	28	29	30	29	29	28	—
Percentage lost at 36 months	29	32	30	30	31	30	28	30	33	35	36	34	34	—	—
Air Force															
Percentage lost at 6 months	9	11	10	9	9	10	11	9	12	12	13	12	13	12	12
Percentage lost at 12 months	12	13	12	11	12	12	13	12	14	14	16	15	15	14	15
Percentage lost at 24 months	17	18	17	16	17	19	20	18	20	20	22	21	21	20	—
Percentage lost at 36 months	22	22	22	20	22	25	25	23	26	26	27	26	25	—	—
Marine Corps															
Percentage lost at 6 months	14	16	13	13	14	16	14	13	14	13	15	13	15	16	13
Percentage lost at 12 months	18	20	17	16	18	20	19	17	18	18	19	17	19	19	17
Percentage lost at 24 months	25	26	24	22	23	26	25	23	24	24	25	23	25	24	—
Percentage lost at 36 months	32	34	31	28	29	33	31	29	29	30	31	28	29	—	—

SOURCE: Correspondence from Vice Admiral P.A. Tracey, U.S. Department of Defense, Washington, DC (2001).

less crippling than those caused by defects in material and machines. Lest the lesson be lost: the Apache helicopters grounded in Bosnia for lack of experienced pilots portend an increasingly critical dimension to the military recruitment/retention challenge.

Why Members of the Enlisted Force Leave the Service

Although there are indications that the Services may currently be experiencing a welcomed resurgence in retention and recruitment rates, in recent years the military has had difficulty meeting its retention goals. This section examines the reasons for attrition and voluntary losses—why military members leave the Services—from several perspectives. With strong caveats about the quality of the available data, we examine the Service reports on the reasons personnel leave the military as indicated by broad separation codes. We also peruse Service career intention surveys, which monitor the attitudes and plans of troops under contract, and we review exit interviews with personnel separating from service for the years 1999, 2000, and 2001.

First-Term

The Services employ eight broad separation codes as reasons personnel leave the military:

Behavior	Failure to meet performance standards, misconduct, in lieu of trial by courts martial, drug/alcohol rehabilitation failure, etc.
Medical	Disability/injury, failed medical physical procurement standards, etc.
Hardship	Hardship, parenthood, custody
Other	Personality disorder, conscientious objector, fraudulent enlistment, defective enlistment agreement, insufficient retainability, military security program, etc.
Homosexual	As stated
Weight	Weight control failure
Unknown	As stated
Pregnancy	As stated

Note that each Service can interpret these codes somewhat differently, which may contribute to some of the differences.

Table 2-8 reports percentages of first-term attrition for men and women by service over the years 1993–1995. Overwhelmingly, men tend to leave the Services for behavior reasons defined as "failure to meet

TABLE 2-8 Reasons for First Term Attritions, Men and Women by Service, 1993–1995 (percentage)

	Army		Navy		Marine		Air Force	
	Men	Women	Men	Women	Men	Women	Men	Women
Behavior	82	48	42	11	60	34	47	21
Medical	12	17	10	16	15	18	15	22
Hardship	0	7	0	4	0	8	3	3
Other	2	2	48	44	21	6	23	24
Homosexual	0	0	0	0	0	0	2	2
Weight	2	1	0	0	5	4	1	1
Unknown	2	0	0	0	0	0	10	10
Pregnancy	0	25	0	24	0	31	0	17

SOURCE: Correspondence from Vice Admiral P.A. Tracey, U. S. Department of Defense, Washington, DC (2000).

performance standards, misconduct, in lieu of trial by courts martial, drugs/alcohol rehab failure, etc." Attrition for behavior reasons is highest in the Army, 82 percent. The rate for the Marines is 60 percent, followed by the Air Force at 47 percent. The reason for most separations in the Navy is "other," defined as "personality disorder, conscientious objector, fraudulent enlistment, defective enlistment agreement, insufficient retainability, military security program, etc." "Other" is the second-highest percentage reason that men separate from the Marines and the Air Force, which report rates of 21 and 23 percent, respectively. The second-highest rate for separation from the Navy is behavior, 42 percent. Medical reasons are the third-highest percentage reason. Percentages for other reasons for separation vary from 10 to 15 percent across the Services. Except for "unknown" reasons reported in the Air Force (10 percent), no other reason accounts for more than 5 percent of men's separations.

Attrition rates for women are more varied. Most women who leave the Army do so for behavior reasons, as do their male colleagues, although the percentage is sharply lower for women, 48 compared with 82 percent. The same pattern holds for women who leave the Marines. Most separate for behavior reasons, although the percentage is sharply lower than it is for men, 34 compared with 60 percent. Women who leave the Navy do so primarily for "other" reasons. Women who separate from the Air Force do so for the most varied reasons. One of four women who leave the Army and the Navy leave for pregnancy, as do nearly one of three who leave the Marines. In summary, men tend to leave the military for behavior reasons, followed by "other" and medical reasons. Women also leave for behavior reasons, but the pattern of separations for women is more varied, less concentrated on behavior, and includes pregnancy.

Separation After the First Term

Each of the Services conducts career intention or exit surveys, which they use to monitor the career intent of troops under contract. These include

- Army: Sample Survey of Military Personnel (SSMP);
- Navy: Quality of Life Domain Survey (NQLDS);
- Air Force: Air Force New Directions Survey;
- Marine Corps: Marine Corps Retention Survey and Marine Corps Exit Survey.

Data on intentions and attitudes toward the Service can be extremely useful in identifying broad domains of factors that enlistees report as influencing either their career intentions or their decision to leave. Classes of variables examined in one or another survey include the attitudes of spouses and families toward staying in the Service, satisfaction with job characteristics, compensation, the availability of civilian jobs, promotion opportunities, pride in the Service, time away from home, and other quality-of-life issues. However, the data currently collected from these surveys cannot support any strong general conclusions regarding which factors have the largest effect on intentions or on decisions, due to the low response rates and differing data-gathering strategies across the Services.

Our review of the survey documentation led the committee to a number of observations: Better data are needed to understand why some individuals choose to separate from the Services, while others reenlist; the quality of the Service retention and exit surveys is suspect; and technical characteristics of the data are often not reported. Furthermore, the surveys tend to focus on interviewing large numbers of respondents rather than selecting smaller representative samples. Due to low response rates, the data that are collected may not adequately represent the attitudes of those reaching the end of a particular enlistment term, and inappropriate conclusions may be drawn. In addition, tests for bias, reliability, and validity are not used, and analyses are typically limited to reporting simple percentages. While in our view a systematic approach to investigating retention would be very valuable to the Services, a detailed consideration of this topic is beyond the scope of this study.

Hidden Costs of Attrition

In addition to the immediate pressures that unacceptable attrition contributes to recruitment goals, there are hidden costs. Four are particularly noteworthy. The first is that attrition in labor-intensive industries including the military is expensive. High attrition costs the taxpayer and requires reallocation of scarce resources within the Services. Two studies provide rough estimates of the costs per attrition. Bowman (2001) reports that for each soldier who leaves, the Army has to spend $31,000 to train a new one. In addition, in 1996 the General Accounting Office reported that the investment in each enlistee separating in the first 6 months was $23,000 per enlistee (U.S. General Accounting Office, 1997).

A second hidden cost issues from the need to build unit cohesion within the military, to ensure such intangibles as collective memory and a shared culture that enhances unit performance. The carriers and articulators of unit cultures are those retained in active service and, for that reason, too, maintaining a critical mass of active-duty troops is a necessity.

Another impact is closely related. Attrition rates may be interpreted as an unobtrusive measure of quality of life in the military, which is affected by long-standing personnel policies. The litany of service personnel grievances includes low pay; decaying infrastructure, including poor housing and dated base facilities; outdated pension and retirement policies; family-unfriendly relocation policies and inadequate health care services; and such unglamorous workplace deficits as deficient spare parts inventories (Dao, 2001). To the extent that these affect morale and esprit de corps, personnel efficiencies and effectiveness suffer.

Unacceptable attrition rates also affect the care and nurturance of the national recruitment environment. Recruits on delayed entry status, trainees, and soldiers return to civilian life with stories to tell, be they positive or negative. Experiences related by active personnel and personnel separating from service define the viability of military service for the nation's youth. The expressed quality of the military experience articulated by those who have "been there, done that" frames the attractiveness and colors the perceptions of the military career option.

The costs of attrition are broader and deeper than merely adding pressure to the military recruitment apparatus—the most significant being in the area of operational readiness. Because of the interrelated aspects of recruiting and retention, all efforts to more fully integrate them into complementary efforts through the planning and budgeting process could hold significant benefits.

SUMMARY

We support the DoD efforts defined by the QDR to "institute programs to . . . encourage talent to enter and stay in the military and civilian service. . . . The Department must forge a new compact with its warfighters and those who support them—one that honors their service, understands their needs, and encourages them to make national defense a lifelong career" (U.S. Department of Defense, 2001:50).

In spite of changing missions and requirements for the military, indications are that the size of the military will not increase significantly over the next 20 years. Although as priorities change, the force mix among military occupational specialties and among the active-duty, Reserve and National Guard structures will also change with increasing emphasis on the Reserves and the National Guard because of the newly emphasized responsibilities associated with homeland defense. The recruitment and retention of both the Reserve force and the National Guard must receive appropriate attention.

Given current trends, it seems extremely doubtful that the force size will decrease, easing pressures on recruiting. In addition, physical and moral requirements for military service are unlikely to change over the next 20 years. The Services are currently accessing recruits who have sufficient aptitude and can be trained to perform military tasks adequately. Recruits satisfying current quality levels can be trained to meet future demands.

There have been few major changes in the occupational distribution of first-term personnel in the past 10 years, but future military missions coupled with advances in technology are expected to require military personnel to make greater use of technology. Technological changes will make some jobs in the future easier and others more difficult, but overall minimum aptitude requirements are unlikely to change much over the next 20 years.

Changes in national security strategy and national military strategy have the potential to influence the size and shape of the Services. The current administration's emphasis on transformation, as well as recent new emphasis on homeland security, could also argue for changes in force size and shape, to include balancing the forces between active-duty, reserve, and National Guard forces. The QDR Report provides some of these decisions and also requires studies to provide the rationale for future decisions that could affect military force structure. In addition, changing technology can affect the numbers and types of people needed in the military of the future.

It is the required military force structure and the difference between that and the current population less anticipated annual losses that determine the numbers required to be recruited. Controlling losses can mitigate recruiting requirements to some extent.

Assuming the validity and the urgency of attrition rates and retention expectations, the task at hand is to bring knowledge and information to bear that informs policies and practices designed to enhance retention and to expand the recruitment pool. In fact, the QDR specifically states, "DoD can no longer solely rely on such 'lagging' indicators as retention and recruiting rates to detect personnel problems; by the time those indicators highlight a problem, it is too late" (U.S. Department of Defense, 2001:59).

3

Demographic Trends

Recruitment to the armed forces is constrained by the broader social environment in which young people grow up. A key part of that environment is the demography of the youth population, which changes substantially over time as a result of long-term trends in fertility and the activities that structure the lives of young people. Demographic processes constrain armed forces recruitment in a number of ways. First, secular trends (i.e., trends over time) in fertility determine the sizes of future cohorts of youth and thus affect the ease with which recruiting targets can be met. The absolute sizes of cohorts of women of childbearing age and the rates at which those women bear children determine the numbers of children in each year. These numbers are also augmented by immigration of children and teenagers and depleted by mortality and emigration of these groups, but these effects on the youth population are small relative to fertility.

Second, trends and levels of *differential* fertility affect each youth cohorts' average propensity to enlist in the armed forces. The family environments in which young persons are raised affect their propensities to enlist or to engage in other activities that may complement or substitute for military service. These environments are affected by the socioeconomic characteristics of families, their size and structure, and other aspects of parents' social backgrounds. Adults from different kinds of families bear children at different rates, with these differentials often varying over time. Other things being equal, therefore, fluctuations in these patterns of differential fertility will affect the average enlistment propensities of later cohorts of offspring when they reach the appropriate age for military service.

Third, demography is concerned with the activities of young persons themselves, without regard to the characteristics of the families in which they were raised. The propensity to enlist in the military is connected in a complex way to the other activities and opportunities faced by youths. These include schooling, work, marriage, childbearing, living arrangements, leisure, and possibly crime and incarceration. As these activities and opportunities have changed over time, so has the attractiveness of military service.

As we consider the impact of demographic trends on recruitment into the armed forces, we take two perspectives on the youth population. One perspective, which we term "young adult demography," focuses on recent trends in the activities and statuses of the youth population. This provides background for understanding recent trends in enlistment behavior. The second perspective, which we term "child demography," looks at recent trends in fertility, including overall levels as well as differentials across social groups. These trends, which correspond to the first two ways in which demographic trends may affect military recruitment, provide a window on the future youth population over the next 15 to 20 years and thus a way of forecasting the size and makeup of the target population for recruitment. Because mortality and emigration rates are very low during the first 20 years of life, we can use recent trends in numbers of births to forecast the size and makeup of the youth population when recent birth cohorts will reach an appropriate age for service in the armed forces.

In this chapter, we first describe key recent trends in the activities of young persons, focusing particularly on the school enrollment, employment, and enlistment patterns of youth and rates of college attendance of civilian youth. Then we describe the numbers and demographic composition of cohorts of children who will be reaching the eligibility age for service in the Armed Forces over the next two decades. We focus on characteristics of children and their parents that may affect their likelihood of entry into the armed forces and develop a rudimentary statistical model for predicting the likelihood of enlisting among demographic groups. Using this model and known trends in demographic characteristics, we can assess the implications of demographic trends for future numbers and rates of enlistment.

YOUNG ADULT DEMOGRAPHY

Upon high school graduation, youth typically face a menu of options about the major activities in which they will engage. These activities include further schooling, participation in the civilian workforce, service in the military, marriage, childbearing, and possibly other activities as well.

At least some of these activities may be pursued simultaneously, and others may be pursued in sequence. However, although decisions about what to do after leaving high school are not irreversible and considerable diversity and flexibility of choices are available to youth, the pursuit of nonmilitary options necessarily diminishes their availability for military service. The principal activities that "compete" with military service for recent high school graduates are further schooling and civilian employment. The trade-offs among these alternatives shift substantially over time because of both fluctuations in the civilian economy and secular trends in the aspirations of youth. In this section, we describe recent trends in schooling, work, and enlistment in the armed forces for young persons, focusing primarily on recent high school graduates, typically the target population for armed forces recruitment.

Trends in Educational Attainment

Perhaps the most dramatic secular trend in the youth population during the 20th century has been the growth in school enrollment and educational attainment, a trend that has continued unabated during the past several decades. Figures 3-1 and 3-2 show the trends in proportions of young men and women (ages 25–29) who have completed selected levels of schooling from 1970 to 1999.[1] These trends show substantial increases in educational attainment for cohorts reaching adulthood during this period. For example, in 1970 only about 40 percent of men ages 25–29 had completed at least some college, whereas by 1999, more than 55 percent had done so.

In the 1990s, the fraction of new cohorts of adults who completed four years of college began to accelerate, a reflection of increasing rates of college attendance throughout the decade (discussed below). The upward trend in educational attainment is even sharper for women than for men, showing a steady secular increase to a point such that now the average attainment level for women exceeds that of men. The difference in the trends for men and women is largely a result of unusually high rates of college attendance for men during the late 1960s and early 1970s, in all likelihood a response to the Vietnam-era draft and the availability of draft deferments for college students. Following the Vietnam era, the trend in men's educational attainment reverted to what would have been expected on the basis of pre-Vietnam-era education trends.

---

[1]This age group is young enough to provide some indication of the education levels of recent labor force entrants but old enough that most members of the birth cohort have completed their education. These data include both the civilian noninstitutional and armed forces populations in this age group.

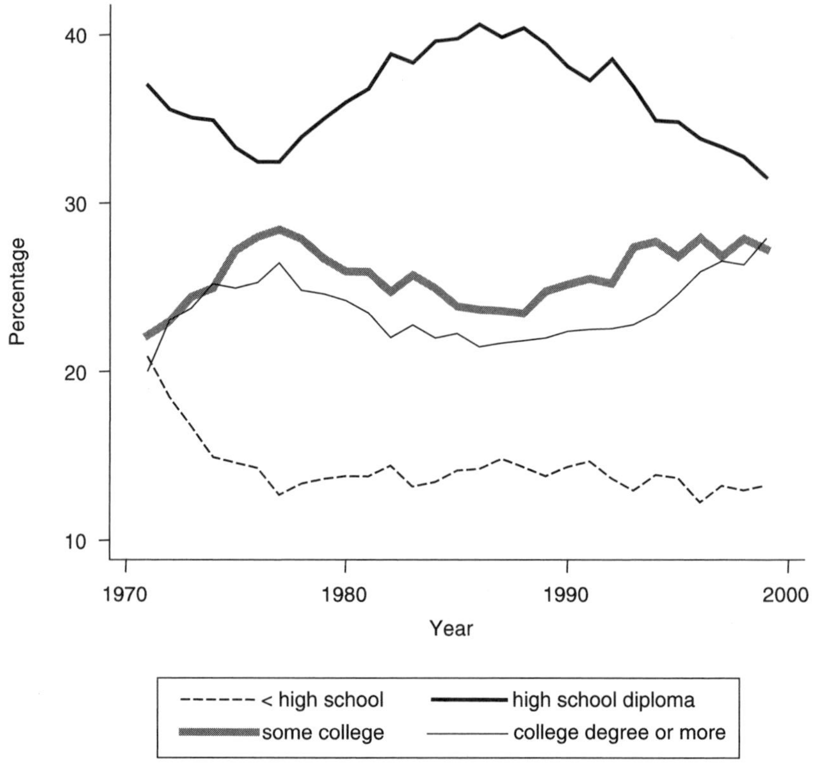

FIGURE 3-1 Trends in educational attainment for men ages 25–29.
SOURCES: Adapted from October Current Population Surveys (U.S. Bureau of the Census, 2002b) and unpublished U.S. Department of Defense data.

As we discuss throughout this report, the expectations and aspirations of youth about their educational attainment have an important impact on their propensity to enlist in the armed forces as well as their actual enlistment behavior. Trends in educational attainment also, to some degree, affect the makeup of armed forces population itself, although trends in educational attainment of the youth population as a whole are not closely mirrored in the enlisted population.

Figures 3-3a and 3-3b show trends in the educational makeup of male enlisted members of the armed forces for two age groups, ages 18–24 and 25–29. Figures 3-4a and 3-4b show the same trends for women. These figures show that persons with a high school degree and no higher educational credential have been a large plurality of the enlisted population throughout the past 30 years, although persons in other education groups

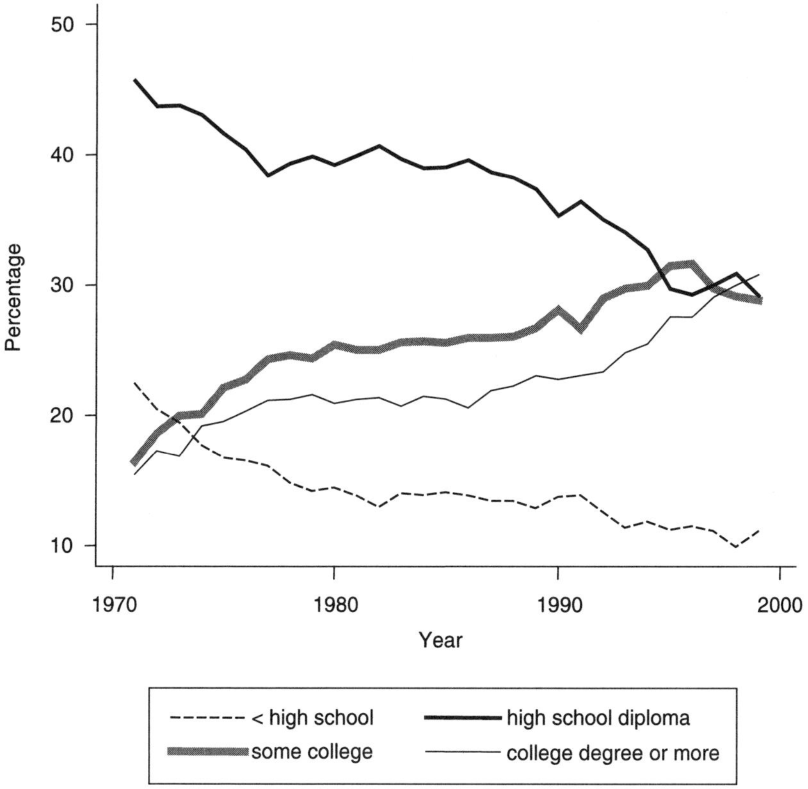

FIGURE 3-2 Trends in educational attainment for women ages 25–29.
SOURCES: Adapted from October Current Population Surveys (U.S. Bureau of the Census, 2002b) and unpublished U.S. Department of Defense data.

have increased in representation in recent years. Most importantly, a substantially increasing proportion of enlisted men and women have some postsecondary education. This results from trends in the educational makeup of persons who enlist in the armed forces, as well as increases in the numbers of enlistees who earn college credit while serving in the military.

Activities of Recent High School Graduates

The principal activities of youth are school enrollment and market work. In this section we briefly review trends in enrollment and employment of persons who have recently left high school. The discussion is

a

b

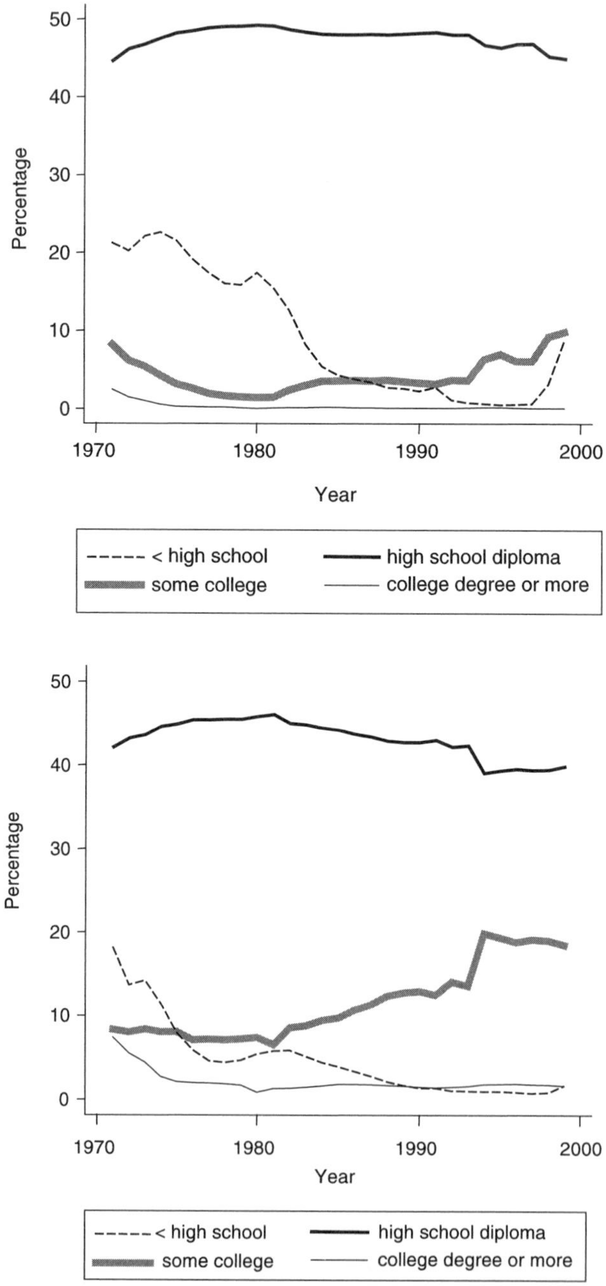

FIGURE 3-3 Trends in educational attainment for male members of the armed forces: **a**, ages 18–24; **b**, ages 25–29.
SOURCE: Adapted from unpublished U.S. Department of Defense data.

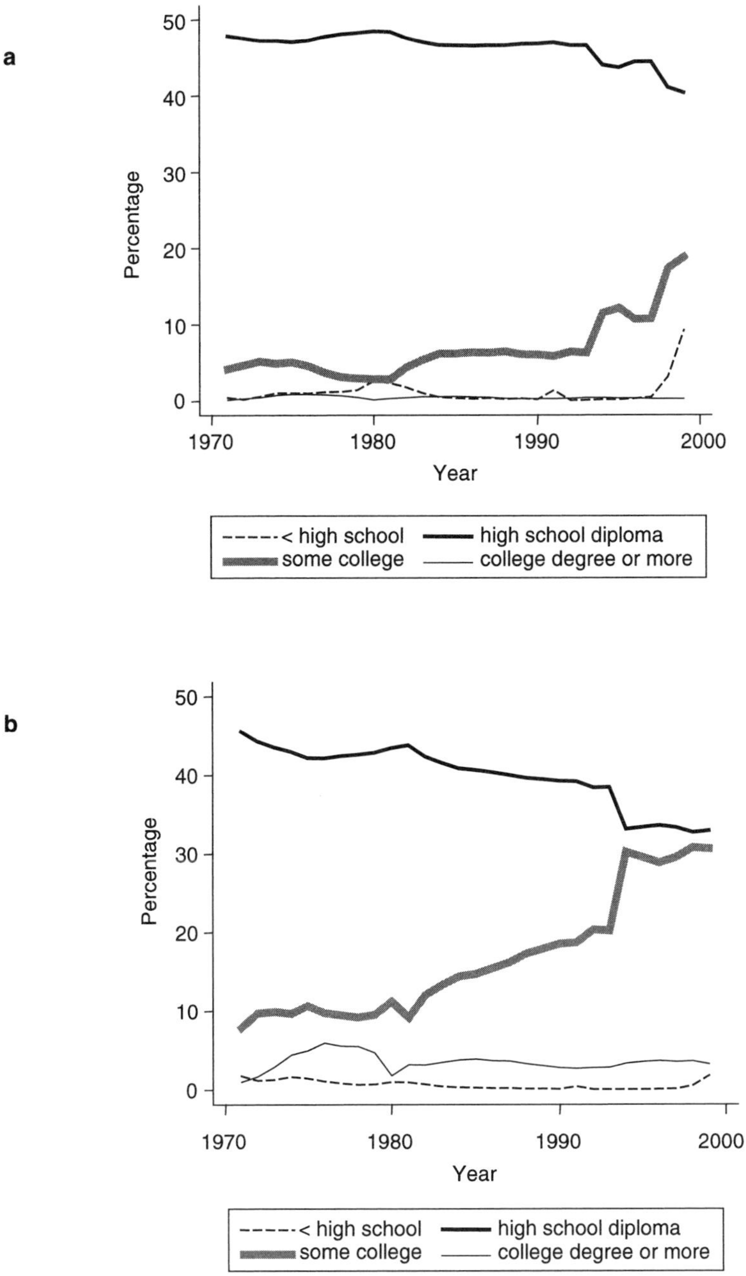

FIGURE 3-4 Trends in educational attainment for female members of the armed forces: **a**, ages 18–24; **b**, ages 25–29.
SOURCE: Adapted from unpublished U.S. Department of Defense data.

based on data from the October Current Population Survey, an annual survey of the civilian noninstitutional population of the U.S. Bureau of the Census that focuses on school enrollment and employment, as well as unpublished Department of Defense data.

Trends in College Enrollment

Corresponding to the large increases in educational attainment shown in Figures 3-1 and 3-2 are corresponding increases in rates of college enrollment by recent high school graduates. Figure 3-5 shows college enrollment rates of new high school graduates over the past three decades. These rates, based on the civilian noninstitutional population, apply to persons who graduated from high school during the previous 12

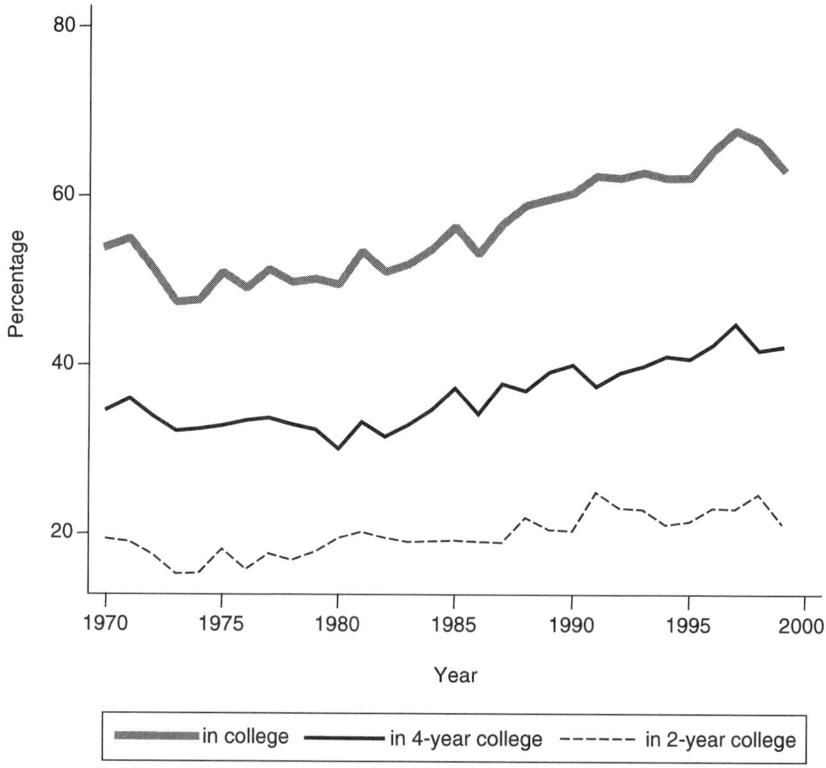

FIGURE 3-5 Trends in college enrollment of recent high school graduates.
SOURCE: Adapted from October Current Population Surveys (U.S. Bureau of the Census, 2002b).

months. Overall fractions of recent high school graduates who attend college have grown dramatically during this period, ranging from less than half in the early 1970s to approximately two-thirds in the late 1990s.[2] Although increases in attendance rates have occurred for both two- and four-year institutions, the bulk of the increase has been for four-year enrollments. In 1970, approximately half of college entrants entered two-year colleges. By 2000, only about one-third of new entrants went to two-year colleges.

Although college enrollment rates have risen for the youth population as a whole, the increases in enrollment have varied among demographic groups. As shown in Table 3-1, enrollment rates for women have been more substantial than for men. Whereas men's enrollments have increased by about 7 percentage points over the past 30 years, women's rates have increased from less than 40 to almost 70 percent, reversing a traditional pattern of higher enrollment levels for men. Enrollment trends have also varied across major race-ethnic groups. Black enrollment rates did not keep up with the growth in white rates during the 1980s and indeed declined for several years (Figure 3-6). During the 1990s, however, the gap between black and white enrollment rates has closed considerably. The trend for Hispanic youth is unclear because of considerable sampling variability in the Current Population Survey data for this group. So far as it is possible to tell, Hispanic youth have not experienced the same increases in college enrollment rates that are seen for non-Hispanic whites and blacks.

Trends in Employment and Enrollment

Figures 3-7 and 3-8 show trends in school enrollment, civilian employment, and participation in the armed forces for persons ages 18 and 19 over the past three decades for men and women, respectively. The fraction of 18- and 19-year-old men who participate in the armed forces has declined steadily over much of the past 30 years, although this fraction was stable in the late 1990s. Participation in the armed forces is extremely low for women throughout this period when considered in the context of the entire 18- and 19-year-old female population.

The long-term decline in armed forces participation for men implies that they have increasingly pursued civilian activities. We have already noted the substantial increase in proportions of recent high school graduates who attend college. In addition to the growth in school enrollment, there have been significant changes in rates of employment for youth. The

---

[2]Attendance rates were unusually high prior to 1973 because of the draft exemption of college students during that period.

TABLE 3-1 College Enrollment of Recent High School Graduates: 1970–1999

	Number of High School Graduates					Percentage Enrolled in College[a]				
Year	Total[b]	Male	Female	White	Black	Total[b]	Male	Female	White	Black
1970	2,757	1,343	1,414	2,461	(NA)	51.8	55.2	48.5	52.0	(NA)
1975	3,186	1,513	1,673	2,825	(NA)	50.7	52.6	49.0	51.2	(NA)
1980	3,089	1,500	1,589	2,682	361	49.3	46.7	51.8	49.9	41.8
1982	3,100	1,508	1,592	2,644	384	50.6	49.0	52.1	52.0	36.5
1983	2,964	1,390	1,574	2,496	392	52.7	51.9	53.4	55.0	38.5
1984	3,012	1,429	1,583	2,514	438	55.2	56.0	54.5	57.9	40.2
1985	2,666	1,286	1,380	2,241	333	57.7	58.6	56.9	59.4	42.3
1986	2,786	1,331	1,455	2,307	386	53.8	55.9	51.9	56.0	36.5
1987	2,647	1,278	1,369	2,207	337	56.8	58.4	55.3	56.6	51.9
1988	2,673	1,334	1,339	2,187	382	58.9	57.0	60.8	60.7	45.0
1989	2,454	1,208	1,245	2,051	337	59.6	57.6	61.6	60.4	52.8
1990	2,355	1,169	1,185	1,921	341	59.9	57.8	62.0	61.5	46.3
1991	2,276	1,139	1,137	1,867	320	62.4	57.6	67.1	64.6	45.6
1992	2,398	1,216	1,182	1,900	353	61.7	59.6	63.8	63.4	47.9
1993	2,338	1,118	1,219	1,910	302	62.6	59.7	65.4	62.8	55.6
1994	2,517	1,244	1,273	2,065	318	61.9	60.6	63.2	63.6	50.9
1995	2,599	1,238	1,361	2,088	356	61.9	62.6	61.4	62.6	51.4
1996	2,660	1,297	1,363	2,092	416	65.0	60.1	69.7	65.8	55.3
1997	2,769	1,354	1,415	2,228	394	67.0	63.5	70.3	67.5	59.6
1998	2,810	1,452	1,358	2,227	393	65.6	62.4	69.1	65.8	62.1
1999	2,897	1,474	1,423	2,287	453	62.9	61.4	64.4	62.8	59.2

SOURCE: U.S. Bureau of the Census (2001).
NOTE: High school graduates in thousands. For persons 16 to 24 who graduated from high school in the proceeding 12 months. Includes persons receiving GEDs.

NA=not available. [a]As of October. [b]Includes other races, not shown separately.

proportion of young men and women who adopt a "work-only" pattern shortly after leaving high school has dropped substantially over this period, a trend that is offset by both an increase in the proportions who adopt a "school-only" pattern and who combine school enrollment with employment. Young persons are increasingly using work as a means of financing further education, as well as pursuing additional schooling while earning a living. A key feature of these trends is that these developments have occurred in tandem for men and women and, in general, women's activity patterns have come increasingly to resemble those of men.[3]

[3]In Figures 3-7 and 3-8 individuals must work 20 or more hours a week to be considered "employed." Using a less stringent definition of employment would result in higher estimated proportions of persons who are employed or both employed and enrolled. The trends in these proportions, however, are similar to those reported in Figures 3-7 and 3-8.

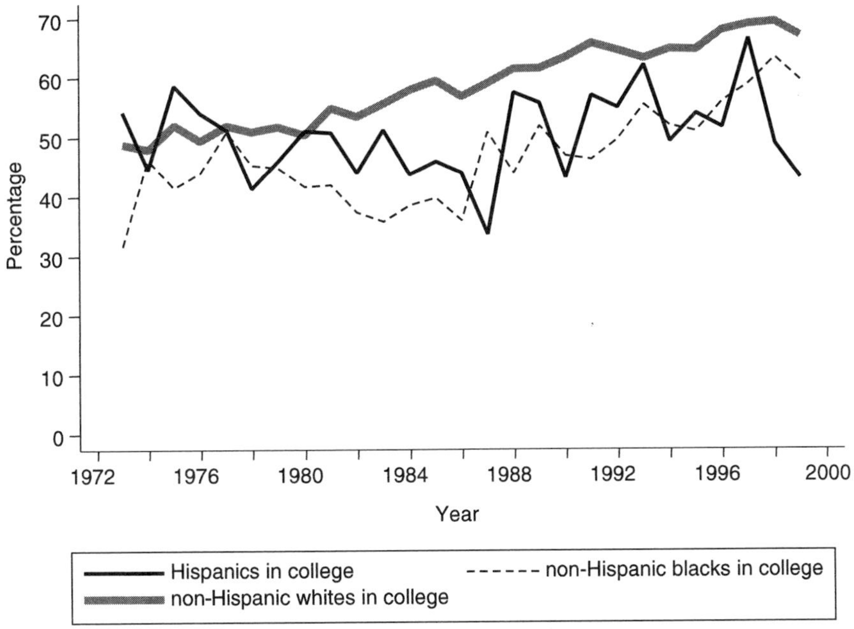

FIGURE 3-6 College enrollment rates of high school graduates, by race-ethnicity.
SOURCE: Adapted from October Current Population Surveys (U.S. Bureau of the Census, 2002b).

FUTURE SIZE AND COMPOSITION OF THE YOUTH POPULATION

Trends in the Size of the Youth Population

We can estimate the size of the youth population with reasonable reliability for the next two decades. Because all persons who will reach age 18 over the next 18 years have already been born, recent fertility trends provide the core data for future estimates of the size of this population. The Census Bureau routinely calculates projections of the size of the population specific to age, sex, race-ethnicity, and nativity (native-born versus foreign-born). Although the Census Bureau makes projections as far as 100 years in the future, from the standpoint of estimating the number of persons of a suitable age to enter the armed forces, we can be most confident of the estimates for cohorts that are already born. For example, because the cohort that will be age 18 in 2015 was born in 1997, its initial size is already known from 1997 birth records, and it is thus necessary to forecast only future mortality, immigration, and emigration. In contrast, the cohort that will reach age 18 in 2025 is not yet born and

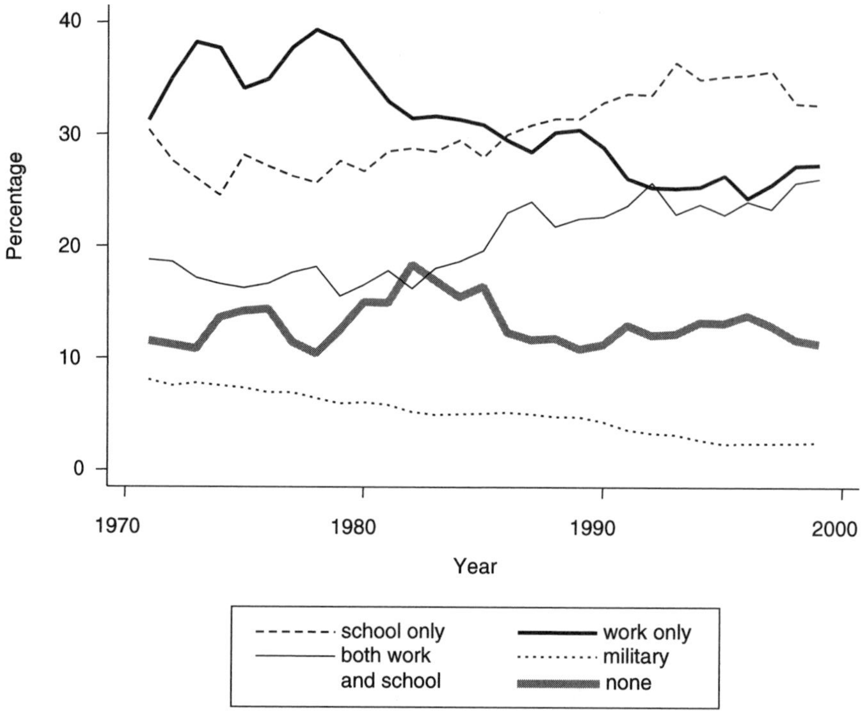

FIGURE 3-7 Enlistment, enrollment, and employment of 18-19 year old men.
SOURCES: Adapted from October Current Population Surveys (U.S. Bureau of the Census, 2002b) and unpublished U.S. Department of Defense data.

thus it is necessary to forecast future fertility as well as mortality, immigration, and emigration.

In addition to fertility, Census Bureau projections of the size of a population incorporate estimates of immigration, emigration, and mortality, which augment and deplete each birth cohort as it ages. Of these three components, the most important is immigration, which more than offsets the negative impacts of emigration and mortality, thus enlarging the size of youth cohorts beyond what would be expected on the basis of fertility alone. Past and future immigration also affects the contributions of fertility and emigration to future population because of the relatively high rates of fertility *and* emigration experienced by immigrants relative to native-born persons.

Figure 3-9 shows the Census Bureau's "medium" projection for the 18-year-old population from 1999 to 2020 (U.S. Bureau of the Census, 2000a). Approximately 3.9 million residents of the United States were 18

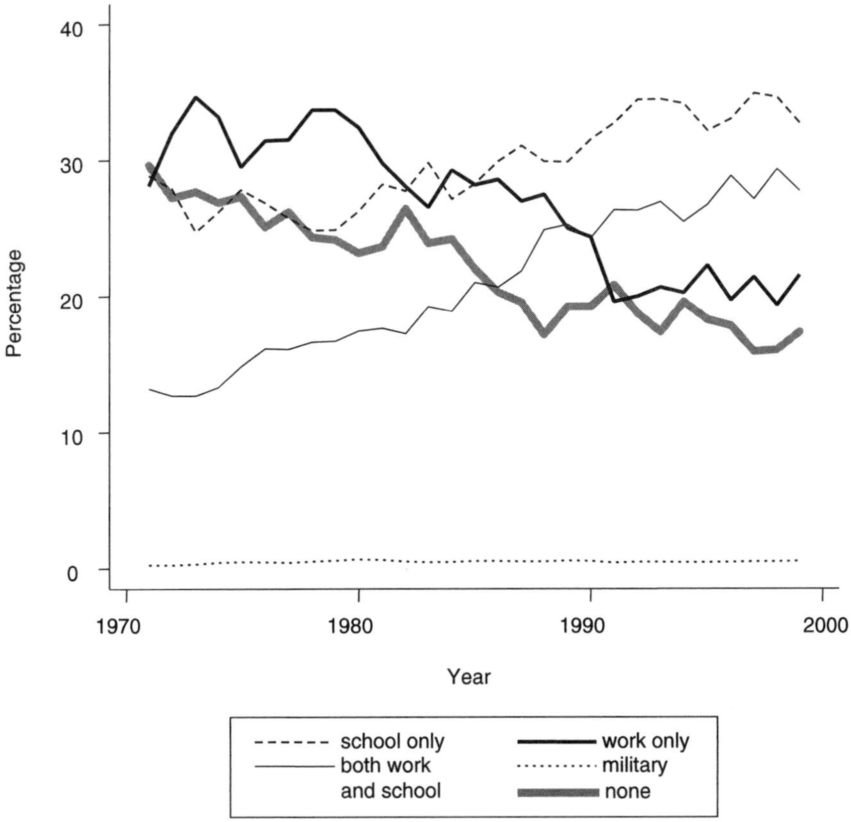

FIGURE 3-8 Enlistment, enrollment, and employment of 18-19 year old women.
SOURCES: Adapted from October Current Population Surveys (U.S. Bureau of the Census, 2002b) and unpublished U.S. Department of Defense data.

years old in 1999, a number that is expected to grow over the next 10 years to a maximum of about 4.4 million in 2009. This increase is an "echo" of the baby boom of the 1950s, which, a generation later, produced a relatively large number of births in the 1980s. After 2009, the 18-year-old population will subside to some degree and reach a plateau between 4.1 and 4.2 million persons in the years from 2015 to 2020.

We can get some idea of the future trend in the immigrant makeup of children and teenagers from the recent past. Table 3-2 shows the proportions of persons who are foreign born for several recent years by 5-year age groups for persons under age 25. The proportion of those ages 20–24 who are foreign born appears to be holding stable at about 10 percent,

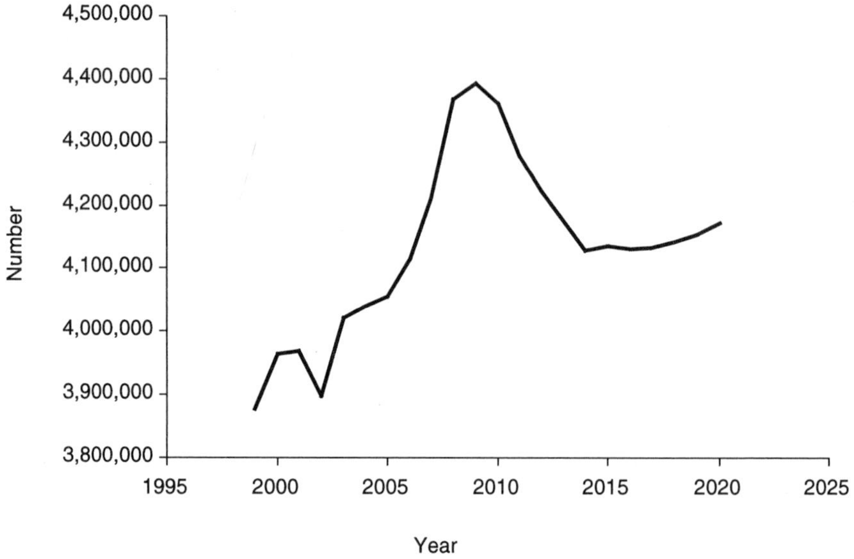

FIGURE 3-9 Projected 18-year-old population.
SOURCE: U.S. Bureau of the Census (2000a).

whereas the proportion ages 15–19 is approximately 7 percent. Although these proportions for 1995 and 1999 are slightly higher than for 1990, the fractions at younger ages have dropped slightly over the 1990s. This suggests that, barring increases in immigration by teenagers that are large enough to more than offset the small declines in proportions of young children who are foreign born, the proportions of persons in their late teens who are foreign born will remain approximately constant.

This conclusion is consistent with Census Bureau projections of the nativity composition of the youth population over the next two decades, which are based on data such as those shown in Table 3-2, plus forecasts of the future streams of immigration and emigration for children and teenagers. Figure 3-10 shows that the percentage of 18-year-olds who are born outside the United States will vary between about 6 and 8 percent between now and 2020 and suggests that, if any change is likely to occur, it will be a reduction in the fraction of young persons who are born outside the United States.

TABLE 3-2 Native and Foreign-Born Populations by Year and Age (in thousands)

	Native			Foreign-Born			Proportion Foreign-Born		
Age	1990	1995	1999	1990	1995	1999	1990	1995	1999
Total	228,945	239,826	246,859	19,846	22,978	25,831	0.080	0.087	0.095
<5	18,495	19,372	18,766	270	160	176	0.014	0.008	0.009
5-9	17,555	18,580	19,481	488	517	466	0.027	0.027	0.023
10-14	16,334	18,001	18,645	733	852	904	0.043	0.045	0.046
15-19	16,687	16,886	18,317	1,206	1,317	1,431	0.067	0.072	0.072
20-24	17,260	16,146	16,077	1,883	1,837	1,949	0.098	0.102	0.108

SOURCE: U.S. Bureau of the Census (2001).

NOTE: For 1995–1999, as of July; for 1990, as of April. Foreign-born residents are those people born outside the United States to noncitizen parents, while native residents are those people born inside the United States or born abroad to United States citizen parents. One notable difference between the two populations concerns children. Any child born to foreign-born parents after entering the United States, by definition, becomes part of the native population. The foreign-born child population, therefore, is quite small, while the native child population (and the overall native population) are inflated by births to foreign-born parents after migrating to the United States.

Race-Ethnic Composition

We can also use Census Bureau projections to examine the race and ethnic makeup of future cohorts of 18-year-olds. Given that race-ethnic groups may vary in their propensity to enlist in the armed forces, it is instructive to examine the implications of recent fertility trends for the race-ethnic composition of successive birth cohorts. The proportions of births in successive cohorts that are identified with one or another race-ethnic group are affected by the relative sizes of the race-ethnic groups in the parental generation, by differential fertility across groups, and by group-specific patterns of immigration, emigration, and mortality.

Figure 3-11 charts the projected trends in the relative numbers of Hispanics and non-Hispanic whites, blacks, and others in the 18-year-old population over the next two decades. The largest change is in the Hispanic population, which is expected to grow from about 14 to about 22 percent of 18-year-olds over this period. This growth is mirrored by a corresponding decline in the relative size of the non-Hispanic white population, from about 66 to about 57 percent of the population. Non-Hispanic

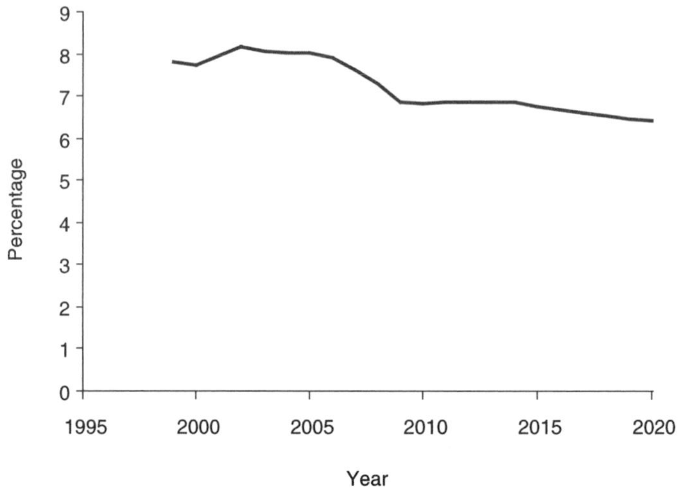

FIGURE 3-10 Projected percentage of the 18-year-old population who are foreign born.
SOURCE: U.S. Bureau of the Census (2000a).

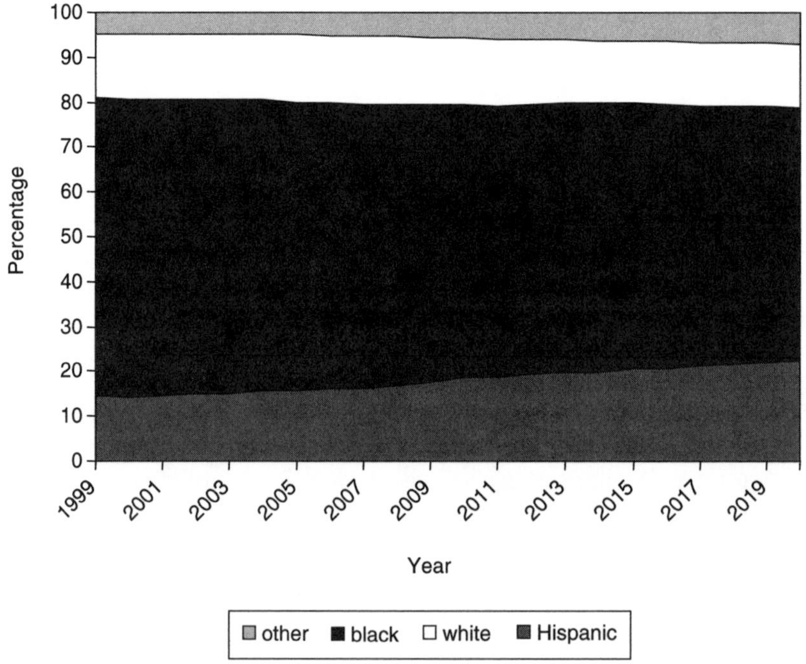

FIGURE 3-11 Projected race-ethnic makeup of the 18-year-old population.
SOURCE: U.S. Bureau of the Census (2000a).

blacks will remain at approximately 14 percent over this period, and other groups will grow slightly in relative terms from 5 to 7 percent.

Immigration and Nativity

Although the proportion of immigrants among the youth population is likely to remain stable over the next two decades, immigration may affect the makeup of youth in another way. Past trends in immigration, which have brought substantial numbers of foreign-born adults to the United States, undoubtedly have increased the proportion of children and youth who are born in the United States to foreign-born parents. Trends in immigration have longer-term consequences in that they alter both the ethnic and generational makeup of the population. Increases in immigration among women of childbearing age will increase the proportion of each cohort of youths who were raised by immigrant (foreign-born) mothers and, in a large proportion of cases, two immigrant parents. These trends are further reinforced by the higher fertility of immigrants compared to native-born women. Figure 3-12 shows the trend in the proportion of births that are to foreign-born mothers from 1970 to 1999, which are the cohorts that will reach age 18 from 1988 through 2017. A much higher fraction of cohorts born in the late 1990s have foreign-born mothers than earlier cohorts. This fraction has approximately tripled over the past 30 years from about 7 to over 20 percent. The trend is remarkably linear and, as yet, shows no sign of abating. This trend indicates that in that the next two decades the youth population will increasingly be made up of persons who were raised by foreign-born mothers.

Parents' Educational Attainment

Among the many ways that parents affect offspring is by influencing their aspirations for higher education and creating environments that vary in the intellectual and motivation resources for children to achieve in school. These effects are often revealed in a positive association between parents' educational attainment and the attainment of their offspring, which summarizes the diverse pathways through which more highly educated parents provide advantages to their children. If higher levels of parents' schooling raise the educational aspirations of youths, then trends in parental education may affect young persons' propensities to enlist in the armed forces.

Throughout the 20th century, average educational attainment levels increased, a trend that shows no sign of abatement in recent decades. Secular increases in overall education are mirrored in the average educational attainments of parents and in the distribution of education for recently born cohorts. Figure 3-13 illustrates this trend by showing the

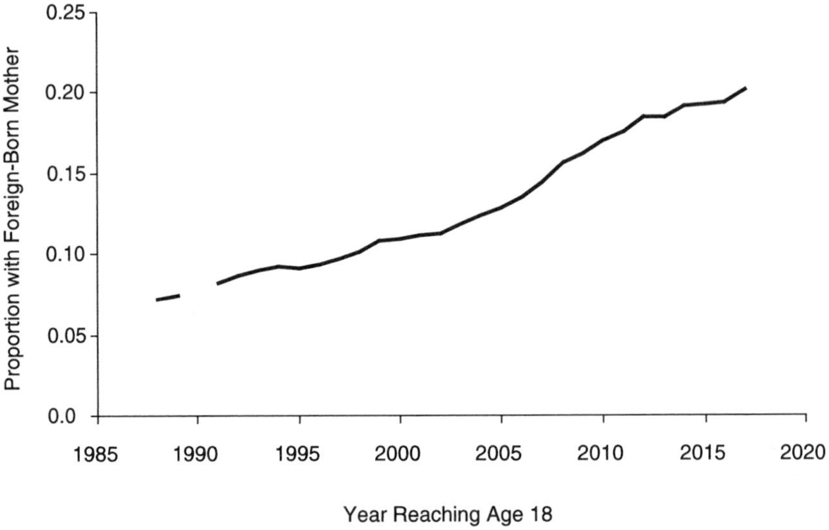

FIGURE 3-12 Proportion born to foreign-born women by year reaching age 18.
SOURCE: National Center for Health Statistics (2002). Birth records from natality statistics microdata files.
NOTE: Data are not available from 1972 on whether mothers were native versus foreign born. Thus, there is a gap in 1990 when that birth cohort was 18 years old.

proportion of persons born between 1980 and 2000 by educational attainment of their mothers. These birth cohorts will reach age 18 between 1998 and 2018. This shows a dramatic increase in the proportions of children to mothers with at least some postsecondary education over this period. For cohorts born in the early 1980s and reaching age 18 in the late 1990s, approximately 30 percent had mothers with some college education, but in cohorts born in the late 1990s and reaching 18 after 2015, more than half have mothers with this level of schooling. These trends are offset by large declines in the shares of births to women with a high school diploma or less education, although the largest decline is for women with a high school diploma. The proportion of births to high school dropout women has not declined as fast as the decline in the proportion of *women* who are high school dropouts because of the relatively high levels of fertility to immigrant women with low levels of educational attainment (Mare, 1995).

The secular increase in proportions of births to women with a college education may be particularly significant for future trends in aspirations for higher education of their children. The educational attainment of offspring is affected not only by the level of parents' schooling in number of years completed but also by whether the parent has completed particular

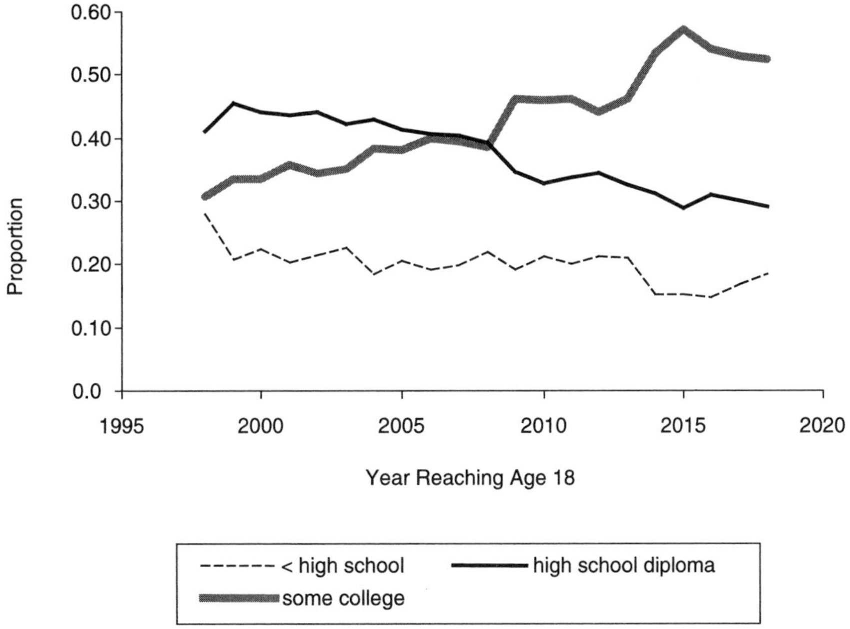

FIGURE 3-13 Proportion of births by education of mother and year reaching age 18.
SOURCE: Adapted from June Current Population Surveys (U.S. Bureau of the Census, 2002a).
NOTE: These estimates are based on birth history data from microdata from the June Current Population Surveys of 1975, 1980, 1985, 1990, 1995, and 2000. Estimates are based on births occurring up to five years prior to the survey date. Similar trends would be observed for births classified by levels of father's education (not shown here).

milestones in the educational process (e.g., high school degree, some college, college degree). That is, parental educational attainments tend to set a "floor" beneath the educational attainments of their children. In particular, whether parents have completed at least some college has a particularly large impact on whether their offspring complete at least some college (Mare, 1995). As shown in Figure 3-13, the proportion of births to women with at least some college has jumped sharply in recent years and, despite considerable short-term fluctuations, shows no sign of declining. This suggests that even in the absence of changes in the economic incentives that affect the relative desirability of alternative levels of schooling, youth will increasingly seek higher education during the next two decades.

Parents' Occupations

Another aspect of the family environment that affects young persons' aspirations and their propensities to engage in various activities after high school is the work roles of their parents. Youths' occupational aspirations are, in part, a reflection of the occupations and socioeconomic levels of their parents. Although both upward and downward mobility are common in the United States, parents' socioeconomic levels tend to set a floor on the aspirations of their offspring. Across cohorts, young persons' family environments change because of secular transformations of the occupational structure and changing rates of adult labor force participation. Figures 3-14 and 3-15 present trends from 1970 to 2000 in the proportions of births to fathers and mothers in broad occupational groups. Over this period, an increasing proportion of births are to couples in which fathers and mothers have managerial, professional, or technical occupations. This reflects the general drift in the occupational structure toward a professional, nonmanual workforce and a relative decline in manual occupations. In addition, a shrinking proportion of births are to mothers who have never worked. These trends, which illustrate the family backgrounds of cohorts that will reach adulthood over the next two decades, show that young persons are increasingly raised in families in which at least one parent has a managerial or professional job. It suggests that future cohorts of potential enlistees may have higher occupational aspirations than their counterparts in the past.

Parents' Military Service

A further element of the family environment that may affect propensity to serve in the armed forces is whether children are raised by parents who have had military experience. Insofar as parents are the primary influencers of their children, and nonfamily influences, such as schools and peer groups, provide no direct contact with the military, children whose parents served in the armed forces may receive unique information about the possibility of service. Over time, the proportion of children who are raised in families with experience in the armed forces varies considerably, because cohorts of adult Americans have varied in whether they entered adulthood during times of peace or war and during times in which policies of conscription or volunteer forces were in effect.

The proportion of a cohort of children who had one or more parents who served in the armed forces is affected primarily by how common military service was when the parents' generation was in its late teens and early twenties, and secondarily by the differential fertility of men and women who served in the military at some point in their lives compared

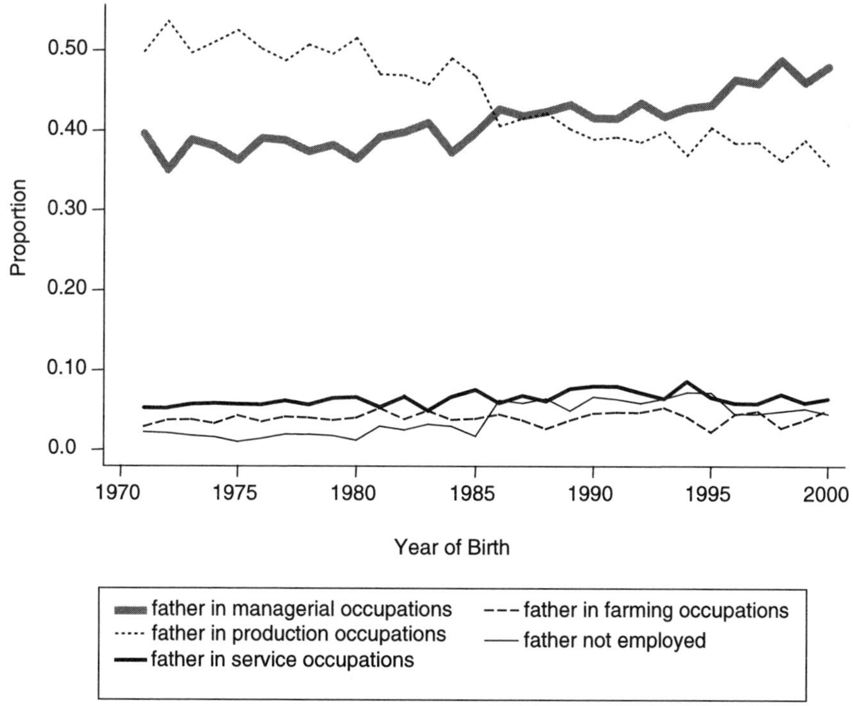

FIGURE 3-14 Proportion of births to families by occupation of father and year.
SOURCE: Adapted from June Current Population Surveys (U.S. Bureau of the Census, 2002a).
NOTE: These estimates are based on birth history data from microdata from the June Current Population Surveys of 1975, 1980, 1985, 1990, 1995, and 2000. Estimates are based on births occurring up to five years prior to the survey date. Estimates for fathers are based on couples in which the father is present in the household.

with those who never served. If we consider cohorts born since 1970, we see a dramatic decline in the proportion of persons in each birth cohort whose fathers or mothers were veterans of the armed forces. Figure 3-16 reports estimates of the proportion of each birth cohort who had at least one parent who was a veteran at the time of the cohort member's birth. These estimates show a dramatic decline, from approximately 40 percent of births occurring in 1970 to couples in which at least one parent was a veteran to approximately 8 percent of births in 2000. Of the cohort of youths reaching age 18 in 2000, approximately 18 percent had a father or a mother who was a veteran of the armed forces. Between 1982 and 2000, therefore, the fraction of cohorts of 18-year-olds with veteran parents

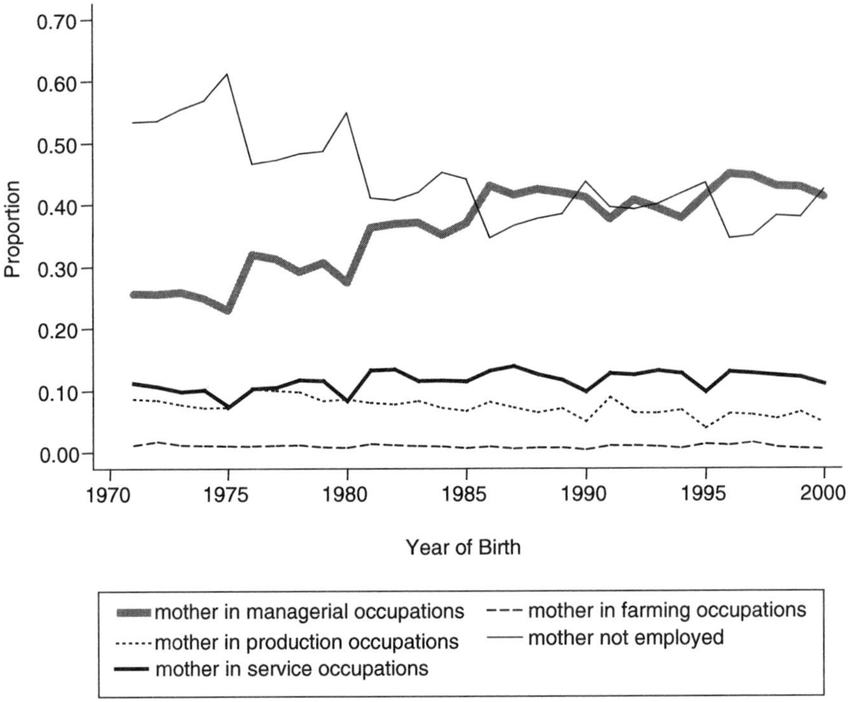

FIGURE 3-15 Proportion of births to families by occupation of mother and year.
SOURCE: Adapted from June Current Population Surveys (U.S. Bureau of the Census, 2002a).
NOTE: These estimates are based on birth history data from microdata from the June Current Population Surveys of 1975, 1980, 1985, 1990, 1995, and 2000. Estimates are based on births occurring up to five years prior to the survey date. Estimates for fathers are based on couples in which the father is present in the household.

declined by more than 50 percent. Between 2000 and 2018, this fraction will decline again by more than 50 percent. This constitutes a dramatic reduction in the proportion of young persons who are raised by parents with military experience.

A SIMPLE FORECASTING MODEL FOR MILITARY ENLISTMENT

The trends in child demography shown in the previous section suggest that demographic influences may have offsetting effects on the propensity of future cohorts of youths to enlist in the armed forces. On one

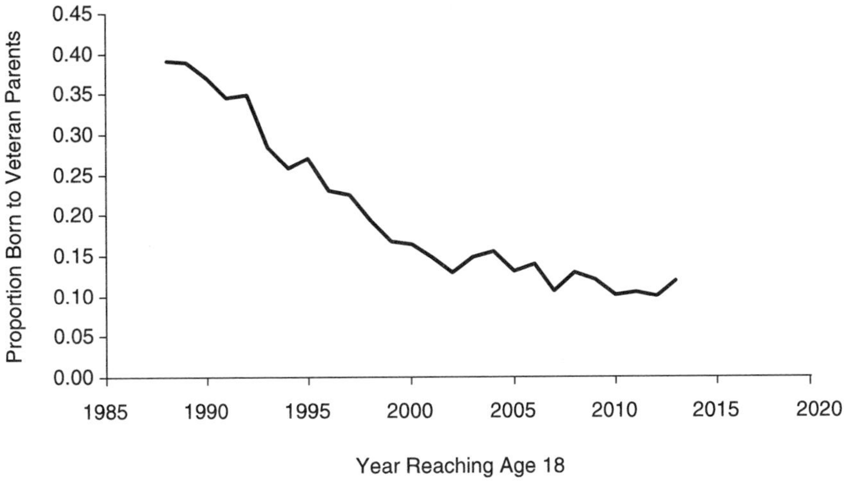

FIGURE 3-16 Proportion of 18-year-olds born to one or more veteran parents.
SOURCE: Adapted from June Current Population Surveys (U.S. Bureau of the Census, 2002a).
NOTE: These estimates are based on birth history data from microdata from the June Current Population Surveys of 1975, 1980, 1985, 1990, 1995, and 2000. Estimates are based on births occurring up to five years prior to the survey date and use information on whether the mother or the resident father had served in the armed forces prior to the survey date.

hand, successive cohorts of youths will increase during the next decade, thereby enlarging the pool of potential enlistees. On the other hand, future cohorts of youths may increasingly opt for higher education at the expense of other postsecondary options, including direct entry into the civilian workforce or service in the armed forces. To appraise the net effect of these potentially offsetting forces requires a model of how they work together. Our approach to this is (1) to estimate a simple logistic regression of the effects of a small number of family background factors on the probability of enlistment in the armed forces, using panel data for a single cohort of youths and (2) to use the estimated coefficients, in combination with the annual series of birth statistics based on the June Current Population Surveys from 1970 to 2000, to compute the predicted probability of enlistment and total number of enlistees in the armed forces for each birth cohort represented in the June Current Population Survey.

The strength of this procedure is that is it a useful way to combine disparate influences and see their net effects. Moreover, it makes clear the logical steps needed to extrapolate from recent trends. Of course, the drawback to this procedure is that it makes several strong assumptions that may not be accurate. In particular, it assumes that the relationships between the demographic factors and the probability of enlistment that are estimated for a single birth cohort will hold over the entire forecasting period. It also ignores economic, political, and cultural factors that may deflect enlistment rates from the long-term trajectory implied by recent fertility patterns. On balance, this method should be viewed as a way of seeing the effects of known demographic trends, rather than an accurate forecast of enlistment rates and numbers.

We estimate the model for the effects of demographic factors on enlistment using the National Longitudinal Survey of Youth (NLSY79), a panel of youths ages 14–22 in 1979 who were followed annually until 1994 and biennially thereafter (U.S. Bureau of Labor Statistics, 2002c). Although these data provide a rich array of potential proximate determinants of military enlistment, including civilian work experience, aptitude test scores, living arrangements, etc., we focus only on demographic characteristics that are known at birth. Only the latter, of course, are known for cohorts that are still too young to serve in the armed forces. These demographic characteristics include mother's educational attainment, race-ethnicity, whether one or more parents served in the armed forces, and region of birth.[4] Separate models are estimated for men and women. Table 3-3 presents the estimated logit coefficients from these models.

Figure 3-17 shows the implied trend in the probability of enlistment for cohorts reaching age 18 between 1989 and 2017. The implied cohort enlistment rates decline by approximately 10 percent over this period, with a roughly constant decline per year. It is important to recognize that these overall enlistment rates are considerably higher than the actual rates during the 1990s, a result of relying on the NLSY79 cohort for the underlying prediction model. Provided that the effects of the demographic characteristics have not changed much over time, however, the secular trend in predicted cohort enlistment rates would not change even were we to

[4]Father's educational attainment is not included in the model because it is highly correlated with mother's educational attainment and thus adds little to the predictive power of the model. Region of birth is a dichotomous variable denoting whether or not an individual was born in the South. Persons of Southern origin are somewhat more likely to enlist in the armed forces than persons born elsewhere. Over cohorts, the proportion of persons born in the South increased slightly between 1970 and 2000. Parents' occupations are not included in the model because the NLSY79 does not include measures of parents' occupation at the time of birth of respondents, which are the measures available in the birth data. Although persons whose parents have professional and managerial occupations are somewhat less likely to enlist than persons whose parents have other occupations, their estimated net effects are small once parental education is taken into account.

TABLE 3-3 Logit Coefficient Estimates for Model Predicting Probability of Enlistment in the Armed Forces

Independent Variables	Logit B	Robust S.E.(B)	z	P>z
Men				
Parent in armed forces	.185	.087	2.1	.03
South	.229	.092	2.5	.01
Mother high school diploma (vs. HS dropout)	–.098	.098	–1.0	.31
Mother some college (vs. HS dropout)	–.230	.154	–1.5	.14
Mother college degree or more (vs. HS dropout)	–.533	.170	–3.1	.00
Black (vs. non-Hispanic white)	.400	.098	4.1	.00
Hispanic (vs. non-Hispanic white)	–.136	.138	–1.0	.33
Other (vs. non-Hispanic white)	.117	.143	.8	.41
Intercept	–1.577	.104	–15.2	.00
Log likelihood = –2666.1; N = 5663				
Women				
Parent in armed forces	.444	.192	2.3	.02
South	–.177	.195	–.9	.36
Mother high school diploma (vs. HS dropout)	.246	.202	1.2	.23
Mother some college (vs. HS dropout)	.281	.309	.9	.36
Mother college degree or more (vs. HS dropout)	.091	.389	.2	.82
Black (vs. non-Hispanic white)	.919	.197	4.7	.00
Hispanic (vs. non-Hispanic white)	.298	.358	.8	.41
Other (vs. non-Hispanic white)	–.203	.380	–.5	.59
Intercept	–3.938	.211	–18.7	.00
Log likelihood = –774.1; N = 5623				

SOURCE: Center for Human Resource Research (2000).

adjust the levels of the rates downward to agree with those observed in the 1990s.

To see the impact of these effects on overall expected numbers of enlistees, we multiply each predicted probability of enlistment by the size of the 18-year-old population as projected by the Census Bureau.[5] These estimates, however, drastically overstate recent and probably future enlistment levels because they are based on a model estimated in an earlier period when enlistment levels were much higher. To obtain a somewhat more realistic view of enlistment levels, we calibrate the estimates to agree

[5]Male and female enlistment trends are combined to yield the total numbers of enlistees presented in Figure 3-17.

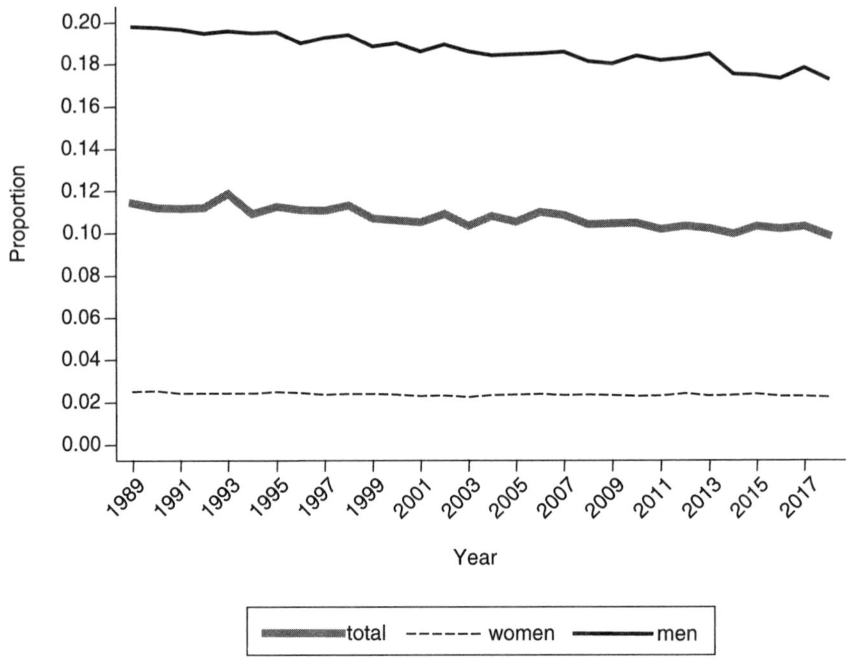

FIGURE 3-17 Trend in proportions ever enlisting implied by trends in mother's education, race-ethnicity, parents' veteran status, and region of birth.
SOURCES Authors' calculations from National Longitudinal Survey of Youth (Center for Human Resource Research, 2000); June Current Population Surveys (U.S. Bureau of the Census, 2002a); birth records from National Center for Health Statistics natality statistics microdata files (National Center for Health Statistics, 2002).

with the number of first-time enlistments in 1997 but otherwise preserve the trend implied by the model (U.S. Bureau of the Census, 2002a, 2002b).[6] Figure 3-18 shows these estimates for the period from 1989 to 2017. Our

[6]This calibration procedure consists of multiplying the predicted number of enlistees in each year by the ratio of the actual number of first-time enlistees in 1997 by the number of enlistees predicted by the model for 1997. This ratio is approximately 0.46. This procedure is likely to yield an overestimate of the number of 18-year-old enlistees in each year because the published number of first-time enlistees is not confined to 18-year-olds. Under the assumption that the age distribution of new enlistees does not change appreciably over time, however, this method does yield a plausible estimate of the number of new enlistees of any age, rather than just at age 18.

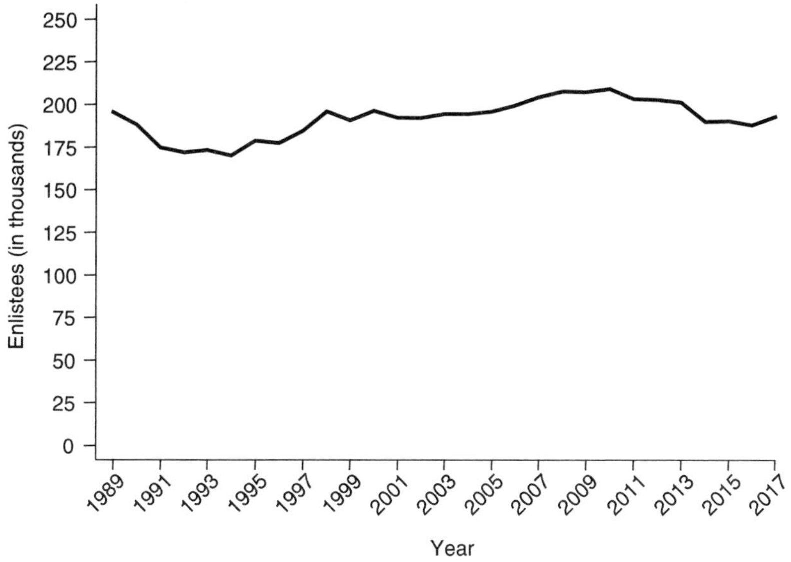

FIGURE 3-18 Expected number of potential enlistees at age 18 based on trends in births and child demographic characteristics.
SOURCES Authors' calculations from National Longitudinal Survey of Youth (Center for Human Resource Research, 2000); June Current Population Surveys (U.S. Bureau of the Census, 2002a); birth records from National Center for Health Statistics natality statistics microdata files (National Center for Health Statistics, 2002).

estimates, however, nonetheless illustrate the trend in enlistments implied by recent demographic trends. These estimates strongly suggest that, despite the downward drift in enlistment rates that is implied by the changing family backgrounds of future youth cohorts, there will be somewhat of an increase in expected numbers of enlistees. Secular growth in the size of youth cohorts almost completely offsets the downward trend in enlistment rates implied by trends in parental education and parents' experience with the armed forces.

SUMMARY

Demographic trends in the youth population place fundamental constraints on armed forces recruitment. The size and composition of the youth population will affect future recruitment efforts. We can forecast the size and some aspects of the composition of the youth population for the next 15 to 20 years with some accuracy because these persons are already born. Cohorts of persons reaching 18 years of age are expected to grow significantly over the next 10 years and then remain approximately at a plateau during the following decade. Approximately 4 million youth reached age 18 in 2000, a number that will increase to approximately 4.4 million by 2008 and trail off to between 4.1 and 4.2 million during the subsequent decade.

The ethnic composition of the youth population will change significantly over the next 15 to 20 years (even in the absence of changes in immigration patterns) because of recent changes in the ethnic makeup of women of childbearing age and ethnic differences in fertility rates. In particular, based on recent fertility and immigration patterns, the percentage of young adults who are Hispanic will increase substantially. A growing percentage of youth will be raised by parents who are immigrants to the United States, a result of high rates of recent immigration and relatively high fertility levels of foreign-born women.

In addition to the size and composition of the youth population, trends in parental characteristics and in the activities of youth affect armed forces recruitment. During the 1990s, rates of college enrollment and levels of education completed increased dramatically as a result of three broad trends: (1) secular changes in parental characteristics, especially parents' educational attainment, which increased youths' resources and aspirations for education; (2) the greater inclusion in higher education of women and some ethnic minorities; and (3) increased economic incentives to attend and complete college, a result of changes in the labor market for college- and non-college-educated workers (as discussed in Chapter 5, this volume). Recent trends also show that youth are increasingly using work as a means to finance additional schooling and that they do so by pursuing an education while earning a living.

Increases in educational attainment in the population will continue over the next two decades. Average levels of maternal education have increased markedly and, in the future, the majority of youths will be raised by mothers who have completed at least some college. Parents' educational attainment has a large effect on the aspirations and decisions of youths, especially concerning higher education.

While youth are expected to have at least one parent with at least some college in the coming years, the proportion of young adults who have had at least one parent with military experience has fallen dramatically and will continue to fall in the coming years. This represents a large decline in exposure to military experience within the nuclear family.

In sum, trends in the numbers of births and in the composition of the child population have offsetting effects on potential enlistment trends. Although the annual number of births has increased in recent years, children are increasingly raised by highly educated parents and by parents who have no direct experience with the armed forces. The net impact of these offsetting trends is a small increase in expected numbers of enlistees in the next decade, implying that the supply of young persons will be large enough to meet recruitment goals. This conclusion refers to the expected impact of demographic trends alone. Thus, demographic trends do not emerge as factors that will contribute to increasing difficulty in meeting enlistment goals. Other factors discussed in this report, including advertising and recruitment practices, will determine whether potential enlistees actually enlist at a rate necessary to meet goals.

4

Trends in Youth Qualifications and Enlistment Standards

The discussion of military requirements in Chapter 2 distinguished two different aspects of manpower requirements, one having to do with the quantity of personnel and the other related to the qualifications of personnel (sometimes abbreviated "quality" by the military services). While force structure dictates the number of people needed to fill military units, the qualifications of those people in terms of knowledge, aptitudes, skills, and motivation determines the effectiveness of those units.

Manpower qualifications include a cluster of human attributes that influence how well a new recruit can adjust to military life and how well the recruit can perform in military jobs. Based on many years of research and experience, the two most important qualifications for military service are aptitudes (as measured by the AFQT—the Armed Forces Qualification Test) and a high school diploma. Other qualifications include good physical health and moral character (e.g., no criminal record). Recognizing the importance of these qualifications for effective military performance, the Department of Defense (DoD) sets various standards that specify minimum levels of these qualifications in order to be eligible for enlistment. These levels comprise what are known as "enlistment standards."

When evaluating the problem of recruiting shortfalls, both quantity and qualifications are involved. If manpower qualifications were of no concern, there would be far less likelihood of recruiting shortfalls, simply because the number of youth available in a single birth cohort is 20 times the number of youth needed in any given year. When enlistment standards are factored in, the number of qualified youth contracts appreciably.

Since enlistment standards and the supply of qualified youth can change over time, present or future recruiting shortfalls can arise either from higher enlistment standards or from declining qualifications in the youth population. The primary purpose of this chapter is to examine trends in the supply of qualified youth in relation to enlistment standards and to assess whether gaps exist now or might exist in the future. The chapter begins with a summary of current military enlistment standards, the rationale for those standards, and the possibility of future changes in requirements that would increase or decrease these standards. The chapter then moves to the supply side and examines trends in youth attributes, including education, aptitudes, physical characteristics, and moral character.

There are two demographic considerations that complicate the supply of youth qualifications, one relating to gender and one relating to race and ethnicity. Regarding gender, not all military jobs or units are open to women, particularly ground combat units in the Army and the Marine Corps;[1] furthermore, women do not have the same propensity to enlist as men. Accordingly, the military has to recruit more men than women, which necessarily reduces supply. Regarding race and ethnicity, DoD has always desired reasonable representation of all racial and ethnic groups, although it does not impose any type of arbitrary targets or quotas. Since different racial and ethnic groups can have different rates of qualifying characteristics and different propensities to enlist, a representation goal creates further constraints on the supply of qualified youth. For these reasons, some of the trends in youth qualifications are examined within these demographic categories.

MANPOWER REQUIREMENTS AND ENLISTMENT STANDARDS

Current Enlistment Standards

This section outlines various enlistment standards (or requirements) used by DoD to screen volunteers for military service. Some standards are common across all Services, such as education and certain aptitude measures, while others vary somewhat by Service. A few standards (e.g., physical fitness) are unique to a single Service. In most cases one can calculate the effects of these standards on reducing the eligible recruit

[1]Approximately 24 percent of Air Force enlisted personnel are women; virtually all jobs in the Air Force are open to women except special operations helicopter personnel, including pilots and para rescue.

population; this is generally true for education and aptitude requirements. In other cases, such as certain physical and moral standards, there are no precise population measures, so one can only estimate their impact on the potential recruit population.

The enlistment standards reviewed here include education, aptitudes, physical or medical attributes, moral character, and certain demographic characteristics. Our primary focus is on DoD-wide standards; Service-specific standards are discussed as appropriate. There are two types of DoD enlistment standards. One type consists of absolute minimums or maximums set by statute or by DoD policy directives. The other type comes from Defense Guidance, which provides DoD policy benchmarks used during the budgeting process.[2] While Defense Guidance benchmarks are not rigid requirements, the secretary of defense monitors Services budgets for compliance and may require budget reallocations in order to meet the benchmarks.

1. *Education:* Defense Guidance says that at least 90 percent of non-prior-service accessions must have a high school diploma. Youth with GED certificates are considered nondiploma graduates.

2. *Aptitudes:* Minimum aptitude standards are expressed in terms of categories of the AFQT, as follows: Category I is the 93rd–99th percentile; Category II is 65th–92nd percentile; Category IIIA is 50th–64th percentile; Category IIIB is 31st–49th percentile; Category IV is 10th–30th percentile; and Category V is below the 10th percentile.

Minimum aptitude standards:

- Youth who score in Category V are ineligible to enlist by statute.
- No Service may enlist more than 20 percent Category IV recruits by statute.

Defense Guidance:

- At least 60 percent of accessions in each Service should be Category I-IIIA.
- No Service should enlist more than 4 percent Category IV, and all should be high school diploma graduates.

[2]Defense Planning Guidance is issued annually by the secretary of defense to all military departments and agencies during the early stages of the budgeting process. It provides goals and priorities for preparing the budget and consequently defines DoD policy in many areas. For example, the education and aptitude benchmarks allow the Services to establish sufficient resources in their recruiting budgets to meet these benchmarks. Service budget submissions are monitored for compliance during the year, and should achievement of the benchmarks be in doubt, the secretary of defense may require reallocation of resources in the final version of the budget submitted to Congress.

3. *Physical and medical requirements* (DoD Directive 6130.3, December 15, 2000):

   To be eligible for enlistment, applicants must be:
   - Free of contagious diseases that probably will endanger the health of other personnel;
   - Free of medical conditions or physical defects that may require excessive time lost from duty for necessary treatment or hospitalization or probably will result in separation from the Service for medical unfitness;
   - Medically capable of satisfactorily completing required training;
   - Medically adaptable to the military environment without the necessity of geographical area limitations;
   - Medically capable of performing duties without aggravation of existing physical defects or medical conditions.
4. *Moral character* (DoD Directive 1304.26, December 21, 1993):

   Moral standards are designed to disqualify:
   - Individuals under any form of judicial restraint (e.g., bond, probation, imprisonment, or parole);
   - Individuals with significant criminal records (i.e., a felony conviction);
   - Individuals who have been previously separated from Military Service under conditions other than honorable or for the good of the Service;
   - Individuals who have exhibited antisocial behavior or other traits of character that would render them unfit to associate with military personnel.
5. Demographic (DoD Directive 1304.26, December 21, 1993):
   - Age: At least 17, not more than 35.
   - Citizenship: Citizen or legal permanent resident.
   - Dependency: If married, no more than 2 dependents under 18; if unmarried, ineligible with custody of dependents under 18.
   - Sexual orientation: Irrelevant; however, homosexual conduct is disqualifying.

Each branch of Service may set enlistment standards above these minimums, and in fact each Service has a separate set of aptitude requirements for each major type of job in that Service. These job-specific aptitude requirements are made up from various subtests of the Armed Services Vocational Aptitude Battery (ASVAB), and they differ by Service. For the purpose of this report, it is not necessary to go into the details for each Service's aptitude requirements for each job, but there is a recent report by the RAND Corporation that provides such detail (Levy et al., 2001).

Rationale for Enlistment Standards

Since enlistment standards affect the size of the population eligible for enlistment, during a time of recruiting shortages there must be a clear rationale for these manpower requirements, especially for those requirements that reduce the size of the eligible population to a significant degree. In this regard, the enlistment standards involving education and aptitudes are perhaps the two most important.

Education standards have been justified primarily on the basis of first-term attrition rates. Research conducted over many decades has demonstrated repeatedly that non-high school graduates have very high attrition rates during their first term of enlistment, rates that are nearly twice as high as for high school diploma graduates. High attrition rates impose a substantial manpower cost, driving up the number of accessions needed to maintain force size and increasing training costs substantially.

Table 4-1 shows enlisted attrition rates calculated by the Defense Manpower Data Center for cohorts entering military service during the 10-year period from 1988 to 1998. The average 24-month attrition rate for high school diploma graduates is about 23 percent, compared with 44 percent for non-high school graduates. It is important to note that attrition rates are nearly as high for GED graduates (the largest number of high school equivalency certificates) as for nongraduates, thereby justifying education standards based on having a high school diploma.

Aptitude standards have been justified on several grounds. Historically, aptitude standards were justified by passing rates in training schools and other commonsense criteria, such as reading skills. For example, a Category IV recruit reads at only the 3rd or 4th grade level, which means that even the most basic training manuals for the easiest jobs are beyond their reading comprehension ability.

More recently, DoD requested a comprehensive study of job perfor-

TABLE 4-1 24-Month Enlisted Attrition Rates by Education, 1988–1998

Education	Number	24-month Attrition (%)
HS diploma graduate	2,027,546	23.4
College	95,628	25.5
Adult education	32,330	36.8
HS GED	73,371	41.2
HS certificate of completion	6,798	35.4
Other Equivalent	1,149	34.5
Non-HS graduate	26,440	43.7
Total DoD (less missing data)	2,263,262	24.5

SOURCE: Data from Defense Manpower Data Center (2001).

mance that found substantial correlations between AFQT scores and performance in a wide range of enlisted jobs, including combat specialties (National Research Council, 1991, 1994). Not only did these correlations validate the use of AFQT for setting aptitude standards, but also the correlations were used in special cost-benefit studies to establish optimal education and aptitude cutoff scores under various assumptions about the cost of high-quality accessions. These cost-benefit studies were used to help establish the minimum aptitude standards currently used by the DoD.

The Defense Guidance criteria recommends a maximum of 4 percent Category IV accessions, which is considerably lower than the 20 percent maximum set by minimum aptitude standards. Since the cost-performance model did not establish 60 percent Category I-IIIA as an absolute minimum, it is possible that, in the face of a shortfall problem, a modest increase in Category IIIB recruits in place of Category IIIA (with high school diplomas) would not appreciably lower military job performance.

The justification for this conclusion is the relationship between AFQT scores and job performance scores, as shown in Figure 4-1 (National Research Council, 1994:20). At all levels of job experience, Category I-II personnel have much higher hands-on job performance scores than Category

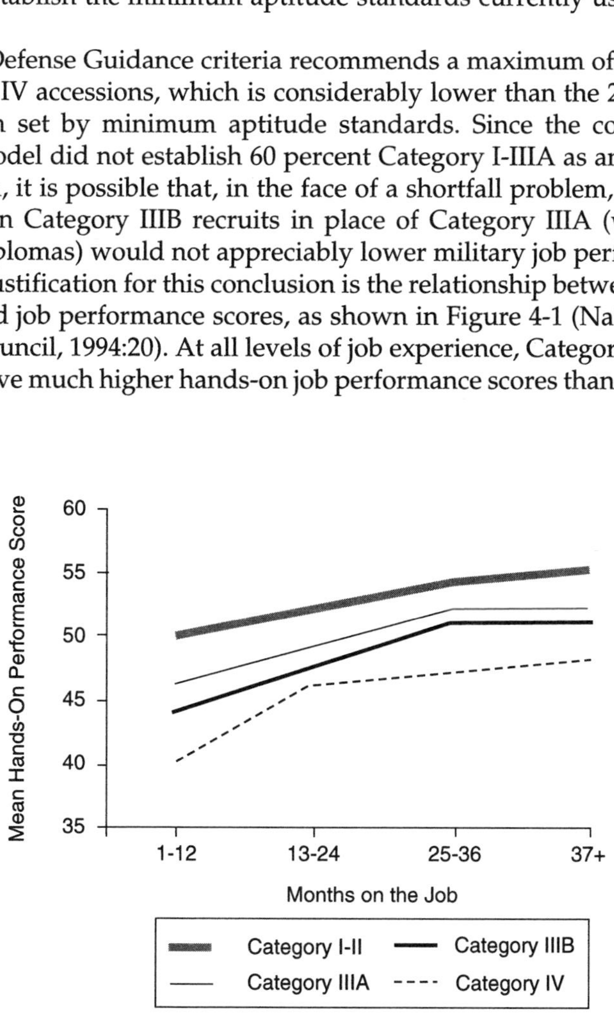

FIGURE 4-1 Job performance and AFQT (from the JPM Project, by AFQT and job experience).
SOURCE: National Research Council (1994).

III personnel, while Category IV personnel score much lower. In contrast, however, the performance scores between Category IIIA and Category IIIB personnel are very close, with an average difference of only about 2 points.

TRENDS IN APPLICANT AND RECRUIT QUALIFICATIONS

One way to evaluate the role of enlistment standards in recruiting shortages is to consider historical trends in recruiting success with respect to the major standards. Historical trends can be examined for pools of applicants as well as for cohorts actually enlisting each year. In this chapter, an applicant is defined as a person who has contacted a recruiting office and gets at least as far as taking the ASVAB.[3] The applicant pool includes those who go on to enlist and those who do not, for whatever reason. On one hand, trends in applicant numbers give some idea of potential supply, although since applicants have made contact with the recruiting station, they are not necessarily representative of the total youth population. On the other hand, trends in accession cohorts tell us about the qualifications of those actually enlisting, thereby revealing the extent to which the Services are meeting or exceeding enlistment standards.

Education Levels

Figure 4-2 shows trends in education levels for applicants (enlisted force) between 1980 and 2000. Three levels of education are identified: those who have graduated from high school or have had some years of college, those with a high school equivalency certificate of some type such as a GED, and nongraduates. Those who have attended some college (not shown in the figure) comprised about 10 percent of applicants during the 1980s, but those with some college declined to less than 4 percent after 1990.

The peak in applicants with high school diplomas occurred in 1992 at 97 percent, and it has been declining since that time. In 1999 it fell below 90 percent for the first time since 1982, and it remained below 90 percent in 2000. This downward trend in applicants with high school diplomas is offset by a slight increase in the percent of applicants with a GED or other high school credential, which rose steadily during the 1990s and reached 10 percent or more in 1999 and 2000. The percent of applicants who did

[3]The applicant pool excludes persons (1) who take the ASVAB in high school but do not contact recruiting offices or (2) who contact a recruiting office but are screened out during the preliminary interview for various reasons without taking the ASVAB.

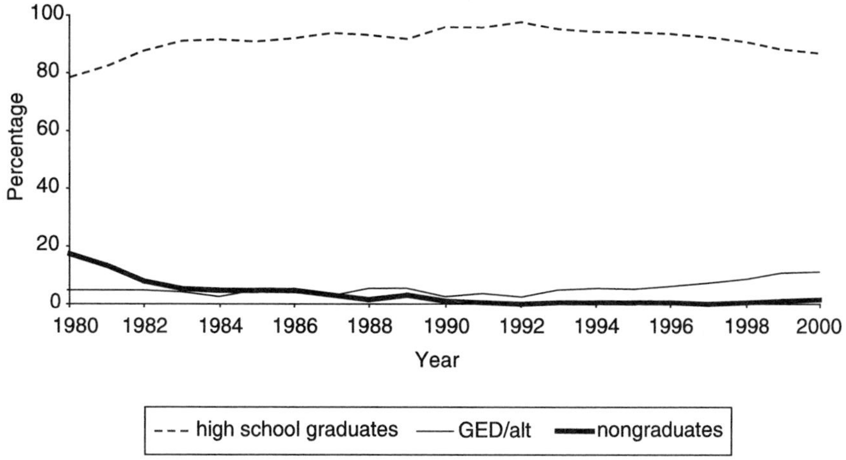

FIGURE 4-2 Trends in education for military applicants (enlisted force).
SOURCE: Data from Defense Manpower Data Center (2001).
NOTE: GED/alt = high school equivalency certification.

not graduate from high school declined rapidly during the early 1980s, and throughout the 1990s it has comprised less than 2 percent of all applicants. Clearly, the decline in diploma graduates is being filled by GED certificates and it appears that nongraduates are not being encouraged to apply for military service.

The applicant trends can be compared to the accession trends in Figure 4-3, which are virtually identical to the trends in Figure 4-2. The rate of accessions with high school diplomas rose during the 1980s, peaked in 1992 at 98 percent, and then gradually declined until 2000. In 1999 the diploma rate fell below 90 percent for the first time since 1982. As with applicants, the rate of accessions with GEDs or other high school certificates has risen to just below 10 percent, while the percent of new recruits who are non-high school graduates has remained below 2 percent since 1987.

Between the mid-1980s and the late 1990s, all Services were able to meet their recruiting targets while sustaining rates of 90 percent or more accessions with high school diplomas. Since 1995 the rate of high school diploma accessions began dropping, and in 1999 and 2000 it finally fell below 90 percent for the first time in many years. Interestingly, the Services have made up the gap in nondiploma graduates by recruiting mostly persons holding GEDs and other high school equivalency credentials rather than recruiting nongraduates, even though their attrition profiles

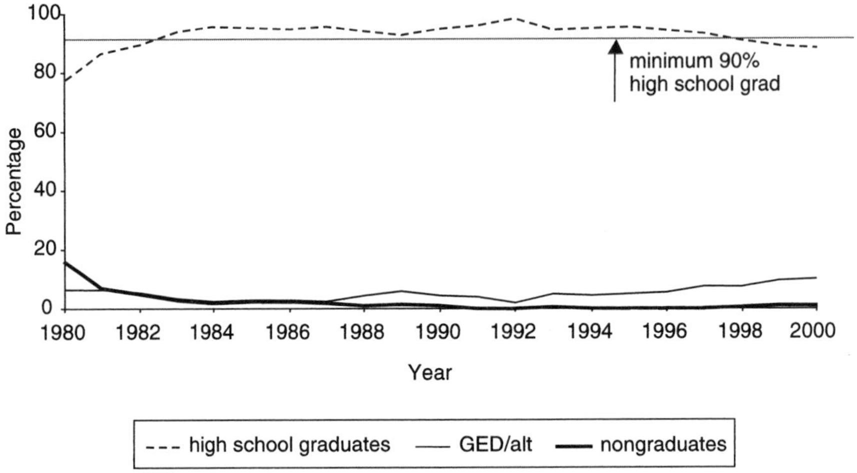

FIGURE 4-3 Trends in education for military accessions (enlisted force).
SOURCE: Data from Defense Manpower Data Center (2001).
NOTE: GED/alt = high school equivalency certification.

are not that different. Like the applicant rate, the rate of accessions who are nongraduates has remained below 2 percent since 1991.

It is noted that the Army and the Navy have the greatest difficulty meeting the Defense Guidance minimum of 90 percent high school diploma graduates. In 1999 both the Army and the Navy fell to 88 percent high school graduates, while the Marine Corps and the Air Force recruited 95 and 98 percent high school graduates, respectively.

This downward trend in the accession of high school diploma graduates is especially noteworthy given the population trends in high school graduation and dropout rates. As we discuss later, high school dropout rates have continued to fall for the past 10 years, so the trends in military applicants and accessions appear to reflect youth propensity rather than the supply of qualified youth.

Aptitude Levels

The trends in applicant aptitudes, as measured by the AFQT, are shown in Figure 4-4. Like the education trends, the percentage of Category I–IIIA applicants rose steadily during the 1980s and early 1990s, peaking in 1992 at 61 percent, and then showing a gradual decline up to 2000 when it fell to about 53 percent. Correspondingly, the percent of

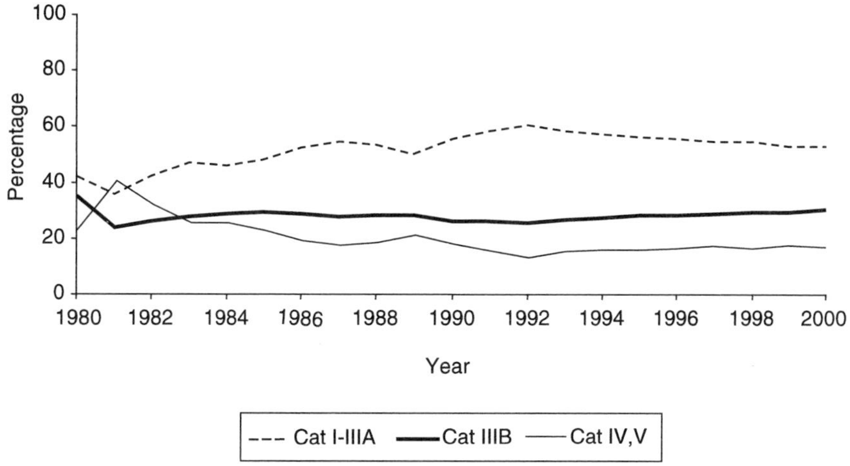

FIGURE 4-4 Trends in aptitudes for military applicants (enlisted force).
SOURCE: Data from Defense Manpower Data Center (2001).

Category IV applicants declined during the 1980s and then stabilized at between 15 and 17 percent. The loss of above-average applicants has been largely replaced by a modest increase in the percentage of Category IIIB applicants, which rose from 25 percent in 1992 to 30 percent in 2000.

Again, the trends in accession aptitudes are quite similar to applicants, with one major exception: the percentage of Category IV accessions dropped throughout the 1980s and then continued to drop throughout the 1990s (Figure 4-5).[4] Category IV accessions have not been above 2 percent since 1990, and their highest level in the past decade was 1.4 percent in 1999 (the most difficult recruiting year in recent times). Thus the Services have had no difficulty meeting the DoD maximum of 20 percent Category IV accessions, and they remained well under the Defense Guidance maximum of 4 percent even during the most difficult recruiting years.

With respect to Category I–IIIA goal, while the Services no longer reach levels of 70 percent or more that were common in the early 1990s, they have remained well above the Defense Guidance minimum of 60 percent. Even in the most challenging recruiting year, 1999, the Services recruited 65 percent high-aptitude enlistees overall. The Army had the

[4]The sharp increase in the percentage of Category IV accessions in 1981 followed by a sharp drop was due to an ASVAB misnorming problem and subsequent recruiting changes in the aftermath. The misnorming problem was a technical test scoring error such that persons classified as Category IIIB were, in actuality, Category IV.

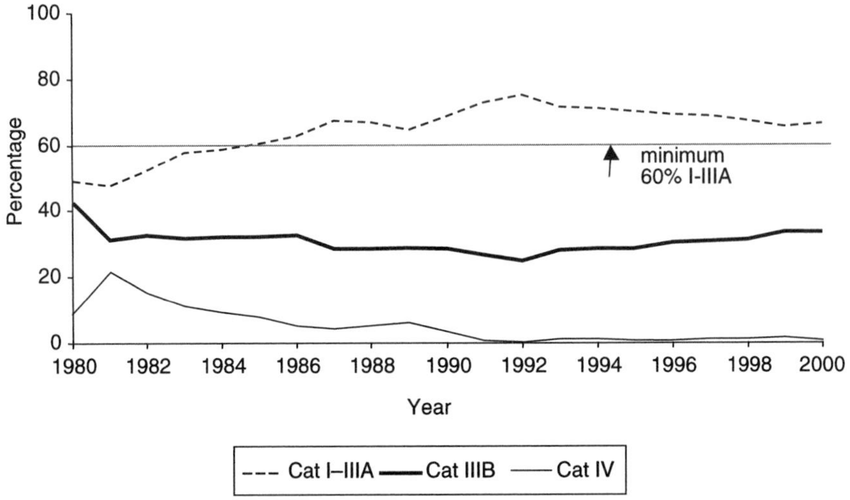

FIGURE 4-5 Trends in aptitudes for military accessions (enlisted force).
SOURCE: Data from Defense Manpower Data Center (2001).

fewest but still attained 62 percent. As the recruiting environment became more difficult during the late 1990s, it appears that in the cost and quality trade-offs between education and aptitude, the Services were more willing to accept a limited number of non-high school diploma graduates (with GEDs) rather than Category IV recruits.

A summary of education and aptitude trends is shown in Figure 4-6, which provides trends in the percentage of "highly qualified" applicants and accessions. A highly qualified youth is a young person in Category I–IIIA who also has a high school diploma. After reaching all-time lows in 1980 because of the ASVAB misnorming problem, the percent of highly qualified applicants and accessions reached historic highs of 61 and 75 percent, respectively, in 1992. After those years the percent of highly qualified applicants and accessions began a steady decline, falling to lows of 52 and 64 percent, respectively, in 1999. Highly qualified accessions rose about 0.5 percent in 2000.

Although the rate of highly qualified accessions is still high by historical standards, the downward trend would be a matter of concern if it continues. With the increasing use of high-technology equipment in all aspects of military operations, most military manpower experts believe that aptitude needs will be greater in coming years. At the very least, military planners would like to stop the downward trend. An important question is the extent to which the downward trend is driven by a decline

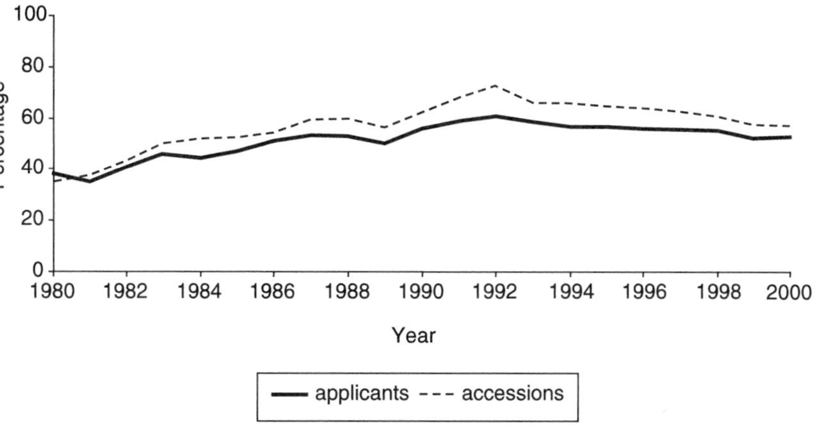

FIGURE 4-6 Trends in highly qualified applicants and accessions.
SOURCE: Data from Defense Manpower Data Center (2001).

in the supply of highly qualified youth or rather a decline in the propensity of highly qualified youth to join the military. The supply question is discussed in a later section, and the propensity question is addressed in subsequent chapters.

Trends by Gender and Race/Ethnicity

How do education and aptitude trends differ by various demographic subgroups, such as gender and race or ethnicity? Do subgroup trends offer any information about special recruiting problems given the DoD goal of maintaining a fair social representation of all population subgroups? Education and aptitude trends for gender and racial subgroups can be summarized using the percentage of highly qualified applicants and accessions.

Figure 4-7 shows the percentage of highly qualified applicants and accessions by gender. The rates of highly qualified applicants and accessions are virtually identical for both men and women. Before 1993, female accessions tended to be more highly qualified than male accessions, but since that time the rates have become very similar for both genders. This may reflect the fact that women are being recruited in higher numbers, and the Services cannot be as selective as they once were. The most important implication of the small differences, however, is that women show the same rate of qualification as men with respect to education and aptitude. Accordingly, women continue to offer an alternative potential source of recruits when the recruiting climate for men turns difficult.

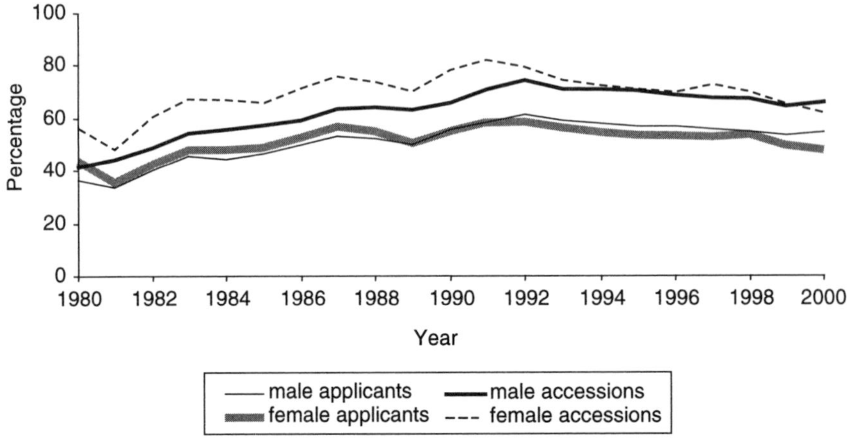

FIGURE 4-7 Highly qualified applicants and accessions by gender.
SOURCE: Data from Defense Manpower Data Center (2001).

Figures 4-8a and 4-8b show the rate of highly qualified applicants and accessions for white, black, and Hispanic subgroups. With respect to the applicants in Figure 4-8a, it is interesting that all three groups showed similar increases in highly qualified rates during the 1980s and a leveling off during the 1990s. Both black and Hispanic groups show lower rates of highly qualified youth than whites, with Hispanics showing higher rates than blacks.

Figure 4-8b shows about the same patterns for accessions, but the rate of highly qualified white accessions has remained fairly constant while the rates of highly qualified black and Hispanic accessions has dropped off, particularly Hispanics. This may reflect more aggressive recruiting of Hispanic youth, a group that has been somewhat underrepresented in the U.S. military compared with their population proportions. In this respect, it should be emphasized that the share of Hispanic recruits has risen steadily during the 1990s, reaching 11 percent in 2000, only a few percentage points shy of their representation in the youth population. In addition, the propensity to enlist of Hispanic youth (both male and female) is slightly higher that of white youth (Segal et al., 1999).[5] One important reason for the overall underrepresentation of Hispanics is that they have a substantially lower high school graduation rate than either white or black youth (see below).

[5]Based on the responses of males and females who responded that they definitely will or probably will to the question: "are you likely to enter or want to enter the military." From Monitoring the Future (1976-1997).

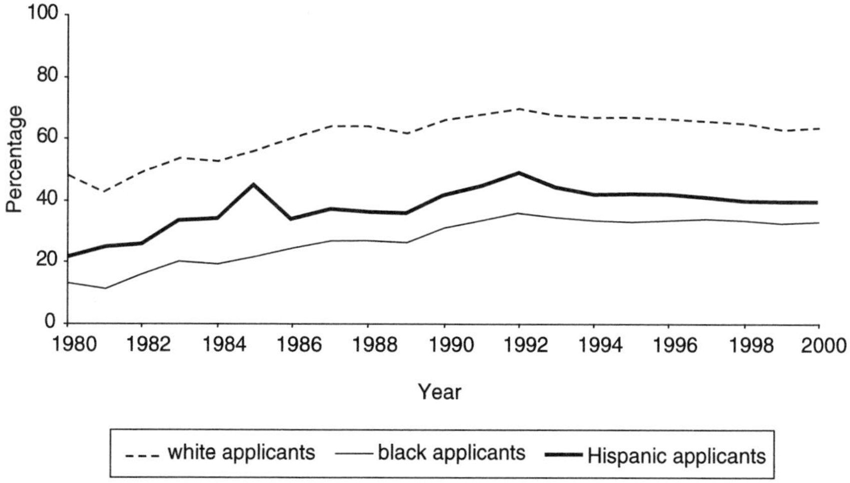

FIGURE 4-8a Highly qualified applicants by race/ethnicity.
SOURCE: Data from Defense Manpower Data Center (2001).

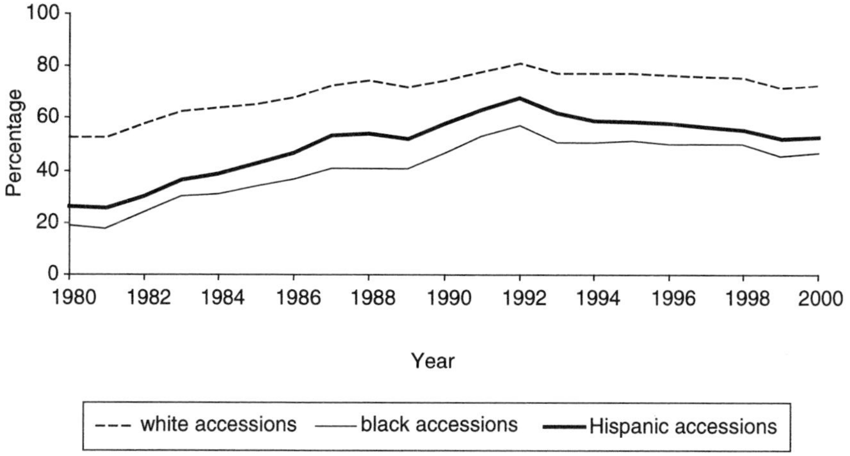

FIGURE 4-8b Highly qualified accessions by race/ethnicity.
SOURCE: Data from Defense Manpower Data Center (2001).

The fact that black youth have a lower rate of highly qualified applicants and accessions compared with whites is offset by their higher propensity to enlist. Thus the Services as a whole continue to recruit black youth at a rate somewhat higher than their population representation.

Other Enlistment Standards

The other enlistment standards include various physical, moral, and demographic characteristics as outlined in the previous requirements section. There are very few systematic data available on military applicants who have various characteristics that might make them ineligible for military service. The data are more complete for accessions, however, for which waivers (of standards) must be given for a variety of health, moral character, or demographic conditions to enable enlistment. For example, waivers have to be issued for such medical conditions as obesity and respiratory problems and for such moral conditions as drug abuse or felonies.

Figure 4-9 summarizes total DoD waivers given for moral character, physical problems, and a variety of other conditions grouped into "other." The three most important categories of "other" waivers are dependents, education levels, and aptitude requirements (such as low scores on ASVAB tests for specific job areas). The rate of waivers for moral character, which include drug use and various criminal behaviors, was quite high between 1980 and 1991, averaging between 16 and 18 percent of all accessions. The most frequent moral waiver was for pre-Service marijuana use. After 1991 the moral waiver rate declined steadily, dropping to about 5 percent by 2000. One reason for this decline is that the Services

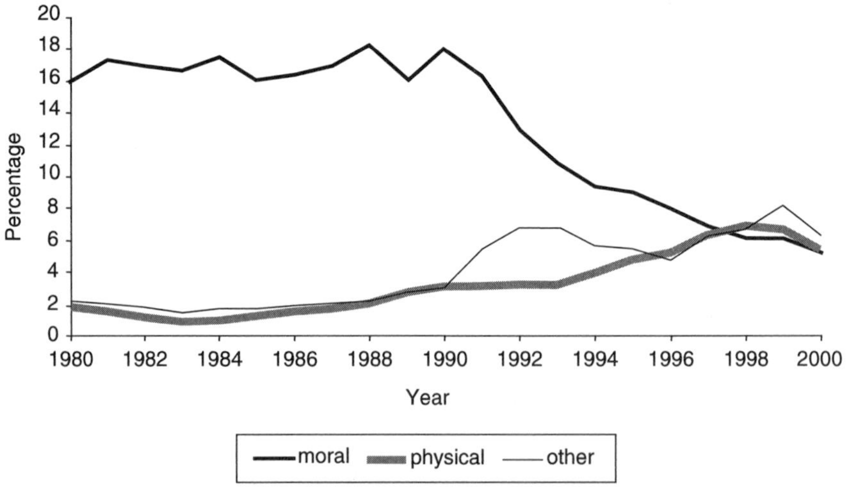

FIGURE 4-9 Trends in waivers for military accessions (enlisted force).
SOURCE: Data from Defense Manpower Data Center (2001).

phased out the requirement of a waiver for pre-Service use of marijuana, assuming the applicant tested negative for the drug during the physical.[6]

In contrast to moral character, the rates of waivers for physical problems and other conditions both increased during the 1990s, although their overall rates remain relatively low at about 5 or 6 percent, respectively. The modest increases in waivers for physical and other problems perhaps reflects the more difficult recruiting climate, in which potential shortfalls are avoided by enlisting persons with mild and remediable physical, aptitude, and education deficiencies.

TRENDS IN YOUTH QUALIFICATIONS

The historical trends for military manpower characteristics reveal declining rates of high-quality applicants and accessions, and recent data show that accession cohorts are slightly below Defense Guidance targets for education, approaching the minimum targets for aptitude levels. While the potential shortages of highly qualified recruits have not yet reached crisis proportions, military performance could be adversely effected if these trends were to continue to worsen in the future.

As mentioned earlier, shortages of more highly qualified recruits could be caused by a declining supply of highly qualified youths in the general population, or it could be ascribed to a reduced propensity to enlist for these highly qualified youths. The goal of this section is to assess the extent to which the supply of highly qualified youth might affect current or future trends in manpower quality. That is, are American youth characteristics with respect to education, aptitudes, and other quality indicators increasing, decreasing, or remaining relatively stable?

The quality indicators used here come from various sources. Data on education trends, including graduation rates, come from Current Population Surveys. Data on youth aptitudes and skill levels come from achievement tests administered by the National Assessment of Educational Progress (NAEP). Data on youth physical, moral, and various demographic characteristics come from a variety of sources.

Education Levels

The U.S. trends in education attainment have been rising for many decades, if not for the entire 20th century. Although rising educational attainment is manifested at all levels of schooling, the most important

---

[6]For the Army, the Navy, and the Marine Corps, applicants who test positive for marijuana must wait 6 months for a retest, and then if negative may be enlisted with a waiver (Burnfield et al., 1999).

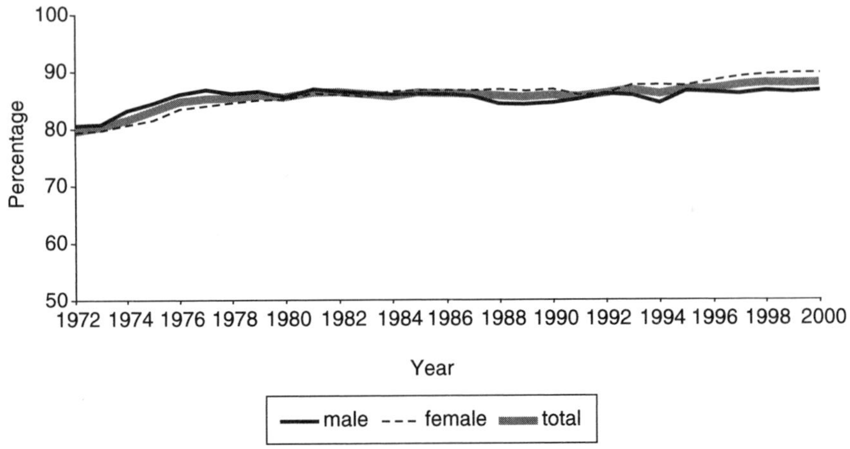

FIGURE 4-10 Trends in high school graduation rates by gender (ages 25–29).
SOURCE: Data from Current Population Surveys (various years).

marker for the enlisted force is high school graduation. Figure 4-10 shows trends in high school graduation rates broken down by gender. For the nation as a whole, high school graduation rates have risen from 80 percent in 1972 to 88 percent in 2000. During the 1970s, male graduation rates were two or three points higher than female rates, but since the late 1980s this pattern has reversed, and females have had slightly higher graduation rates than males. In the year 2000, graduation rates were 89 percent for females and 87 percent for males.

Figure 4-11 examines youth graduation rates by race or ethnicity. For white youth, graduation rates have remained relatively constant since the late 1970s, rising slightly from 86 to 88 percent by 2000. The most significant change has occurred for black youth, whose graduation rates rose dramatically from about 65 percent in 1972 to over 85 percent by 1995. Since 1995, graduation rates for black students have remained very close to white rates. In contrast, the graduation rate for Hispanic youth has not improved much and remains substantially below the rates for black and white youth. Hispanic graduation rates have risen just 5 points in the past 25 years, from about 58 percent in 1976 to only 63 percent in 2000. In fact, the Hispanic graduation rate in 2000 was not as high as the black rate was in 1972.

It appears, then, that over the past few decades the supply of youth with high school diplomas has improved slightly for whites and very substantially for blacks, and thus there is not likely to be a problem with the future supply of these two groups. There must be some concern about

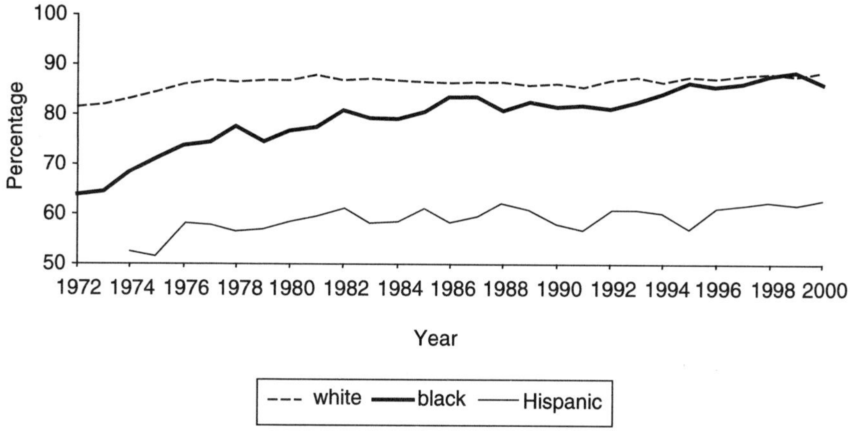

FIGURE 4-11 Trends in high school graduation rates by race/ethnicity (ages 25–29).
SOURCE: Data from Current Population Surveys (various years).

Hispanic youth, however, who had very low graduation rates even in 2000. This low graduation rate will make it more difficult for the Services to increase their Hispanic representation.

Aptitude Levels

The best long-term trend data on academic aptitudes and skills comes from the NAEP project, which has been administering achievement tests to national samples of students since 1970 (Campbell et al., 2000). These achievement tests are administered every two to four years to 4th, 8th, and 12th graders (ages 9, 13, and 17), and the subject matter covers reading comprehension, mathematics computation and concepts, and science. Although these tests are not identical to the subtests used in the AFQT, which is comprised of subtests in word knowledge, paragraph meaning, arithmetic reasoning, and math knowledge, the NAEP reading and mathematics content areas are similar. NAEP contains two separate measures: the NAEP trend data, which is the focus here, and a second assessment known as national NAEP. National NAEP is more closely linked to curriculum, and its content changes to stay abreast of curriculum changes. This report focuses on the NAEP trend data, as the measures used are more similar than the national NAEP to AFQT content. It is worth noting that the national NAEP results indicate that there is some substantial growth in mathematics achievement for the nation's students as a whole, as well as for all groups of students.

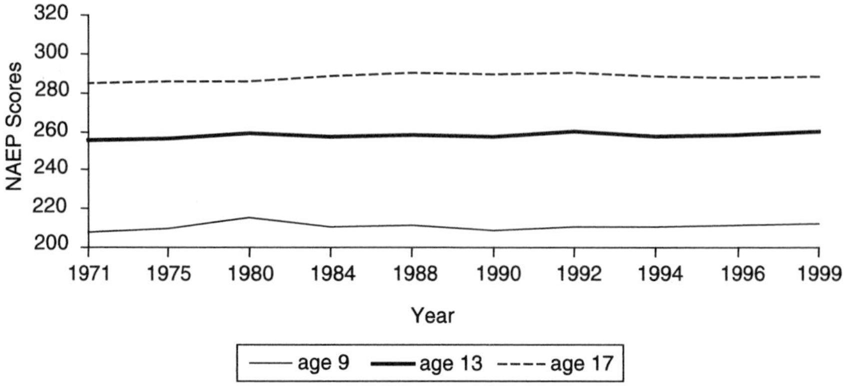

FIGURE 4-12 National trends in reading achievement.
SOURCE: Data from National Assessment of Educational Progress.

Note that the use of the labels "aptitude" and "achievement" reflect the intended use of the different tests. The Services use the term "aptitude as the AFQT is being used to forecast future job performance; NAEP uses the term "achievement" as the test is being used as a description of current levels of accomplishment. Different users assign the labels of ability, achievement, and aptitude to very similar tests.

Figures 4-12 through 4-14 present trends in reading, math, and science scores for the three NAEP age groups. For the reading scores shown in Figure 4-12, the trends are just about flat; there are no significant increases or decreases in reading skills for any of the three age groups between 1971 and 1999.

For the math scores shown in Figure 4-13, a different pattern emerges. All three age groups show upward trends in math scores starting in 1982. Age 17 students experienced a slight decline of about 5 points in math between 1973 and 1982, but then they show a steady increase up to 1992. After 1992 the math scores for age 17 level off, but they end the period several points higher than they were in 1973. After 1978 the age 13 math scores show gradual increases until 1999; altogether 13-year-olds gained more than 10 points, which is about one-third of a standard deviation.[7] But the largest gains occurred for age 9 students, whose math scores rose 13 points between 1973 and 1999.

---

[7]On most standardized tests, students gain about one standard deviation per year in the early grades.

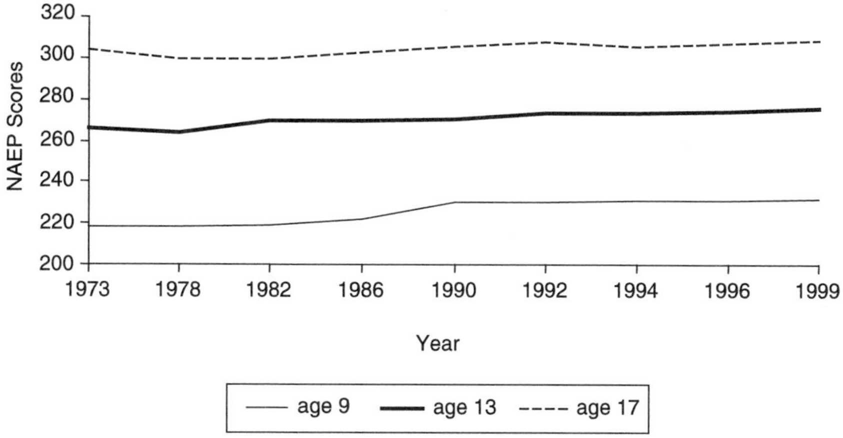

FIGURE 4-13 National trends in math achievement.
SOURCE: Data from National Assessment of Educational Progress.

Finally, the science achievement trends shown in Figure 4-14 are similar to math scores in that after 1982 all age groups have rising science scores. All three groups show declines in science scores during the 1970s, and the drop was especially bad for age 17 students, whose scores dropped by more than 20 points between 1970 and 1982. After 1982 all science scores were on the rise, although age 17 students have not re-

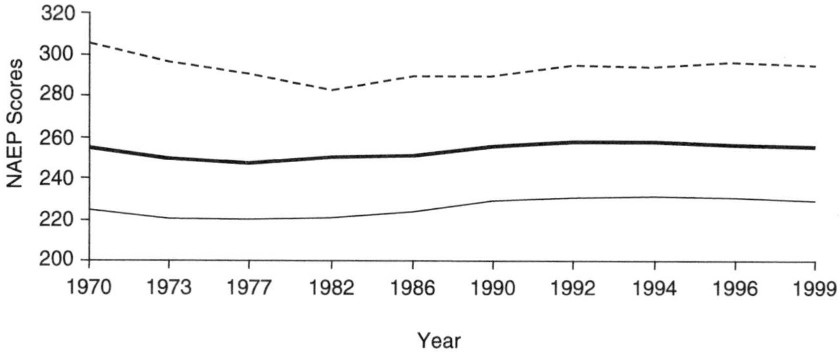

FIGURE 4-14 National trends in science achievement.
SOURCE: Data from National Assessment of Educational Progress.

gained their 1970 highs; they are still 10 points behind. The other two age groups have been able to surpass their 1970 scores, age 13 by 1 point and age 9 by 4 points. It is interesting that the turnaround in both math and science scores occurred the same year that *A Nation at Risk* (National Commission on Excellence in Education, 1983) appeared, a report that proclaimed a crisis in American education and which prescribed a series of reforms aimed at reversing these declining test score trends.

For the purpose of projecting future levels in aptitudes and skills for the youth population, the age 9 grades scores are most important since they will become the 18-year-olds of the future. The cohorts of age 9 students between 1992 and 1999 will become the age 18 cohorts between the years 2000 and 2007, so it is especially important to consider those scores. In addition, since the military uses the 50th percentile to define highly qualified persons, it would be worthwhile to convert the NAEP scores to 50th percentile scores before discussing the future supply of youth with higher skills and aptitudes. One would expect the trends to be about the same, but if changes have occurred at the top or bottom of the score distribution, it is possible that trends in median scores might differ somewhat from trends of mean scores.

Figure 4-15 shows the trends in age 9 NAEP scores at the 50th percentile (median score trends). Just as for the trends in mean scores, the reading median scores are relatively flat, while math and science show significant improvements over the past 30 years. It is also interesting that most of the improvements in age 9 math and science scores took place

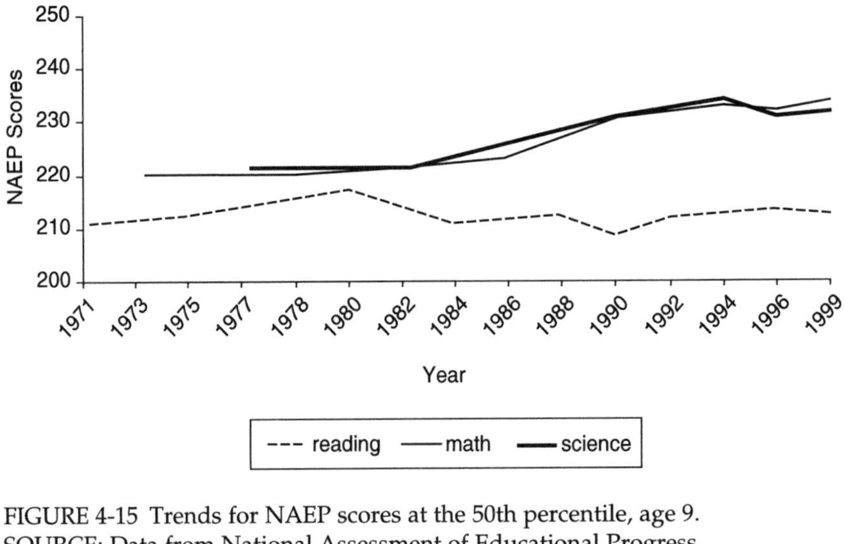

FIGURE 4-15 Trends for NAEP scores at the 50th percentile, age 9.
SOURCE: Data from National Assessment of Educational Progress.

between 1970 and 1990, and these cohorts have already passed through age 18, a prime year for enlisting in the military. Since there are no significant changes between 1992 and 1999, one would not expect to see significant changes in aptitudes for these late teen youth between 2000 and 2007.

Of course, there is no guarantee that improvements in test scores at age 9 will be carried forward to later ages, and in fact, considering the math trends for age 17 students in Figure 4-13, this does not seem to have happened. That is, the sharp gains in age 9 math scores between 1982 and 1990 are not replicated by age 17 gains between 1990 and 1999.

Do these aptitude trends differ by gender or race/ethnicity in ways that might affect recruiting targets? With respect to gender, the trends for both male and female age 17 reading scores are generally flat, but females outscore males by more than 10 points. That situation is reversed with math scores, for which age 17 males outscore females by several points, but both groups show gains. Females have gained slightly more than males, and by 1999 the male advantage is only 3 points. If one averages reading and math scores (to approximate the AFQT content), one would find females scoring somewhat higher than males overall. Thus we expect the supply of highly qualified females to be on the same order of magnitude as that for males.

Unlike gender, there are significant differences in aptitudes according to race and ethnicity. Generally, white youth have the highest scores, Hispanics second, and black youth have the lowest scores. One interesting development has been a reduction of the achievement gaps between white and both black and Hispanic students, mostly between 1970 and 1990. During this period, black students reduced the age 17 reading gap by about 20 points and the age 17 math gap by about 10 points. Hispanic students reduced the gaps by about 10 points each, but they were not as large to begin with. Since 1990 there has not been much change in the relative scores of each group. There is no consensus among social scientists about the causes of these narrowing achievement gaps; various explanations include school desegregation, compensatory education, and improved socioeconomic status of black families (see Jencks and Phillips, 1998, for a discussion of alternative explanations).

The reduced gap means that more minority youth should fall into the highly qualified category, and indeed this is supported by the trends in highly qualified applicants (Figure 4-8b), which show increasing proportions of highly qualified black and Hispanic applicants between 1980 and 1990. The fact that a large gap still exists, however, is a matter of some concern when recruiting shortfalls occur. During times of recruit shortages, there is a tendency to enlist black youth at a higher rate than their proportionate share of the population, thereby making it harder to maintain the social representation goals of the DoD.

Taken together, the trends in youth aptitudes offer no reason to expect changes in reading skills over the next 10 years or so, and there might be increases in math and science skills. Most important, analysis of the NAEP data yields no support for past declines, or a prediction of future declines, in the types of cognitive aptitudes and skills measured by the AFQT. Accordingly, to the extent that there has been some decline in the enlistment of higher-aptitude youth in recent years, we must conclude that it is due to falling propensity of these youth to enlist rather than to their supply in the youth population.

With respect to the gains in math and science, it should be noted that there is a well-documented phenomenon called "the Flynn effect," in which IQ test scores in the general population have risen about 3 points per decade over the past 80 years or so (Flynn, 1998). While experts do not agree on the precise causes for these increases in intellectual ability, it is likely that the rise in NAEP scores is a part of this phenomenon. The practical implication is that each succeeding generation has greater cognitive skill than the previous generation. When IQ and other tests are "renormed" so that the average remains constant (e.g., 100), the renormed scores mask actual improvement in underlying cognitive abilities.

Like the NAEP tests, the AFQT is also a standardized, "normed" test. By construction, 50 percent of the youth population on which the test was normalized will score in Categories I to IIIA. To the extent that the improvement in NAEP scores reflects the Flynn effect, it is possible that over time a greater proportion of the youth population will score above the 50th percentile on the AFQT due to improved aptitudes, especially in math and science skills. For this reason, DoD periodically renorms the AFQT. If any future AFQT renorming occurs due to rising aptitudes, DoD should also consider readjusting its recruiting goals in light of the renormed test so that it does not inadvertently increase aptitude targets.

Physical and Moral Attributes

Population data are more limited with regard to some of the enlistment standards that involve physical and moral attributes. In this section we present and discuss information on four youth attributes that constitute a sizable proportion of military waivers: criminal behavior, drug use, obesity, and asthma.

Figure 4-16 shows the rates of illegal drug use among high school seniors from 1979 to 2000. The data are taken from the Monitoring the Future surveys conducted by the Institute for Social Research, University of Michigan (adapted from Johnston et al., 2001). All rates of illicit drug use by seniors decline substantially between 1979 and 1992, from just under 40 percent for all types of drugs to about 15 percent. After 1992 that

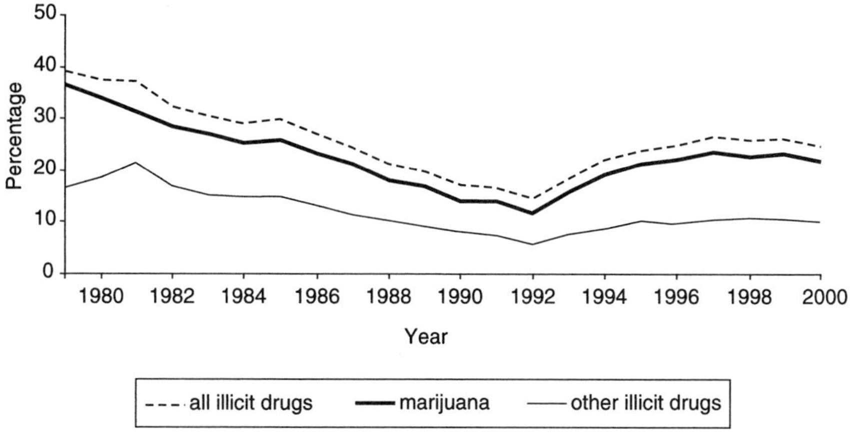

FIGURE 4-16 Trends in illicit drug use for high school seniors, past 30 days.
SOURCE: Data from Monitoring the Future surveys.

the rates begin moving upward again, and then they flatten out at about 25 percent starting in 1997, still far lower than their high levels in the late 1970s. By far the dominant drug abused is marijuana, with over 20 percent usage in 2000, compared with only 10 percent for all other drugs. Given the fact that the Services no longer require a moral waiver for preservice marijuana use, it is unlikely that the modest increase in marijuana use after 1992 has had a significant adverse affect on military recruiting.

Trends in arrest rates for juveniles, shown in Figure 4-17, likewise seem to suggest little impact on recruiting (U.S. Department of Justice, 1999). These figures are the percentages of offenses known to police that were cleared by arrests of youth under 18. Although there was some increase in both violent and property crimes between 1989 and 1994, both rates began falling in 1995 and have continued to fall through 1998. By 1998, property crime rates were slightly lower than their levels during the 1980s, and violent crime rates were only a couple of percentage points higher (at about 12 percent).

The youth population data for medical problems present a somewhat different picture, in that the rates for two common medical problems that require waivers are increasing. Figures 4-18 and 4-19 show youth rates for asthma and obesity, respectively, which can be grounds for physical disqualification if they are serious. If the conditions are viewed as manageable, youth with these conditions can be enlisted with a waiver.

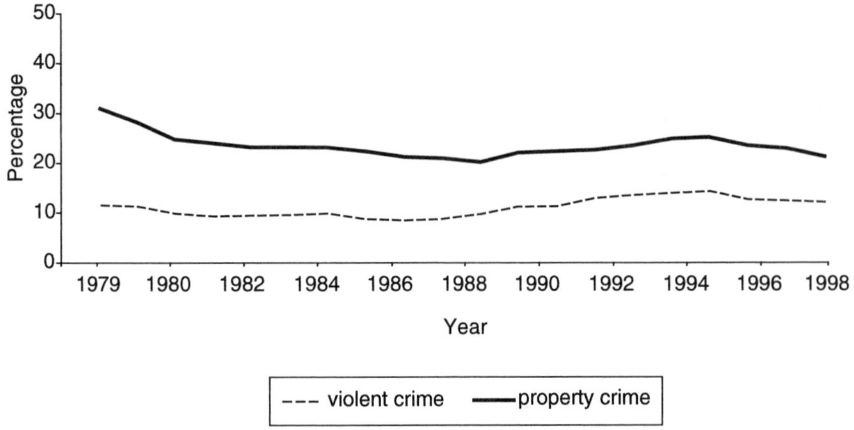

FIGURE 4-17 Trends in arrest rates for persons under 18.
SOURCE: U.S. Department of Justice (1999).

The rates of asthma doubled between 1980 and 1995, and even though the definition of asthma was changed in 1997 (from a condition of asthma to an asthma attack in the past 12 months), it is believed to represent a trend that continues to rise (Mannino et al., 2002). Likewise, obesity rates for youth ages 12 to 17 have nearly tripled between 1980 and 1999 (National Center for Health Statistics, 1999). In view of these trends, it is perhaps not surprising that the rate of military waivers for physical problems has risen markedly in the past 10 years (see Figure 4-9). If these

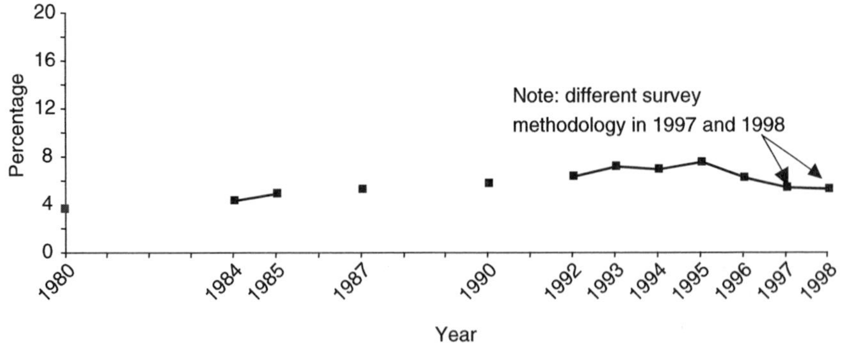

FIGURE 4-18 Trends in asthma rates for persons under 18.
SOURCE: Mannino et al. (2002).

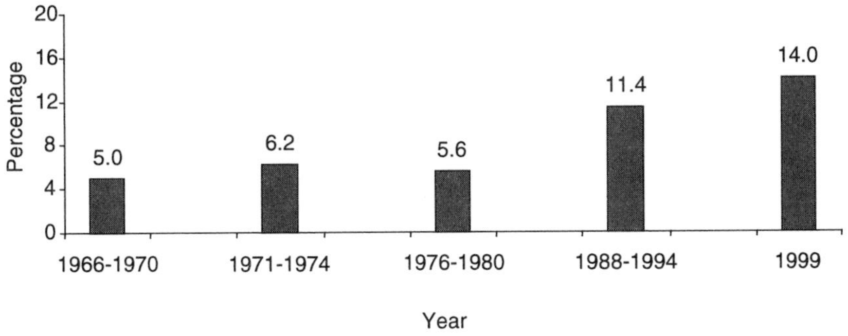

FIGURE 4-19 Trends in obesity rates for youth ages 12–17.
SOURCE: National Center for Health Statistics (1999).

trends continue, they could adversely impact recruiting success or, with waivers, possibly have negative effects on military job performance.

SUMMARY

Research shows that there is a significant correlation between recruit qualifications (especially education and aptitudes) and military performance, and hence current statutory and policy requirements call for somewhat higher rates of highly qualified recruits than their proportions in the general youth population.

While the demand for highly qualified personnel remains strong, the actual enlistment of highly qualified recruits has declined in the past 10 years after reaching all-time highs during the early 1990s. In spite of this decline, the rate of highly qualified youth recruited for the enlisted force continues to be above DoD targets, even in the most difficult recruiting year of 1999. However, if the downward trend continues, there could be shortfalls of recruits with high school diplomas or with higher aptitudes or both over the next 20 years.

The potential supply of highly qualified youth in the U.S. population, in terms of education and aptitude, will remain fairly stable over the next 10 years, and there is no reason to expect any declines over the next 20 years. If anything, the proportion of highly qualified youth may increase slightly, particularly if the high school graduation rate remains high while the math aptitude scores continue to rise. If AFQT scores rise to the point at which renorming is contemplated, DoD should also consider lowering aptitude targets to avoid an inadvertent reduction in the supply of high-aptitude youth.

On the basis of recent population trends, there may be further increases in certain youth population characteristics over the next 20 years that require waivers for accession, particularly physical health conditions. The trends in obesity and asthma are among the most worrisome.

Since the supply of highly qualified youth has not declined, but there has been a decline in highly qualified personnel applying for and enlisting in the military, it is quite possible that these declines reflect lower propensity for military service rather than shortages of supply in the youth population. This possibility is examined extensively in subsequent chapters.

5

Trends in Employment and Educational Opportunities for Youth

As they approach the completion of primary and secondary schooling, eligible youth confront the choice of entering military service. Competing with this choice are two primary alternatives: (1) entering the civilian labor market and (2) continuing education by entering college.[1] In this chapter, we examine these choices, focusing on understanding the aspects of the choice that may affect the decisions of recruit-eligible youth (e.g., high school graduates).

THE DECISION TO ENLIST: A CONCEPTUAL MODEL

Underlying the rationale for examining these choices is a simple model of occupational choice: individuals compare pecuniary and nonpecuniary benefits of enlisting and conditions of service relative to their next best alternative and choose the one that provides the greatest net benefits. In this model, we hypothesize that individuals, at the completion of high school, choose a time path of jobs, training, and education that maximizes their expected welfare or utility over their lifetime. Elementary labor economics suggests that the utility of a job is a function of earnings and deferred compensation, benefits, working conditions, and hours of work (or its complement, leisure.) However, the earnings the individual can command, as well as the other aspects of the job, are a function of education, training, and experience. Hence, the individual has an incentive to

[1]While there are other options available after finishing high school (e.g., marriage or leisure), this chapter focuses on the primary choices that compete for youth.

invest in additional education by attending college and to seek jobs that provide training that is generally valued in the labor market because this improves future job opportunities and earnings. The path chosen by any particular individual will depend on the individual's tastes, innate abilities, information, and resources.

This choice model emphasizes the notion of a path of activities, rather than a single choice. In making a choice (or a series of choices), the individual invests in gathering information regarding alternatives, which can be a costly process. Furthermore, the information that the individual has with regard to various career paths is imperfect. Military service may be part of a path that also includes additional schooling and eventual entrance into the civilian labor market. Alternatively, military service could include a full career of 20 or more years of service.

In the remainder of this chapter, we outline the key aspects of each of the three major alternatives—military service, additional schooling, and civilian employment—that are likely to be relevant to the individual's choice. We examine the tangible benefits associated with an option, the other conditions associated with that choice, and possible intangible factors.

MILITARY SERVICE

The Enlistment Process

We begin by outlining the enlistment "production" process itself. Though there are differences in the details, the general process is the same for all of the Services (see Figure 5-1). All, or almost all, entrants enter at the "bottom" of a closed personnel system; there is little or no lateral entry. Consistent with our conceptual model, each Service competes for the youth population with civilian employers, colleges, and the other Services. Recruiters—the Services' sales force—have quotas or targets for the enlistments in each period, typically one month. The Services also offer enlistment bonuses and education benefits, in addition to the basic education benefit offered to all recruits under the Montgomery GI Bill, which is targeted to qualified recruits who enlist in particular occupational specialties.

In order to qualify for military service, all applicants must first take the Armed Services Vocation Aptitude Battery (ASVAB). In addition, a background check is used to determine if the applicant is morally qualified for service. Once these two areas of qualification are met, the applicant can be offered a military job.

Most qualified applicants who accept the offer of enlistment do not begin military service immediately. Instead, they enter the delayed entry

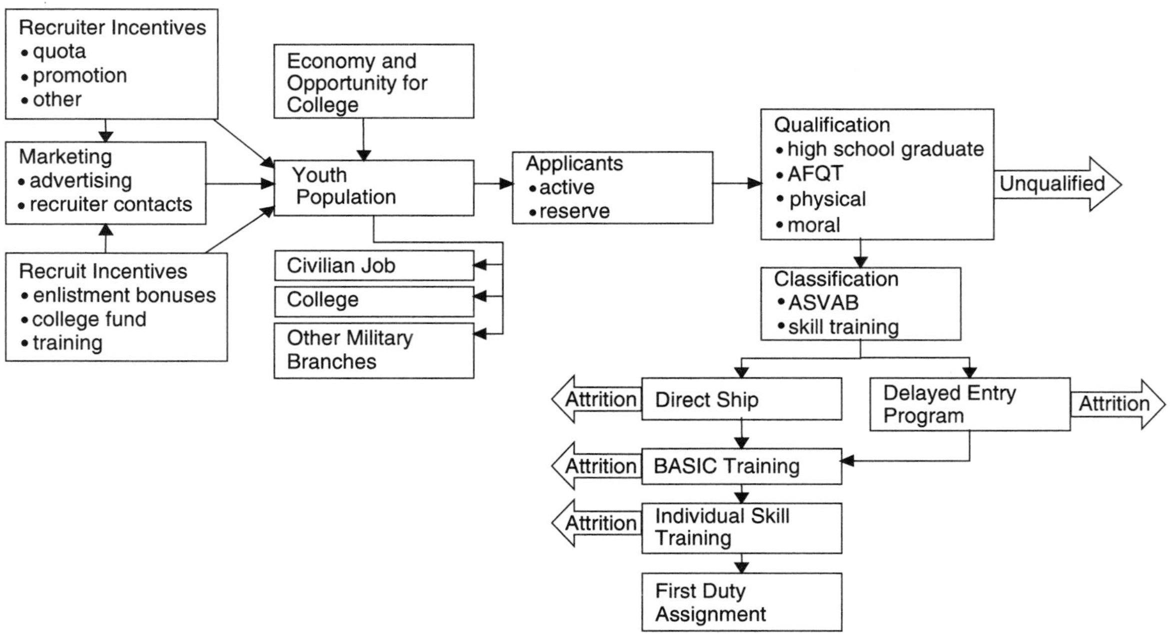

FIGURE 5-1 Simple description of the enlistment production system.

program (DEP). While in the program, the recruit remains at home, continuing life as a civilian. The primary purpose is to schedule the recruit's entry into service to coincide with basic and initial skill training classes and to accept applicants who are high school seniors and have not yet graduated. At some point, the recruit will enter basic training and upon completion will be assigned to a duty station. Typically, the first duty station is overseas for the Army and sea duty for the Navy.

Conditions of Military Service: An Overview

Enlistment in military service is unlike accepting a civilian job or even entering college; the military institution has a much larger influence on the member's life while in the military. Although the service is constrained by laws and regulations and the recruit voluntarily agrees to accept the order and discipline of military life, the degree to which the military service influences the recruit's location, hours of work, food and housing, and even leisure activities vastly exceeds that of any other choice.

The recruit begins by signing an enlistment contract that obligates the individual to serve for a specified period. This both constrains the individual's employment over the term of the agreement and provides a degree of security in employment over the period. The potential recruit must consider this aspect of enlistment when making a decision. The contract itself is potentially enforceable, but in practice has not been strictly enforced in a peacetime, volunteer environment.[2]

All recruits receive basic training and most receive initial and advanced training in a particular skill. The skill training can range from technically intensive training in electronics, computers, nuclear power, or avionics to training in law enforcement, clerical skills, or mess management skills.

Initial and advanced skill training can last from a few weeks to well over a year. During this time in training, the member receives full pay. Most formal training occurs when the Service member is relatively junior and receives only modest pay. Nevertheless, civilian-sector employers do not typically provide training in such general areas as basic and advanced electronics, at no cost to the employee, while paying the employee's full salary. One of the reasons the military Services are able to do this is that

[2]In particular, about 10-20 percent of recruits who enter the delayed entry program do not enter the service. In effect, they break their enlistment contract before they even begin. The Services do not attempt to enforce the contract. Similarly, almost 40 percent of recruits fail to complete their initial term of service, for a variety of reasons. Again, the only adverse consequence is that the member may receive a general discharge instead of an honorable discharge, a distinction that has lost its value.

the recruit is obligated by a potentially enforceable enlistment contract to serve for a specified period of time.

The military member is subject to military discipline and to the Uniform Code of Military Justice, meaning they must obey all lawful orders. On rare occasions, this may mean being deployed for significant periods of time without notice and the possible cancellation of individual or family plans. More typically, it means planned periods of family separation.

Frequent moves are a fact of life in the military, with planned rotations about every three years. The armed forces of the United States are stationed all over the world and on all of its oceans. The typical Service member considers some of these locations very interesting and desirable assignments, but others are not. Members are assigned to positions throughout the world through a process that might be described as "share the pain, share the gain." Typically, this means that the member can anticipate that relatively onerous assignments will be followed by relatively desirable assignments, as judged by the typical member.

For most permanent assignments, the member may choose to bring his or her family. The Service, according to schedules of coverage and allowances, reimburses moving expenses for the member and his or her family. Moreover, there is an infrastructure at most military locations to support the member's family. Some types of assignments, however, are designated as "unaccompanied." The member's family does not accompany the member to the location. These assignments are typically to onerous locations or locations of higher risk. Unaccompanied tours are typically of shorter duration.

The frequency of moves and the overseas assignments present hardships, primarily to married members or members with dependents. First, frequent moves make it difficult for the member's spouse to pursue certain types of careers. Second, at some locations, there are very limited opportunities for any spouse employment.[3] Third, members with school age children can find the frequency of moves disruptive to their education. However, the recruit typically has no dependents. The opportunity to be stationed overseas or even in other parts of the United States in the first term of service may be seen more as an opportunity for travel than a hardship.

Members in all of the military Services who are assigned to combat or combat support units—which includes most initial duty assignments for recruits—will be subject to deployments. Deployments occur when the

---

[3]In many countries overseas, Status of Forces (SOF) agreements limit the opportunity to work in the local civilian economy. The only opportunities for spouses are positions on the military base.

unit leaves its permanent station for a period of time (ranging from several days to many months) for an operational or training mission. During periods of deployment, the member is separated from his or her dependents. In the Navy, deployments typically mean time at sea. In the other Services, humanitarian, peacekeeping, and similar missions are typical deployments.[4]

Working conditions may be onerous or unpleasant, depending on the job, the assignment, and the member's tastes, but they may also be quite pleasant. On sea duty, sailors work and live on a ship for 180 days at a time. Accommodations aboard ship are not to everyone's taste, especially for junior enlisted members.

Military members may be physically at risk from two sources. First, some types of military jobs are inherently risky. Any job that puts one in constant proximity to live ordnance is potentially risky. Jobs requiring sailors to be on the deck of a carrier during launch and recovery operations, operation of high-performance aircraft, airborne operations, diving, and demolition are a few examples. Second, all members are subject to being deployed to hostile fire areas—to be put in harm's way—regardless of their particular jobs. Some units, particularly combat and combat support units, and some types of jobs are more at risk than others.

In principle, active-duty members are "on call" 24 hours per day. Actual hours of work will vary by the nature of the job or skill, the current assignment or duty station, and factors that may be affecting the command at any particular time. It is not unusual for sailors on board ship, for example, to work 16 hours per day. Similarly, soldiers and airmen may work equally long hours in preparing for a deployment, for example. However, normal duty hours under typical circumstances will require a workweek not unlike that in the civilian sector. Unlike the civilian sector, however, members will be called on to take rotations for extra duty, such as watch standing. There is no overtime pay.

Military Compensation System

Because recruits enter voluntarily, rather than through conscription, pay and benefits must remain competitive with alternatives, if the Services are to attract and retain required numbers of qualified personnel. The military compensation system for active-duty members consists of a complex array of basic pay, nontaxable allowances, special and incentives pays, deferred compensation, and in-kind benefits. We briefly review its major elements below.

---

[4]For example, most soldiers stationed in Bosnia are actually deployed from bases in Germany.

Current Cash Compensation

Basic pay is the major element of military compensation. Monthly basic pay is a function of the member's rank or pay grade and years of service. The enlisted pay table in effect as of January 1, 2001, is shown in Table 5-1. Typically, recruits enter at the lowest pay grade, E-1. However, if they have some college or are highly qualified and are entering certain occupational specialties, they may enter at an advanced pay grade, typically E-2 or E-3.

Members progress through the pay table in two ways: length of service or promotion. As they gain longevity, pay increases in the ranges of the pay table. In addition, members may be promoted. Typically, promotions through E-3 are relatively automatic as long as the member is making satisfactory progress. From E-4 through E-9, promotions become increasingly competitive. Typically, the member must reach noncommissioned officer status (E-4) to reenlist beyond the first term of service. Members who remain competitive for promotions will reach senior noncommissioned officer (NCO) or chief petty officer status at about their 12th year of service.

In addition to basic pay, all members receive either housing and food in-kind, or a tax-free allowance for housing (basic allowance for housing or BAH) and a tax-free allowance for food (basic allowance for subsistence or BAS). These allowances vary by pay grade and, unlike the civilian sector, by dependency status. Members with dependents receive higher allowances than those without dependents. In addition, BAH varies by location, reflecting geographic variation in rental prices. BAH rates reflect the cost of renting standardized housing types across different geographic locations. Table 5-2 shows current BAH rates for enlisted members for Virginia.

BAH is substantial—about 33 percent of basic pay. Moreover, it is not subject to federal or state income tax, increasing its value by 15 percent or more for most enlisted members. Unmarried junior enlisted members (E-4 and below) typically live in on-base housing, however, and receive rations in-kind.

Regular military compensation (RMC) is defined as the sum of basic pay, BAS, and BAH (including the tax advantage on each). It is the most frequent way that military cash compensation is defined.

All members are paid from the common basic pay table, regardless of occupation. Similarly, allowances vary only by rank, dependency status, and location. Yet the training and experience offered in some military occupations is much more valuable in the civilian sector than that offered in other occupations. Hence, other things being equal, it is more difficult to retain personnel in some occupations than in others.

TABLE 5-1 Monthly Basic Pay Table for Enlisted Members (effective 1 January 2001)

	Years of Service							
GRADE	<2	2	3	4	6	8	10	12
E-9	$0.00	$0.00	$0.00	$0.00	$0.00	$0.00	$3,126.90	$3,197.40
E-8	0.00	0.00	0.00	0.00	0.00	2,622.00	2,697.90	2,768.40
E-7	1,831.20	1,999.20	2,075.10	2,149.80	2,227.20	2,303.10	2,379.00	2,454.90
E-6	1,575.00	1,740.30	1,817.40	1,891.80	1,969.50	2,046.00	2,122.80	2,196.90
E-5	1,381.80	1,549.20	1,623.90	1,701.00	1,777.80	1,855.80	1,930.50	2,007.90
E-4	1,288.80	1,423.80	1,500.60	1,576.20	1,653.00	1,653.00	1,653.00	1,653.00
E-3	1,214.70	1,307.10	1,383.60	1,385.40	1,385.40	1,385.40	1,385.40	1,385.40
E-2	1,169.10	1,169.10	1,169.10	1,169.10	1,169.10	1,169.10	1,169.10	1,169.10
E-1>4	1,042.80	1,042.80	1,042.80	1,042.80	1,042.80	1,042.80	1,042.80	1,042.80
E-1<4	964.80	0.00	0.00	0.00	0.00	0.00	0.00	0.00

	Years of Service						
GRADE	14	16	18	20	22	24	26
E-9	$3,287.10	$3,392.40	$3,498.00	$3,601.80	$3,742.80	$3,882.60	$4,060.80
E-8	2,853.30	2,945.10	3,041.10	3,138.00	3,278.10	3,417.30	3,612.60
E-7	2,529.60	2,607.00	2,683.80	2,758.80	2,890.80	3,034.50	3,250.50
E-6	2,272.50	2,327.70	2,367.90	2,367.90	2,370.30	2,370.30	2,370.30
E-5	2,007.90	2,007.90	2,007.90	2,007.90	2,007.90	2,007.90	2,007.90
E-4	1,653.00	1,653.00	1,653.00	1,653.00	1,653.00	1,653.00	1,653.00
E-3	1,385.40	1,385.40	1,385.40	1,385.40	1,385.40	1,385.40	1,385.40
E-2	1,169.10	1,169.10	1,169.10	1,169.10	1,169.10	1,169.10	1,169.10
E-1>4	1,042.80	1,042.80	1,042.80	1,042.80	1,042.80	1,042.80	1,042.80
E-1<4	0.00	0.00	0.00	0.00	0.00	0.00	0.00

SOURCE: Available: <http://www.military.com>.
NOTE: Basic pay for 0–7 and above is limited to $11,141.70, Level III of the Executive Schedule. Basic pay for 0–6 and below is limited to $9,800.10, Level V of the Executive Schedule.

An efficient compensation system recognizes differences in civilian opportunities through pay differentials by occupation. Although all enlisted occupational specialties are paid from a common pay table, there are two ways in which pay differentiation is achieved. First, some occupational specialties have more rapid rates of promotion than others. This means that personnel in these occupations will receive greater compensation. These tend to be occupations requiring technical skills that are highly transferable to civilian employment. Moreover, because the promotion system in the Services tends to be at least partially driven by vacant posi-

TABLE 5-2 Enlisted Basic Allowance for Housing (BAH) Rates: Virginia

Location	GRADE E2	E3	E4	E5	E6	E7	E8	E9
Without Dependents								
Woodbridge	$744	$744	$744	$793	$825	$874	$925	$930
Newport News	559	559	559	580	596	636	705	750
Norfolk	626	626	626	645	658	683	733	773
Fort Lee	529	529	529	537	546	580	624	654
Richmond	506	506	506	558	592	649	713	724
With Dependents								
Woodbridge	$817	$817	$858	$922	$938	$1,005	$1,078	$1,150
Newport News	590	590	624	676	822	871	925	997
Norfolk	654	654	675	707	837	901	972	1,044
Fort Lee	541	541	569	613	669	709	763	853
Richmond	583	583	631	705	742	768	797	877

SOURCE: Available: <http://www.military.com>.

tions, it serves as a natural equilibrating mechanism. Low retention increases vacancies, which make promotions more rapid, and increases pay, which increases retention.

The second source of pay differentiation is "special and incentive pays." The purpose of these pays are to compensate for risks, onerous conditions, or responsibilities associated with particular occupational specialties or assignments and to increase retention in areas where there are, or would otherwise be, shortages. Special and incentive pays are the primary way in which pay differentials are introduced. They can be targeted to occupations and experience levels for which increased retention is needed the most. Because an across-the-board pay raise is very expensive,[5] they tend to be an efficient solution to recruiting and retention problems that are skill-specific. For example, the Services may offer enlistment bonuses of $5,000 to $8,000 to qualified recruits in selected occupational specialties.

Special and incentive pays constitute only a small portion of the total cash compensation package. Although they are especially effective in targeting and alleviating specific recruiting and retention problems, because they are such a small proportion of total compensation, their influence is limited by their relatively modest proportion of total compensation. Figure 5-2 shows the distribution of cash compensation (including deferred retirement pay) for enlisted personnel.

[5]A 1 percent increase in basic pay across the board costs about $400,000,000 per year.

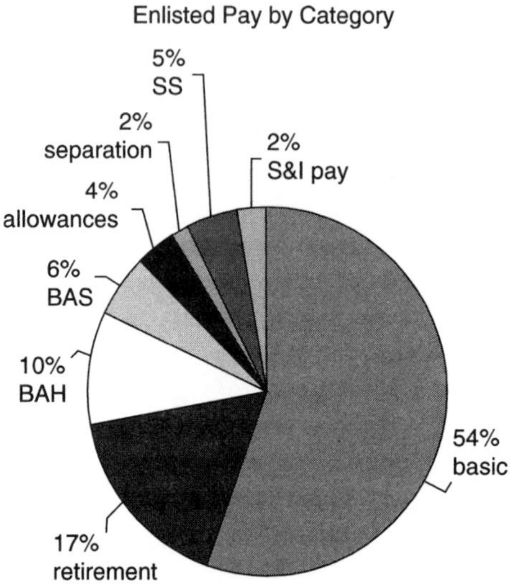

FIGURE 5-2 Distribution of cash compensation: Enlisted personnel.
SOURCE: Paul F. Hogan and Pat Mackin, Briefing to the Ninth Quadrennial Review of Military Compensation Working Group, November, 2000.
NOTE: BAH = Basic Allowance for Housing; BAS = Basic Allowance for Subsistence; SS = Social Security; S&I = Special and Incentive Pay.

The military retirement system is a major component of total compensation. Those members who entered prior to 1980 are eligible to receive 50 percent of basic pay at the time of retirement if they retire with 20 years of service, rising linearly to 75 percent of basic pay with 30 years of service. Those who entered between 1980 and 1986 may retire under similar conditions, except that their annuity is based on an average of their highest three years of pay. Finally, those who entered after 1986 now have a choice to either (1) receive the same annuity as those who entered between 1980 and 1986 or (2) accept a $30,000 lump sum at 15 years of service and receive 40 percent of an average of their three highest years of pay at 20 years of service, rising linearly to 75 percent at 30 years of service.

The military retirement system is relatively generous. It has a dominant effect on retention after the second term of service. However, because of the "cliff" vesting of the current system—the member receives nothing unless he or she stays through 20 years of service—it is not a major factor in recruiting.

Benefits

In addition to housing and subsistence allowances, which may be provided either as cash or an in-kind allowance, there are numerous other in-kind benefits of military service.

Military members themselves receive medical and dental benefits at no cost from military treatment facilities. Dependents receive medical benefits through TRICARE, which offers managed care and fee-for-service (CHAMPUS) plans. The out-of-pocket costs are generally minimal and the overall benefits are competitive with plans offered by civilian employers.

The Montgomery GI Bill, which is offered to all recruits, provides resources for college, primarily to those who have left active duty service. The recruit must agree to have $100 per month deducted from pay for 12 months. In return, the member receives about $20,000 for college over a 36-month enrollment period. In addition, the Service may supplement this amount for qualified recruits in selected occupational specialties. The Army, through the Army College Fund, adds in excess of $30,000 to the benefits for qualified recruits who enlist in selected hard-to-fill occupational specialties. The Navy and the Marine Corps also offer additional education benefits, but on a smaller scale than the Army.

Recently, there have been numerous legislative proposals in both the House and the Senate to enhance the basic benefit of the Montgomery GI Bill and to add new features, such as the ability of the member to transfer all, or a portion, of the benefit for use by dependents. However, no major changes have been enacted as of this writing.

All of the Services offer a tuition assistance program for members who take college courses while on active duty. Typically, this program will reimburse 75 percent of the cost of college credit hours taken by members while on active duty.

In addition, there are numerous other benefits of military service. These include access to recreational facilities ranging from gymnasiums and bowling alleys to the opportunity to fly free on a space-available basis on military flights. Members also receive all of the benefits associated with veteran status, including the ability to obtain a VA mortgage while on active duty.

Finally, many military members serving in particular occupations receive training and experience that is transferable to the civilian sector. While not a benefit in the traditional sense, this training and experience are valued highly in the civilian labor market. The evidence in the research literature is consistent with intuition (e.g., Stafford, 1991; Goldberg and Warner, 1987). On one hand, in occupations that require technical skills and other occupations, such as health care, with a clear counterpart in the civilian sector, military training and experience are about as valuable as training and experience in the occupation in the civilian sector. The advantage to the military members is that they receive this training at no cost to themselves while earning full pay on active duty.[6] On the other hand, in the traditional combat arms occupational specialties, as well as in other occupations that are relatively military-specific, training and experience have a positive effect on potential civilian earnings, but not as large an effect as the equivalent amount of civilian labor market experience.

Competitiveness of Military Compensation and Benefits[7]

The adequacy of the military compensation package is ultimately determined by its ability to attract and retain sufficient qualified staff to maintain readiness levels. Generally, to do this, military pay must at least compensate members for what they could be earning in the civilian sector, given their actual level of education. Moreover, recruiting requires that some proportion of high school graduates who are in the upper half of the distribution nationally on an intellectual qualification test choose to enter the military rather than college. Arguably, because these potential recruits could have gone to college, military pay must compensate them for what they could have earned in the civilian sector, had they gone to college. Because not all would have completed college, at a minimum it should compensate members for what they could have earned with "some college." Indeed, as reported in the Department of Defense (DoD) Personnel Survey for 1999, over half of enlisted personnel and 80 percent of enlisted members with more than 20 years of service report that they have completed at least one year of college (see Asch et al., 2001).

Figure 5-3 compares average enlisted regular military compensation (RMC) to the earnings of civilians with some college. Recall that RMC includes basic pay, allowance for subsistence, the housing allowance, and the tax advantage on the allowances. RMC is the standard measure of military cash pay to compare with civilian cash earnings. Average RMC is about at the same level as median earnings reported by those with some college and comparable experience and below average earnings for that group. Average RMC is significantly below those at the 70th percentile of

[6]Stafford (1991) and Goldberg and Warner (1987) obtain similar results in comparing the effects of military and civilian experience on civilian earnings. A limitation is the age of the studies and that they do not encompass the major information technology revolution of the 1990s.

[7]Much in this section draws from the information provided in Asch et al. (2001).

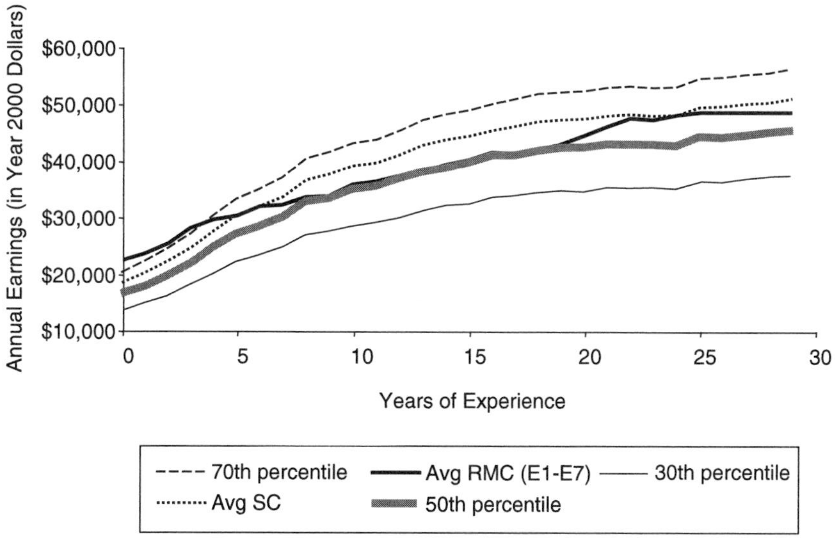

FIGURE 5-3 July 2000 enlisted pay (regular military compensation or RMC) compared with predicted year 2000 earnings of males with some college (SC).
SOURCE: Asch et al. (2001).

civilian earnings for those with some college. Because most military members are in the upper half of the intelligence distribution, as measured by the Armed Forces Qualification Test (AFQT), those in the 70th percentile of civilian earnings are arguably the most relevant comparison group.

Many who enlist could have entered a four-year college and completed an undergraduate degree. Their earnings in the enlisted force would be significantly below the average earnings of college graduates. Figure 5-4 compares average RMC, projected to FY 2005 to average earnings in the civilian sector, at various levels of education, also projected to FY 2005 (Hogan and Mackin, 2000). Note that the difference in earnings between college graduates and high school graduates is quite large, as is the difference between college graduates and those with some college. The returns from completing four years of college appear to be quite substantial. Enlisted pay, as measured by average RMC, does not appear to be competitive with the pay of college graduates.

These pay comparisons suggest that if the Services are competing with the college market, recruiting will become more difficult, given the apparent earnings premium enjoyed by college graduates. Since the mid-1980s, the Services have been relatively successful in recruiting. It is only in recent years, since about FY 1997, that recruiting has become difficult again.

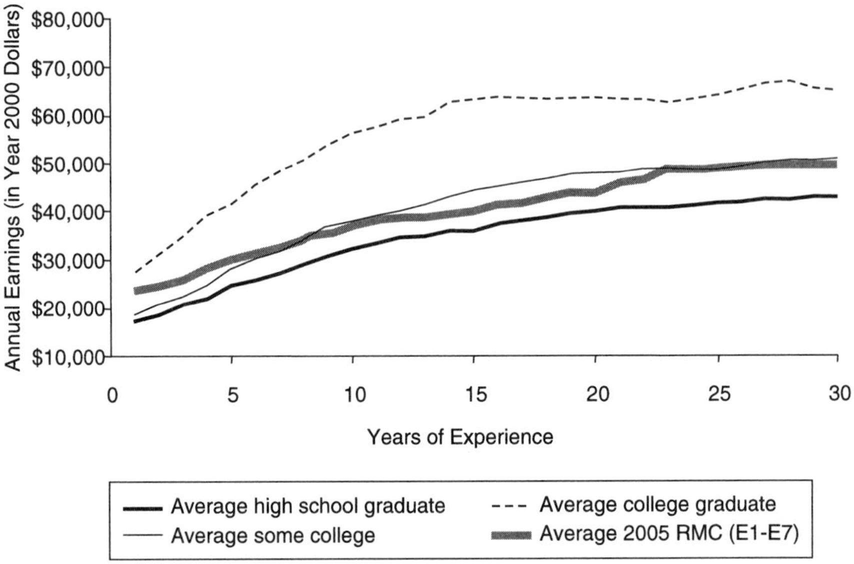

FIGURE 5-4 Average civilian earnings and average FY 2005 enlisted (regular military compensation or RMC) by years of experience.
SOURCE: Asch et al. (2001).

Recruiting success is typically measured not simply by numbers, but by the proportions of recruits who are highly qualified. We present two measures: the proportion of recruits who are high school diploma graduates and the proportion who are high school diploma graduates and also score in the top half of the AFQT. Data on these factors are presented in Chapter 4, this volume. The percentage of high school diploma graduates recruited by the Services has consistently exceeded about 90 percent since about FY 1984, with only the Navy falling below that level, and then only for a few years. However, while the percentage had exceeded 90 percent for the first half of the 1990s, it has fallen to 90 percent or near that for all the Services except the Air Force since about FY 1996. Similarly, while the percentage of highly qualified recruits exceeded 60 percent for each of the Services in the first half of the 1990s, the trend has been downward for all the Services since about FY 1996 and has fallen below 60 percent for the Army and Navy.

While the Services have experienced significant recruiting difficulties since about FY 1996, two related points are important to consider. First, recruiting results were extraordinarily good, as measured by the percentage of highly qualified recruits, in the first half of the 1990s. Hence, the poorer results of the second half of the decade are in comparison to the best recruiting results in the modern volunteer force period. Much of the success of the early 1990s was due simply to the lower demand for recruits during the post-cold war drawdown of forces. Second, compared with the 1970s and most of the 1980s, the results in the second half of the 1990s through FY 2000 are relatively good.

Incentives and Resources Affecting Recruiting

The military environment, military compensation, and in-kind benefits all affect the desirability of enlisting. In addition, the Services and DoD allocate resources directly to improve recruiting. The largest share of the recruiting budget is for military recruiters—the recruiting sales force. In addition, the Services provide targeted incentives to recruits with certain qualifications who enlist in selected occupational specialties. These include enlistment bonuses, which can be as high as $8,000 and are paid on successful completion of initial skill training, and supplements to the Montgomery GI Bill, which can be as much as $30,000. Finally, the Services spend resources on advertising, both to inform potential recruits about opportunities in the military and to convince them to enlist.

The state of the economy has an important effect on the supply of recruits. When the economy is strong, with low unemployment, recruiting becomes more difficult. More recruiting resources are required to achieve a given recruiting goal. Similarly, during economic downturns, recruiting becomes somewhat easier. Fewer resources are required to achieve a given recruiting goal.

The effect of recruiting resources on enlistments is sometimes summarized in a measure called an "elasticity." In the context of recruiting, elasticity indicates the percentage increase in recruits one can expect when a particular recruiting resource or factor increases by 10 percent. If the elasticity of enlistments with respect to recruiters is 0.5, for example, a 10 percent increase in recruiters would result in a 5 percent increase in enlistments.

In Table 5-3, we summarize the range of elasticities found in the recent econometric literature on military recruiting (e.g., Hogan and Dall, 1996; Hogan et al., 1998; Murray and McDonald, 1999; Warner et al., 1998). Although there has not been a large number of studies on recruiting in the 1990s, these studies use methods that build on and, in many cases, improve on the more numerous studies of the 1970s and early 1980s. In almost all cases, the elasticity is measured with respect to highly qualified recruits, a group that is in demand at the margin. The estimates vary by Service, time period of the data, and, in some cases, methods used to estimate effects. The estimated elasticities are statistical estimates ob-

TABLE 5-3 Elasticities of Enlistments with Respect to Various Recruiting Resources

	Factor		
	Army	Navy	Source
Recruiters	0.51	0.57	W
	0.42	0.23	H/M/H
	0.6	0.53	M/M
		0.2–0.32	
College Fund	0.49	0.22	W
	0.01		M/M
Enlistment Bonus	0.13	0.03	W
	0.003		M/M
Advertising-Total	0.319	0.067	W
Advertising-TV	0.089	0.041	W
		0.02–0.08	H/D
Pay	1	1.1	
		0.3–0.8	H/D
Unemployment Rate	0.14	0.11	H/M/H
	0.16		M/M
	0.22	0.29	W
		0.1–0.22	H/D

NOTE: W = Warner (2001). H/M/H = Hogan, Mehay, and Hughes (1998). M/M = Murray and McDonald (1999). H/D = Hogan and Dall (1996).

tained by treating the observed variation in recruiting results and resources as a natural experiment.

An elasticity provides a measure of the importance of the factor on recruiting. However, it is not a cost-benefit analysis. Advertising, for example, has a small elasticity, but the advertising budget itself is small compared with the budget for recruiters. Hence, it is not correct to judge cost-effectiveness of the resource by the elasticity.[8] Table 5-4 provides an estimate of the costs associated with recruiting one additional highly qualified recruit using various incentives and methods (Warner, 2001; Warner et al., 2001). The marginal cost estimates of the major recruiting resources—recruiters, education benefits, enlistment bonuses, and advertising—are generally within a few thousands of dollars of each other. This is what one would expect if all the resources are being used reasonably efficiently.

[8]For example, a key variable in advertising effectiveness is the message strategy, which is not accounted for in econometric models.

TABLE 5-4 Marginal Cost Estimates of Major Recruiting Resources (in thousands)

	Service			
Factor	Army	Navy	Air Force	Marine Corps
Pay	$32.0	$30.0	$40.8	$59.0
Recruiters	$13.2	$8.2	$3.3	$9.3
Educational benefits	$7.4	$12.0		
Enlistment bonuses	$12.0			
Advertising-Total	$9–11.0	$7–8.0		
Advertising-TV	$7.0–9.3	$7.3		
Advertising-non-TV	$3.8–8.4	$2.7–4.7		

SOURCE: Warner et al. (2001).

The table provides a measure of the effect of various resources (and environmental factors) on recruiting. Table 5-5 provides an indication of the budgets and changes in these budgets over time. Note that military pay, the largest single budget item, is largely pay and allowances for recruiters. Since about FY 1997, recruiting resources have increased significantly, particularly for the Army (see Figures 5-5 and 5-6).

Econometric models of enlistment supply, such as those of Warner et al. (2001), can provide insights into an important question: Given the estimated effects of recruiting resources and the economy on enlistment supply, can changes in these resources and changes in the economy explain the decline in enlistments in the late 1990s, or is the decline due to

TABLE 5-5 FY 2001 Recruiting Budget (in millions)

	Army	Navy	Air Force	Marine Corps	DoD
College funds (NCF)	$101.62	$28.18		$29.42	$159.22
Enlistment bonuses	$147.40	$105.12	$116.34	$7.44	$376.30
Loan repayment	$32.89	$0.10	$6.00	—	$38.99
Military pay	$409.60	$252.05	$96.04	$119.87	$877.57
Civilian pay	$47.57	$24.55	$8.76	$7.77	$88.65
Advertising	$102.04	$66.39	$47.86	$38.65	$254.94
Recruiter support	$225.23	$70.31	$35.21	$43.01	$373.75
Total	$1,066.35	$546.70	$310.21	$246.16	$2,169.42

SOURCE: Office of the Assistant Secretary of Defense (2001).

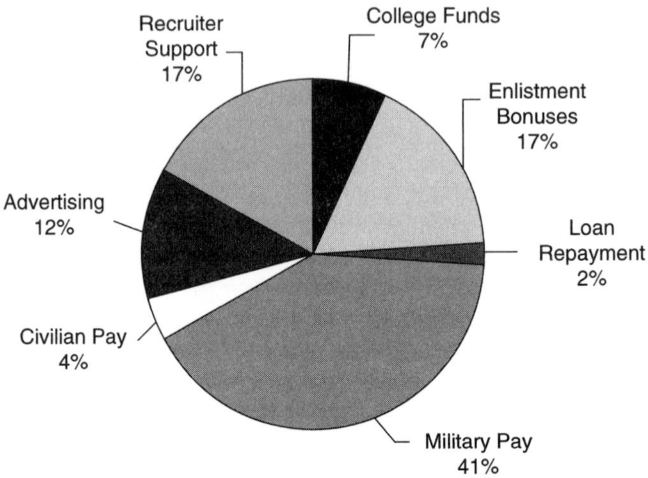

FIGURE 5-5 FY 2001 distribution of DoD recruiting resources.
SOURCE: Office of the Assistant Secretary of Defense (2001).

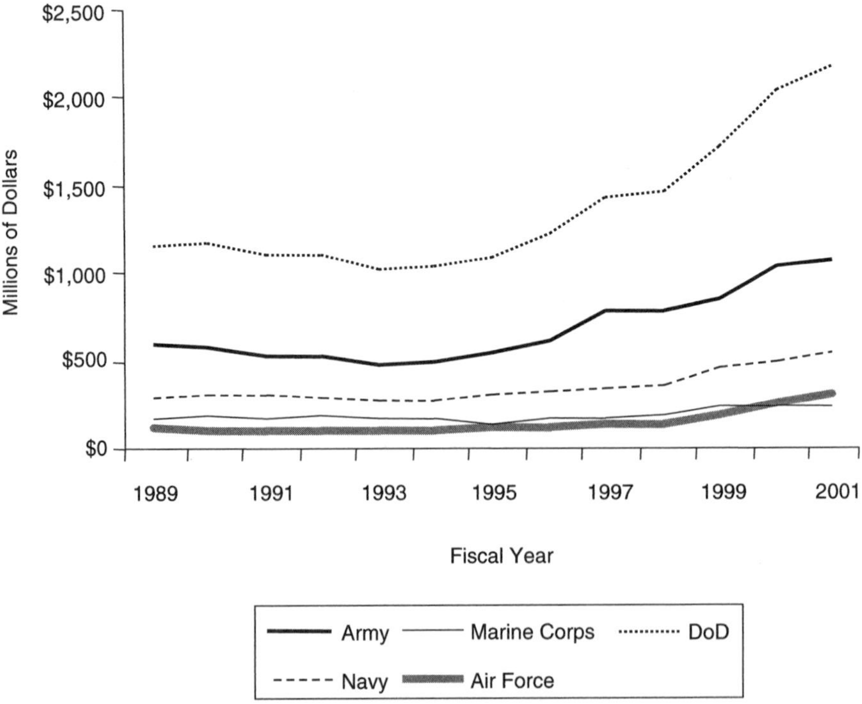

FIGURE 5-6 Recruiting resources: FY 1989–FY 2001.
SOURCE: Office of the Assistant Secretary of Defense (2001).

other, as yet unidentified, factors? Warner et al. accounted for changes in the economy, resources, and the effectiveness of resources. Based on their analyses, Warner and his colleagues conclude the following:

> We do not believe that the downward trend in high-quality enlistment since 1993 can be attributed to shifts from the past in the way recruiting responds to economic factors (pay and unemployment) and it is not explained by diminished effectiveness of recruiting resources. Specifically, we do not find that the responsiveness of enlistment to either recruiters or advertising has declined. Positively, we also find that recruiting is responsive to educational benefits and enlistment bonuses. However, the effectiveness of Army College Fund benefits may have declined due to a large expansion in Army enlistment bonuses since FY 1997. Expansion in college attendance accounts for some of the decline in high-quality recruiting but is not the end of the story. There remains a significant negative residual component to recruiting that appears to be the result of a decline in youth tastes for military service. Understanding the reasons for the decline in youth tastes for service, and policy options for responding to it, would be an important avenue for future research.

In particular, Warner and his colleagues note that the *unexplained* portion of the decline in highly qualified recruits (the portion of the decline that cannot be explained by changes in resources, the economy, or the effects of resources and the economy on enlistments) tracks the decline in youth propensity as measured by the Youth Attitude Tracking Survey (YATS). This is illustrated in Figure 5-7.

Possible sources for youth propensity and its changes over time that are not typically captured in econometric models are altruistic motives to enlist. One form of these motivations, patriotism, is briefly discussed in the next section.

Patriotism and Higher Purpose

We have discussed the financial rewards of military service, in-kind benefits, and the hardships and risks associated with military life. One aspect of military service that differs from most civilian jobs is that there is a higher purpose to military service than simply receiving pay and benefits in return for work and the acceptance of hardships. Patriotism and selfless service to one's country provide a potentially strong motivation to enlist (as discussed in Chapters 7 and 8, this volume). While such sentiments cannot fully compensate for noncompetitive pay and unpleasant working conditions, they do provide a foundation for recruiting and retaining motivated staff that is generally unavailable to civilian-sector employers. Moreover, as suggested in the previous section, changes in

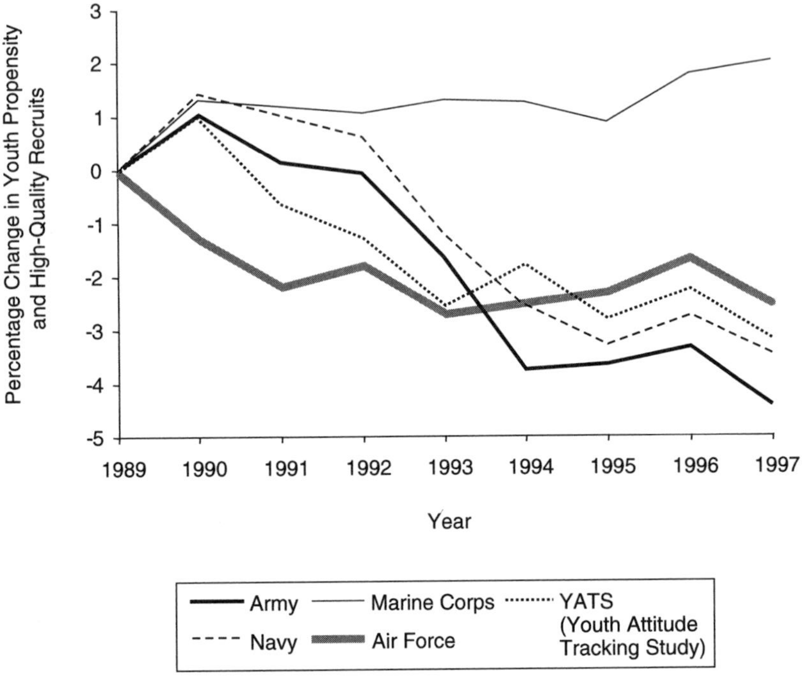

FIGURE 5-7 Normalized male YATS propensity and unexplained recruiting trend.
SOURCE: Warner (2001).

more altruistic motivations for service can have an important influence on recruiting.

Summary

Enlistment in the armed forces is one of three major options eligible youth have upon completion of high school. The armed forces offer a lifestyle that is very different from the one most recent high school graduates experience in college or in civilian employment. Typically, the individual leaves his or her hometown and is subject to military orders and discipline. These orders may, in fact, place the individual in harm's way.

To attract recruits, the armed forces offer a complex but generally competitive system of pay, allowances, bonuses, and benefits to compete with civilian alternatives. In addition, they use a reasonably efficient mix of various recruiting resources, including bonuses, advertising, and recruiters. Finally, the rate at which youth continue to go to college has grown over time.[9] The armed forces offers postservice education benefits (in the form of the Montgomery GI Bill and augmentations to it) to attract those recruits who also have college aspirations.

Recruiting has become increasingly difficult since the second half of the 1990s. This was due, at least in part, to an improving economy through 2000 and to a decline in recruiting resources in the mid-1990s. However, while these factors can explain some of the decline, a portion remains unexplained. It is unclear whether this represents a change in tastes or preferences for military service on the part of American youth or is due to other factors that simply are not captured in econometric equations. Success in the recruiting market now and in the future will require that compensation and benefits remain competitive with those offered in the civilian sector, that other recruiting resources be applied efficiently and in proportion to the recruiting mission, and that innovative ways to combine military service with the pursuit of higher education are offered. Finally, military service offers youth a higher purpose, in the form of duty to country, service to others, and self-sacrifice, that is rare in other alternatives. The armed forces should not neglect this aspect of military service in its recruiting efforts.

THE EMPLOYMENT ALTERNATIVE

Employment in the civilian sector is a major source of competition to the military services. Although a small proportion of the civilian labor force works in the public sector and is employed by a governmental organization, a far greater proportion is employed by private industry. The U.S. Bureau of Labor Statistics (2002b) estimates that 212 million people age 16 and older make up the civilian noninstitutionalized population; in contrast, the U.S. Bureau of the Census (2000b) reported 17 million people employed by federal, state, and local governments. Because the number of job opportunities is far greater in the private sector, our discussion primarily focuses on individuals employed in the private sector. When appropriate, examples from the public sector are included.

In general, the term "civilian" in this section refers to people employed in nonmilitary settings, while "private" and "public" sector refer to private industry or government employment, respectively. This section also provides brief descriptions of the common practices used in the private-sector labor market and compares employment opportunities in private industry to opportunities in the military at general levels. We provide

[9]As we explore more completely in the section on college as an alternative below, it becomes apparent that this effect may be due in part to an increase in the relative earnings of college graduates.

general information regarding entry-level private-sector employment, a description of the current labor market, the perceived benefits and liabilities associated with employment in private industry, and comparisons of various aspects of military service and private-sector employment.

Comparing the private-sector employment option to military service is difficult because of the many variations in civilian employment processes and opportunities. Differences among entrance and hiring practices, compensation strategies, benefit programs, education and training options, and working conditions vary across different industries and organizations. Moreover, such variables as the size of the organization, industry type, job family, presence or absence of labor unions, and geographical location also affect the civilian labor market. Even when the job or type of work is held constant, opportunities in private-sector employment are highly varied. While the military Services offer similar benefits and rewards for the same job level, the civilian sector may offer different compensation and benefits for the same job depending on the organization, industry, location, etc.

It is important to remember that slightly more than half of Americans work in companies with fewer than 500 employees and almost 20 percent work in firms with fewer than 20 employees. The jobs and associated opportunities in these smaller firms may be significantly different from those in larger firms. While the military may be more like the largest companies, only 3,316 of the 5,541,918 firms in the United States had over 2,500 employees (U.S. Bureau of the Census, 1999) and approximately 37 million of the 105 million U.S. employed work for these firms.

Yet another difficulty in comparing occupations across civilian and military settings is defining what jobs are similar across those settings and what jobs are different. For example, while the military may not have Internal Revenue Service (IRS) agents, the military does have personnel who engage in audit functions similar to the duties that might be performed by an IRS agent. Similarly, the military may lack a merger and acquisitions specialist, found in some large organizations, but it does have individuals who use similar skills to analyze situations and organizations. Thus, questions about the degrees of similarity in jobs and tasks as well as environment must be asked if meaningful comparisons are to be made.

Further complicating a comparison of civilian opportunities in the private sector and military opportunities are the continuous changes in the external contexts of work (e.g., markets, technology, and workforce demographics), the organizational contexts of work (e.g., organizational restructuring and changing employment relationships), and the structure and content of work (National Research Council, 1999). The National Research Council report *The Changing Nature of Work* noted many changes that affect the actual content of a job. For example, the increasing use of teams has flattened organizational hierarchies, and teams are often given greater discretion over how work is performed. The report also pointed out that few trends were universal and cited as an example decreasing discretion in some jobs in the service and manufacturing sectors, in which rational processes and information systems are used to control the work.

In sum, there is enormous variability in processes and opportunities both across and within industries, organizations, and jobs. Considering the variability mentioned above, it is impractical to try to describe and compare all possible variations in organizational processes and employment options. Much of the research in these areas is limited to private-sector market research, general polls, and organizational surveys (e.g., American Management Association, 2001; Society for Human Resource Management, 2000). Moreover, the empirical literature on civilian employment is not scientifically sound or readily available. However, when available, we cite relevant sources of information on employment in the civilian sector (e.g., National Research Council, 1999; Office of Management and Budget, 1997; U.S. Bureau of Labor Statistics, 1999a, 2001, 2002a, 2002b).

Demographic Trends: Who Chooses Employment?

Of the 212 million people in the civilian population in 2001 (U.S. Bureau of Labor Statistics, 2002b), over two-thirds (142 million or 70 percent) participated in the civilian labor force. Participation in the labor force varied little among race and ethnic groups. However, participation was higher for men than for women; specifically 74 percent of males age 16 and over participated in the labor force compared with 60 percent of females.

About half of all youth ages 16 to 19 participated in the civilian labor force in 2001 (approximately half male and half female; U.S. Bureau of Labor Statistics, 2002b). A majority (77 percent) of those ages 20–24 participated in the labor force (82 percent male and 73 percent female).

The labor force is projected to increase by 17 million to 155 million people in 2008, a 12 percent increase from 1998 (U.S. Bureau of Labor Statistics, 1999b). Among race-ethnic groups, participation rates for Asians and Hispanics are projected to grow the most compared with blacks and non-Hispanic whites. The rate of participation for men in the labor force is expected to increase, but at a slower pace than in the previous decade. For women, the participation rate will surpass that of men. Older workers ages 45–64 are expected to show the highest growth (spurred by the baby boom generation of workers), but the youth population is also projected to increase over the next several years, leading to rapid growth in the youth labor force.

Participation in the labor force typically increases with level of educational attainment (Table 5-6). In 2001, college graduates (25 and older) participated in the labor force at the highest rates (79 percent) compared with high school graduates (64 percent) and persons who did not complete high school (44 percent). The participation rate for male college graduates was 85 percent compared with 73 percent for female college graduates (U.S. Bureau of Labor Statistics, 2002b). For male and female high school graduates, participation rates were 74 and 56 percent, respectively. Blacks and whites with a high school degree participated in the labor force at lower rates (69 percent and 64 percent, respectively) compared with Hispanic graduates (74 percent).

Together, these data indicate that a number of high school graduates do not participate in the civilian labor force immediately after graduation or in the years thereafter. Instead, they may potentially choose among other options such as time off, postsecondary education, or military service.

Employment Opportunities in the Private Sector: An Overview

The array of jobs in the private sector can be classified into different categories. This overview describes the systems used to classify different occupations and presents a simple analysis of jobs found in the military versus the civilian sector. The National Research Council (1999: 165) provides a detailed discussion of job classification systems and defined occupational analysis as "the tools and methods used to describe and label work, positions, jobs, and occupations". The report reviewed three kinds of systems: (1) descriptive/analytic systems, (2) category and enumerative systems, and (3) systems that combine the two. Of the common enumerative systems, the most important is the Standard Occupational Classification (SOC) system, which is intended to be the primary occupational category system used by federal agencies (Office of Management and Budget, 1997), including the Census Bureau for the 2000 census and the Department of Labor for the Occupational Information Network or O*NET project.

A combination of the descriptive and enumerative approaches is embodied in the O*NET. O*NET was developed by the U.S. Department of Labor to provide employers, job seekers, career counselors, government workers, occupational analysts, researchers, and students computerized access to the database of occupational information. O*NET occupational profiles provide an overview based on 445 variables, such as worker characteristics, worker requirements, experience requirements, occupation

TABLE 5-6 Employment Status of the Civilian Noninstitutionalized Population 25 Years and Over by Educational Attainment, Sex, Race, and Hispanic Origin, 2001 (in thousands)

Household Data Annual Averages Educational Attainment	Total	Men	Women	White	Black	Hispanic Origin
Total						
Civilian noninstitutional population	176,839	84,294	92,546	148,021	20,333	17,850
Percentage in the civilian labor force	67.4	75.9	59.7	67.1	68.2	69.7
Less than a high school diploma						
Civilian noninstitutional population	27,790	13,195	14,595	22,250	4,241	7,736
Percentage in the civilian labor force	43.6	55.6	32.7	44.2	40.0	59.4
High school graduates, no college						
Civilian noninstitutional population	57,367	26,542	30,825	48,277	7,094	4,911
Percentage in the civilian labor force	64.4	74.4	55.8	63.6	69.2	73.9
Some college, no degree						
Civilian noninstitutional population	30,529	14,300	16,229	25,441	3,968	2,281
Percentage in the civilian labor force	71.9	78.9	65.8	70.8	78.0	80.2
College graduates						
Civilian noninstitutional population	46,601	24,002	22,599	39,754	3,411	2,005
Percentage in the civilian labor force	79.0	84.4	73.3	78.7	83.7	82.1

SOURCE: Data from Table 7, Employment Status of the Civilian Noninstitutional Population 25 Years and Over by Educational Attainment, Sex, Race, and Hispanic Origin, 2001 (U.S. Bureau of Labor Statistics, 2002b).

requirements, occupation characteristics, and specific occupational information. Each of these is further broken down into meaningful categories. Detailed information about employment opportunities referenced in the O*NET are available at www.onetcenter.org.

When comparing private-sector jobs and military positions, one finds that not all military jobs are represented in the civilian sector and not all civilian jobs are found in the military. For example, military jobs related to warfare, such as armored assault vehicle crew members, artillery and missile crew members, and infantry, are not found among civilian jobs; civilian jobs like teaching the disabled are not found in the military. In the SOC system, which is linked to O*NET, 1 of the 23 major occupational groups contains occupations specific to the military. The implication of this observation, consistent with the literature briefly reviewed in the previous section on the value of military experience in the civilian sector, is that military service for the purpose of job training that will enhance the chance of successful employment in a civilian position may not be as valuable if the military assignment is in a field that does not have a civilian counterpart.

However, also consistent with the literature, many employers recognize that military service builds many job-related skills even when the actual work performed is not related to the civilian job. Basic skills like teamwork, following directions, and record keeping may be common to most jobs regardless of the nature of the work or the environment in which it is performed.

Many jobs are found in both the military and the private sector. Of the other 22 SOC major groups, all have a military counterpart in some form. The large number of jobs that are found in both the military and the civilian job market indicates that military service may provide job-specific training in many occupations that will facilitate later civilian employment.

The extent to which military and private-sector jobs overlap is difficult to assess without comparing specific jobs in a military service and in a particular organization. However, at a more macro level, comparisons have been made. The current military career guides, available at www.todaysmilitary.com, suggest about an 88 percent overlap of jobs based on analyses from the Defense Manpower Data Center (Today's Military, 2001). These are more likely to be jobs in health care, service and administration, and electrical and machine equipment repair (Gribben, 2001). However, the distribution of personnel in military occupational areas shown in Table 2-5 indicates that almost 17 percent of personnel were in infantry alone in 2000, an occupation with no direct counterpart in the civilian sector. Furthermore, the Army and the Marine Corps have more personnel assigned to infantry positions.

Occupational Trends

Gribben (2001) did an analysis of occupational trends in the civilian sector and in the military; she reported noticeable shifts in the occupa-

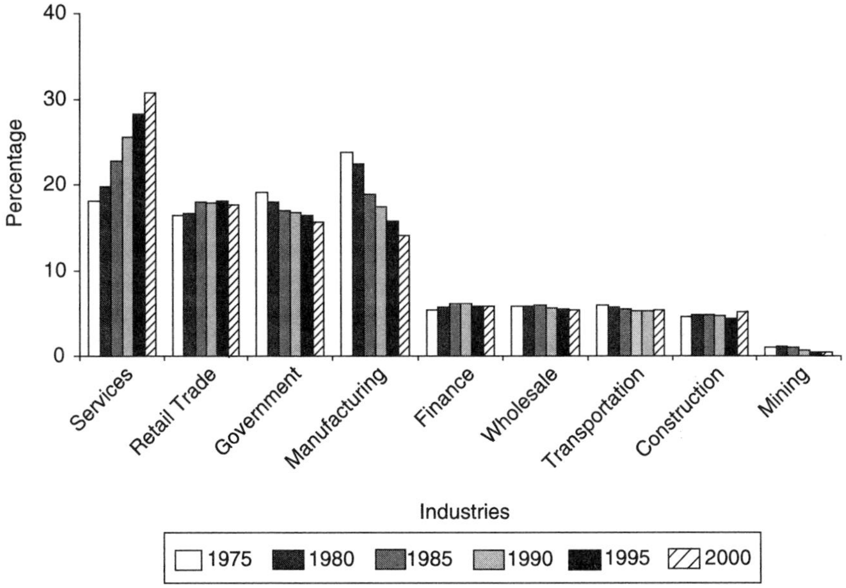

FIGURE 5-8 Civilian industry trends, FY 1975, 1980, 1985, 1990, 1995, and 2000.
SOURCE: U.S. Bureau of Labor Statistics (2001).

tional mix of jobs in civilian employment. Between 1975 and 2000, the proportion of jobs in the service industry grew from 18 to 31 percent of all civilian positions, while manufacturing jobs decreased from 24 to 14 percent of all positions (see Figure 5-8; U.S. Bureau of Labor Statistics, 2001). Government employment also decreased somewhat during the same time period, from more than 19 percent in 1975 to less than 16 percent in 2000. Other industries, such as construction, retail trade, and finance, accounted for about 30 percent of all jobs. These industries did not show appreciable changes in their proportion of the labor market during this time period.

Job growth during the 10-year period 1998–2008 is expected to continue particularly for service employees and professional specialty occupations (U.S. Bureau of Labor Statistics, 1999a). The U.S. Bureau of Labor Statistics projects the following as the 10 fastest-growing occupations for 1998–2008:

- Computer engineers
- Computer support specialists
- Systems analysts
- Database administrators

- Paralegals and legal assistants
- Personal care and home health aides
- Medical assistants
- Social and human service assistants
- Physician assistants.

Demand for Workers

Although there is some military-civilian overlap in high-growth occupations (e.g., health care, service, administration, equipment operation and repair), current and projected trends in the civilian occupational distribution do not precisely mirror the trends in military distributions. Moreover, occupational mix requirements and the demand for workers in different occupations vary between the civilian and military sectors. In the military during the past 25 years, there has been a slight trend toward more technical occupations, such as electronic equipment repair, and communication and intelligence (Gribben, 2001). Service and supply handlers and functional support and administration specialists play a smaller role in the mix of occupations. During the period FY 1976–2000, there was a tremendous need for electrical and mechanical equipment repair specialists in the military along with a large contingent of infantry and related positions.

Military and civilian jobs are categorized by the SOC system. Civilian labor market opportunities depend on the mix of available jobs and demand for specific occupational areas. Though occupational trends have shown significant shifts in certain areas in civilian employment (particularly professional and service occupations), changes have been smaller and more gradual in the military. We have described the opportunities available in the civilian sector. Next, we turn to entry into occupations in each sector and compare the different selection processes and standards.

Labor Market Access

As described earlier in this chapter, the military has clearly defined processes and standards for entry into each service. In contrast, processes and standards in the private sector vary across companies and industries. For example, some employers have no defined processes or standards at all; the employer decides how to hire when a position becomes available and an applicant appears and may use selection methods with little relationship to the requirements of the job (e.g., likeability of candidate in the interview). Such employers may change hiring practices for different jobs and even for candidates for the same positions, setting standards for hiring only after a suitable candidate has been identified. At the other extreme, other employers have specific hiring processes and standards as clearly defined as those of the military (e.g., the U.S. government). In between, there are significant variations in selection processes and standards, even for the same jobs within an organization.

All entrants to military service are subject to a centralized job classification procedure based on the results of the ASVAB. This is not the case in the private sector. Even when required skills are common across jobs and organizations, employers differ in how they assess those skills. For example, one organization may assess the cognitive ability of an applicant by administering a professionally developed and validated paper-and-pencil test, while another may assess cognitive ability by reviewing the applicant's resume and evaluating his or her educational achievement. In 2001, 29 percent of private-sector companies used some form of psychological measurement to assess applicants (American Management Association, 2001). In addition to cognitive ability assessments, which are used most frequently (20 percent of respondents), interest inventories, managerial assessments, personality assessments, and task simulations may be used in the private sector. However, the number of organizations using any form of psychological tests other than basic screening is low and dropping.[10]

Organizations also differ on what knowledge, skills, and abilities they choose to assess. As described in O*NET, they range from basic writing skills to complex problem-solving skills, from knowledge in general domains to specific occupational knowledge and experience, and from physical to cognitive abilities. Even if the identical job with the same requirements exists in two different organizations, finding one organization that measures and selects on cognitive ability and another on "professional appearance" is not uncommon.

The degree of variation and lack of standardization across U.S. companies is reflected in data collected by the American Management Association. It surveyed human resource managers in their member and client companies in January 2001 (American Management Association, 2001). Based on 1,627 usable responses with a margin of error of 2.5 percent, the findings showed that several companies used formal selection processes like literacy testing in 2001 (35 percent), while large numbers did not. Responses also illustrated different testing practices—literacy testing,

---

[10]The American Management Association noted that the drop may be due to skills shortages and the tight labor market that existed at the time the surveys were collected. It is important to note that these data represent the Association's corporate membership and client base, which account for about 25 percent of the U.S. workforce. Smaller firms predominate in the larger U.S. picture and are underrepresented in this survey.

math testing, or both—across industries (e.g., financial services versus wholesale and retail).

In addition to ensuring that recruits have the basic skills and abilities to learn and perform their military occupations, the military imposes other standards related to age, citizenship (or permanent residency), education, physical fitness, dependency status, and moral character. Most private employers avoid use of such standards because of the inherent conflicts with federal, state, and local equal employment and disability laws that protect individuals from discrimination on the basis of race, sex, age, and other variables that define a protected class of individuals. However, some jobs may require the applicant to submit to background checks, meet medical or physical requirements, or obtain certifications or licenses, similar to processes used in the military.

Some characteristics like education and moral character are used in the private sector for selection if they are job relevant. Although the practice might be questioned from a legal defensibility point of view (e.g., most organizations are not able to demonstrate a significant relationship between having a high school diploma and job performance), it is not uncommon for an employer to require a high school diploma. However, it is important to note that the usual question private employers ask is whether a person possesses a high school diploma *or the equivalent*, not whether a person has a GED instead of a high school diploma. Alternatively, organizations sometimes recruit applicants only at places that are likely to provide high school graduates (e.g., job fairs associated with a vocational-technical school). Because some characteristics like having a GED are correlated with other variables that constitute a protected class (e.g., race), most employers avoid making inferences about the kind of individual who earns a high school diploma and the kind who earns a GED.

Comparing the processes for entrance into the military and private-sector employment, there are two apparent differences. First, the standards and processes by which individuals are selected into military occupations are known and widely communicated. In contrast, the standards and processes used to staff jobs in business and industry are highly variable across and within organizations. While personnel in the staffing division of the organization (if they exist) usually know the processes and standards, in many organizations, these processes and standards are often not publicly communicated to other personnel. Although applicants will learn of the process through experience, the standards required are sometimes never explicitly stated.

Second, because of the federal, state, and local laws, civilian employers often may not use personal characteristics like age, physical ability, and dependency status. In fact, some companies avoid any questions regarding these characteristics until after making a job offer. In contrast, these are primary criteria for military qualification.

Training Opportunities

Once an applicant is selected or hired into an entry-level position, the organization may provide training. Organizations vary considerably in their approaches to training. Some companies consider training to be an on-the-job effort and expect employees to learn by doing. So, in an entry-level job, a new hire may only attend a general orientation on the first day of employment and no other training. Other organizations devote considerable resources to providing formal training and evaluating competence (e.g., mentoring, staff development, technical training). Still others sidestep training altogether and hire only experienced employees for positions that require some training.

Organizations also vary on the kinds of training that are offered. While the military offers military training (e.g., basic training and occupational skills training) to all Service members, many private-sector employers provide basic skills training and supervisory training to only a few. The extent to which kind and amount of training are provided is unknown in the private sector as a whole. Similarly, outcome variables like standards and required competency levels are usually not specified.

Many employees are expected to invest in acquiring skills though postsecondary and vocational education and training courses prior to entry. When such general skills training is provided by business or industry, the employee often receives a reduced training wage although the employee may receive reimbursement for tuition. In contrast, the military provides fundamental job training in basic and advanced skills, such as electronics, computer science, and nuclear power, while the recruit continues to earn a salary. Training opportunities are available in some organizations for some occupations; in the military, training is offered to all recruits.

Working Conditions

To attract highly qualified and skilled employees, employers in private industry emphasize various inducements, such as attractive work settings (e.g., location, amenities, firm size), competitive compensation and benefits, advanced technology, career development, and training. For many job applicants, one of the most important aspects considered when choosing an entry-level position is working conditions.

While many positions in the civilian work force do not seem to vary substantially from similar positions in the military on skill or training requirements, there may be substantial differences between working conditions in military and civilian-sector jobs. Some variables that differ include leisure time, level of supervision, autonomy, union membership, personal safety, work/family issues, culture, and commitment.

Unlike serving in the military, working in entry-level jobs may not require a contract with the employer or strict adherence to rules that govern almost all aspects of one's life. A notable exception is certain jobs in local, state, or federal government. However, civilian employees are subject to organizational rules and regulations regarding such things as work hours (e.g., night shifts for security guards), dress code or uniform requirements (e.g., delivery drivers), and personal behavior (e.g., customer service agents). Nevertheless, in most cases, entry-level employees have more freedom to choose where to live and how to spend their free time compared with military personnel.

Not only are military personnel subject to military orders and discipline; they may serve under onerous conditions and in unpleasant places and may be subject to the risk of physical injury or death in combat. In some civilian occupations, conditions may be similarly undesirable or unpleasant. For example, police officers and firefighters risk their lives to ensure the safety of the people in their communities. With the exception of jobs in law enforcement and safety, however, most entry-level jobs are inherently low-risk, and none involves combat in warfare. Other work conditions may be less risky but have undesirable components, such as separation from family and night-shift duty.

The working conditions in the military and civilian employment vary considerably; however, it is important to note that it is the value an individual places on a specific working condition that is important in determining whether it is an advantage or disadvantage in the employment decision. What may be undesirable to one individual may be particularly attractive to another. While some may find the conditions of combat terrifying, others may find the same situation exhilarating. For some, night-shift work may be valued because it allows parental child care that would be precluded if both parents worked day hours.

Compensation

Like other human resource practices in the private sector, compensation strategies vary across and within different organizations, jobs, job families, and labor contracts. While the military has a fairly straightforward pay scale based on enlistment grade and various bonuses, private firms often use a complex array of recurring salaries or wages, merit increases, longevity increases, performance bonuses, and wage credits as well as benefits with direct financial value (e.g., pension plans, savings plan match).

Compensation of entry-level employees typically includes a base salary, incentive pay, awards, and other rewards. In addition, entry-level employees often receive overtime pay for hours worked over 40 in a week or on holidays. Compensation is often determined through job analysis (e.g., comparison with market data, comparison with other jobs, manager interviews, site visits), job description (e.g., the nature and level of the work), job evaluation (e.g., job content, market drivers), and compensation strategy and resources. General wage increases usually occur either annually or semiannually.

Many employees receive incentives in addition to regular pay. Awards and other rewards include cash and noncash prizes for performance, attendance, service, stock plans, and retirement. Wage credits (based on experience), signing bonuses, and referral bonuses are frequent hiring practices. Recent trends that affect compensation are the competitive labor market, retention challenges, and performance-based plans.

Perhaps the most notable difference in compensation for military and private-sector work is the pay practices of each group. Private employers offer wider ranges of pay within smaller groups of individuals, and their practices often indicate a philosophy of rewarding those who make greater contributions (e.g., pay for performance). Pay ranges may vary as a result of wage credits given at entry for previous experience, merit increases based on superior performance, performance bonuses based on individual or team efforts, increases for tenure, credits for high cost-of-living areas, and more generous collective bargaining agreements.

Private employers take into account labor market conditions when setting pay. Most large organizations have compensation plans and a supporting organization to review and set pay scales. There are no laws other than the equal pay laws forcing private employers to pay at a given rate. The implicit limitations on pay rates come from concerns about profits, labor market rates, internal equity, and overall fairness. Private firms are free to change pay rates to meet changes in the marketplace.

In locations where the labor market is tight, such as in major metropolitan areas versus rural areas, organizations may increase pay. According to LinemenOnline, an Internet site that collects pay scales of linemen working for electrical utilities from around the world, linemen for Virginia Power are reported to make $25.62 per hour, while linemen for Tucson Electric Power Company in Arizona make $28.88 per hour (LinemenOnline, 2001). Private employers may also compete and pay more for highly desirable skills, like computer skills, than for common or easily learned skills, like administrative support skills. Although the military also considers labor market conditions, it is on a much more limited basis (it sets one base pay rate that is congressionally approved). Thus, private companies have much more flexibility than the military Services to adapt quickly to the labor market and to reward individuals commensurate with their contributions.

In comparison, military compensation is consistent across large groups of individuals. Service members receive pay and signing bonuses for jobs that require high skills and substantial training (e.g., submarine or aviation personnel), that subject them to a high degree of personal risk, or that have shortages for any number of reasons. Thus, the pay range for everyone in a job at a particular grade in the military is very narrow.

Whether military or private-sector pay is greater overall is not clear. Some evidence regarding this was provided in the previous section. Recall that, in general, the pay of enlisted members compares favorably with the pay of high school graduates of similar age in the civilian sector, but it compares much less favorably to college graduates in the civilian sector (see Figure 5-4). However, like many other points of comparison, specific jobs in each sector should be compared. The military's own information highlights the variability regarding pay competitiveness in a question and answer format on the Internet site, www.todaysmilitary.com:

Is military pay competitive with civilian jobs?

> The answer is yes, it is. In a February 2000 survey, the Army Times Publishing Company compared 40 military positions (officer and enlisted) to their local civilian counterparts. In 19 of them, military pay was estimated to be below the local salary, but in 21 others, military pay was estimated to be equal to or higher than the comparable civilian pay. For purposes of comparison, military pay included base pay, basic allowances (housing and food), and the value of the average tax advantage for that pay grade (Today's Military, 2001).[11]

One of the central questions regarding the role of compensation and military recruitment is the extent to which compensation deters (or aids) enlistments. While salary comparisons can be done for a few jobs, doing these comparisons across large numbers of jobs is quite time-consuming. Also, the data available typically compare wages across industries without accounting for differences in wages within the industry. Findings from the Army Times study mentioned in the quote above suggest that in

[11]The above comparison does not include the value or savings of: (1) the greater security of a military job; (2) the greater purchasing power of the military dollar through the base commissary and exchange outlets; (3) low-cost group life and health insurance; (4) free space-available travel in military aircraft and other similar benefits; and (5) retirement or pension benefits (Today's Military, 2001).

the aggregate there is no clear advantage to civilian employment in terms of pay. This is inconsistent with the aggregate age-education comparisons of earnings presented in the previous section, in which the comparison is with civilians who are high school graduates.

Another belief is that salary levels reach higher levels in the private sector than in the military. Many are willing to accept lower pay on a short-term basis if it leads to higher pay in the long run. Because there is generally greater dispersion in earnings in the private sector than in the military, it is likely that the financial rewards to exceptionally talented employees in the private sector are greater than the financial rewards to exceptionally talented military members.

The real issue may not be whether the actual salaries in the civilian sector are higher than those in the military or whether salaries in the private sector have a higher range. Instead, the important issue may be in how salaries are perceived by young people making decisions about enlistment.

One source of information regarding attitudes on compensation comes from the 1999 Survey of Active Duty Personnel, which measured the attitudes and experience of over 30,000 service members from the Army, the Navy, the Marine Corps, the Air Force, and the Coast Guard with at least 6 months of active-duty service (Defense Manpower Data Center, 2001). Table 5-7 indicates a fairly small proportion of E1-E4 personnel who are satisfied with their compensation. Similar data come from the private sector in the form of a proprietary survey of employees in the Fortune 100 companies (Personnel Research Associates, 1999). Although there are many differences in the way each group was sampled and the percentage favorable are not strictly comparable, these data suggest that nonexempt (from overtime) employees in the private sector are more satisfied than those in the military (Table 5-8). In its magazine, *HR News*, the Society for Human Resource Management reported pay satisfaction of U.S. workers from a study of 1,218 respondents that provides a statistically representative sample of the U.S. private-sector workforce done by Sibson and Co., a global management consulting firm, with WorldatWork, formerly known as the American Compensation Association (Society for Human Resource Management, 2000). According to this study, 70 percent of the respondents reported satisfaction with their pay and benefits, and 65 percent were satisfied with their pay level overall. Interestingly, 43 percent reported that their employer's pay process (i.e., methods the employer uses to determine pay amounts, grades, progress through grades, promotion decisions, and other pay structures) was satisfactory. The survey did not clarify whether employees were reacting to the process itself or the lack of communications about many of the compensation practices.

When asked about which forms of pay were most important to them,

TABLE 5-7 Members by Pay Grade Group Who Indicated Satisfaction with Various Components of Military Life (percentage)

Military Pay and Allowances

Percent Satisfied E1-E3	Percent Satisfied E4	Compensation Element
16.5	16.1	Basic pay
23.4	20.9	Special and incentive pay
27.7	22.8	Reenlistment bonus or continuation pay program
22.4	22.0	Housing allowance
26.4	28.4	Subsistence allowance

Military Benefits

Percent Satisfied E1-E3	Percent Satisfied E4	Benefit
61.0	52.3	Medical care for you
34.7	26.5	Other retirement benefits (e.g., medical care and use of base services)
51.6	46.7	Medical care for your family

SOURCE: Tables 3.7 and 3.8, Defense Manpower Data Center (2001).

19 percent ranked merit pay first; 20 percent ranked overtime pay first; 18 percent ranked cost-of-living allowances first; and 12 percent mentioned individual incentives first. The survey also found that Generation Y (i.e., those born after 1973) were less satisfied than older workers with work in general and less interested in direct financial rewards or work content.

TABLE 5-8 Ratings of Compensation, Benefits, and Working Conditions by Nonexempt Civilian Employees in Fortune 100 Companies

Element and Survey Item	Percentage Favorable
Compensation: "I am paid fairly for my work."	59.0
Benefits: "How do you rate your total benefits package?"	70.9
Working Conditions: "How satisfied are you with your opportunity to get a better job at your company?"	40.1

SOURCE: Personnel Research Associates (1999) unpublished data.

This finding seems to suggest that these young people will not be particularly motivated by financial rewards. Clearly, the entire U.S. workforce is not directly comparable to the E1-E4 sample; however, if the data represent a comparable group of employees in any way, workers in business and industry seem to be more satisfied than military personnel.

Finally, statistical or econometric evidence provided in such studies as Warner, Simon, and Payne (2001) have clearly and consistently indicated that changes in relative compensation are correlated with changes in recruiting. Econometric studies since the inception of the All-Volunteer Force have established that there is an economically and statistically significant relationship between changes in military compensation relative to measures of civilian compensation and changes in recruiting. While the precise magnitude of this relationship has varied over time, from study to study, and by Service, the estimates have centered around an elasticity of about 1.0—that is, a 1 percent rise in military compensation relative to civilian employment earnings is associated with about a 1 percent increase in highly qualified recruits.

Benefits

Civilian employees in the private sector are offered an array of benefits and privileges. Benefits for entry-level employees often include medical and dental insurance, short- and long-term disability insurance, a savings plan (sometimes with a company match), a pension plan, tuition reimbursement, a flexible reimbursement plan, life and other insurance coverage, and personal time off. According to a survey of human resource professionals (N = 745) on employer's use of 160 benefits often used in the private sector (Society for Human Resource Management, 2001), 64 percent of respondents indicated that the value of the benefits provided were worth from 11 to 40 percent of an employee's salary on average.

There are differences in the mix of benefits offered as well as the level of benefit provided and the cost to the employee for the benefit. For example, in some organizations uniforms are required and the employee is expected to pay for them (either on his or her own, at a discount, or through a payroll deduction); in other organizations, the uniforms are free or subsidized. Complicating things further, some organizations offer a set of benefits to all employees, while others offer "cafeteria plans" that allow employees to tailor their benefit package to their current needs to varying degrees.

Simply offering a benefit may not be sufficient to attract and retain employees or military personnel. The quality of the service, which is typically discovered after employment or enlistment begins, and use of the service may affect the perceptions of its value. Also, initial receipt of the benefit may affect its perceived value. For instance, medical insurance programs that take effect after a person has been on the payroll for six months may be less valuable than a program that begins from initial date of employment. Similarly a month of vacation may be highly desirable, but less so if it takes 10 or 20 years to reach that level of benefit.

Furthermore, different employees place different values on the same benefit. For example, a single employee with no dependents may not value a child care assistance program, while an employee who is a single mother with three children may choose employment based on the availability of company-sponsored child care.

Despite the difficulty of objectively comparing benefits across organizations, people still have well-formed opinions about the value of benefits. As Table 5-7 shows, the 1999 Survey of Active Duty Personnel suggests that the percentage of service personnel satisfied with their pay is much lower than the percentage satisfied with their benefits. There is also some evidence that satisfaction with benefits has decreased over time (Defense Manpower Data Center, 2001).

And compared with satisfaction in the private sector, the percentages of military personnel satisfied with pay and benefits are low. Table 5-8 shows survey responses of nonexempt civilian employees in Fortune 100 companies. The results show that more than half are satisfied with their pay and 70 percent with their benefits. Even though these findings may not be comparable with the available data on military personnel, they are reinforced by reports from the Society for Human Resource Management (2000) that 70 percent of private-sector employees are satisfied with their pay.

Higher Purpose

In many, or even most, civilian firms, there is a notion that the goods or services provided by the firm or industry and its employees are valued by society and that providing quality goods or services efficiently is intrinsically rewarding to many employees. In some occupations, such as the health care professions, the belief in this type of intrinsic reward can be particularly strong. However, in private industry, this dedication to serving the public through high-quality goods and services is often tempered by the idea of corporate profits that benefit shareholders rather than the general public. In the general case, the spirit of self-sacrifice and service for a higher purpose is much stronger in the military than in almost any civilian occupation or profession.

Combined Military-Civilian Careers

When considering civilian employment as an alternative to military service, one should recognize that they are not mutually exclusive. The majority of those who enlist serve only one term of service and then enter either the civilian labor market directly, or the civilian labor market after obtaining additional formal education. Moreover, those who serve a full military career of 20 or even 30 years of service typically will have a second career in the civilian sector. For this reason, a strategy by the military Services that attempts to take this into account and offers to prepare members for a return to the civilian labor market may be more successful than one that considers only the direct competition between military service and the civilian labor market at the entry point.

To a large extent, the military Services have followed such a strategy. Military training and experience are valuable in the civilian sector. The Services inform potential recruits of this, through advertising and through recruiters. Moreover, there has been increasing emphasis on providing the opportunity to obtain civilian "credentials" for relevant military training and experience.

Trends in the civilian labor market, in which employees change firms and even industries more frequently than in the past, may make the notion of a combined military and civilian career even more fruitful in the future. In particular, the opportunity for a potential recruit to serve his or her country and to determine, through experience over an initial term of service, whether to pursue a full career of military service is a sound strategy to pursue.

Summary

Almost two-thirds of youth ages 16–24 participate in the civilian labor force today. Over the next several years, the youth population is projected to increase, leading to rapid growth in the youth labor force. There are a large number of job opportunities that exist in the civilian sector, but many opportunities also exist in the military, and many jobs can be found in both. Neither the military sector nor the civilian sector is clearly superior. There are also no significant reasons for preferring opportunities in civilian employment over opportunities in the military. In fact, some specific training and skills obtained in the military are transferable and valued in the civilian sector.

Working conditions and the opportunity to serve a higher purpose may be two areas in which substantial differences are observed. Different individuals may place different values on these characteristics. Thus, while one individual may find the restrictions that come with military life onerous, another may find that service in the military is a source of pride and patriotism that outweighs inevitable restrictions. Similarly, one individual may value the intrinsic reward of serving one's country, while another may value extrinsic rewards (e.g., fame and fortune) more.

EDUCATION AS AN ALTERNATIVE

In this section, we examine the aspects of postsecondary education that attract youth to college and the benefits of obtaining and completing a postsecondary education. Specifically, we address demographic trends in postsecondary education, access and returns to postsecondary education, youth perceptions of the benefits of education, and postsecondary educational opportunities in the military.

As noted in Chapter 3, young people appear to put a very high premium on getting a postsecondary education. Depending on the college and the program, a college education can provide a liberal education, develop the ability to think analytically and solve problems, improve communication skills, and provide the opportunity to learn high-tech skills. Parents and students alike also value the opportunity for students to mature, manage their own lives, and develop a personal value system (Immerwahr, 2000). However, as discussed earlier in this chapter, the greatest benefit may be the economic advantage over individuals with only a high school education; thus, parents, teachers, and high school counselors are increasingly advising students to go to college (U.S. Department of Education, 1999).

In 2000, the unemployment rate was at a 30-year low—3.9 percent (U.S. Bureau of Labor Statistics, 2002a). There was increasing demand for skilled workers in the civilian sector and more opportunities available to college degree holders. In response to this demand, more high school graduates chose to enroll in college and complete their postsecondary education.

By today's standards, a high school degree is often the minimum requirement needed to enter the job market and gain access to various opportunities in the manufacturing and service sectors. However, high school graduates across the United States today can also choose from a complex array of postsecondary education programs. These programs range from short-term certificate programs to public or private two- or four-year degree programs that vary in quality and cost.

Transitions to Postsecondary Education

Vocational and technical preparation can be provided in combination with high school classes. These programs offer a mechanism for young people to make a transition to postsecondary education.

Tech Prep

The Tech Prep Education Act—Title III of the Carl D. Perkins Vocational and Applied Technology Education Act of 1990—emphasizes the development of academic and technical skills of high school and postsecondary students. This federally funded program, also known as 2+2, begins during the final two years of high school (10th–12th grade) and continues through a two-year postsecondary program (e.g., community college, technical college, or apprenticeship program). Participants in the program have the opportunity to obtain an associate's degree or postsecondary certificate in a specific career field, as well as employment in the related area of study. Some schools have expanded the 2+2 concept to a 2+2+2 concept to encourage students to plan a long-term program of education that includes transferring from a community college to a four-year college.

Tech prep programs are prominent across the nation in most secondary schools and community colleges. They are also being implemented in the military. The Navy works with community colleges to recruit graduates with the technical and literacy skills needed to be active-duty sailors and are establishing a training program based on the federal tech prep program (Golfin and Blake, 2000). While participants in tech prep programs have the opportunity to obtain a two- or four-year college degree, many are trained in specialized skills (e.g., computers, advanced technology) that are highly applicable to jobs in either the private sector or the military.

Vocational and Technical Programs

Vocational and technical programs also provide high school graduates with training in specialized skills. However, many more students are completing these programs with the intention of going to college. Between 1982 and 1994, there was an increasing number of public high school students completing college preparatory courses and a decreasing number completing vocational courses (U.S. Department of Education, 1999). Of the students completing vocational coursework, some were completing college preparatory coursework as well (18 percent in 1994 compared with 2 percent in 1982). Public high school graduates who completed both vocational and college preparatory courses in 1992 (89.9 percent) were just as likely as students taking exclusively college preparatory courses (93.2 percent) to enroll in a two- or four-year postsecondary program.

In sum, students in tech prep or in vocational and technical programs develop academic and technical skills that are highly applicable to the military or the civilian employment sector. Yet they are just as likely as other high school students and graduates to prepare for and enroll in a postsecondary institution.

College Enrollment and Completion

High school seniors have a variety of postgraduate choices, but the most popular choice among them is enrollment in college. As discussed in Chapter 3, a majority of high school graduates enroll in college immediately after graduating. Along with the rise in four-year college enrollment, community college enrollment has become increasingly popular during the past 30 years. Community colleges help accommodate the growing demand for a postsecondary education particularly because of their low cost and history of enrollment of low-income and minority students. Community colleges continue to be an important alternative for those in the postsecondary market. Although many high school seniors choose to enroll in two- or four-year colleges right after graduating, others have been choosing nontraditional paths to college, such as waiting one year or more after high school before enrolling, attending school part-time, combining school and work, or working full-time first (Choy, 2002).

Some high school seniors who choose postsecondary education may not remain in school straight through to completion. Students may leave the path toward a college degree at different stages depending on parents' level of education, types of academic courses completed in high school, and availability of financial assistance. Among those high school graduates who enroll in college, some may drop out for a few years and later reenroll in the same program or transfer to another program, while others may drop out before completing their programs and never earn their degrees. Students whose parents completed college, who aspire and prepare for college, and who receive financial support are more likely to complete steps toward a college degree (Choy, 2002). Working full time, enrolling in a community college, and having parents with only a high school education are risk factors for college completion (Choy, 2002).

In 1989–1990, about 30 percent of college students (16 percent of students in four-year colleges and 42 percent of students in two-year colleges) left postsecondary education before beginning their second year (U.S. Department of Education, 1998; Choy, 2002). Of those students who left four-year colleges, 64 percent returned within five years (stopouts) and 36 percent stayed out (stayouts). Stayouts were typically older adults, married, people with children, and full-time workers (Choy, 2002). One-third of the students who returned within five years obtained some degree within that time. About half of the two-year students who stopped out returned within five years. Stopouts were more likely to have attended college full time and been engaged in academic pursuits (e.g., interactions with students and faculty).

By 1998, these trends were relatively the same: 37 percent of first-year students who enrolled in postsecondary education in 1995–1996 left without obtaining a degree (U.S. Department of Education, 2001). Students in four-year colleges were less likely to drop out and stay out of college by their third year than were students in two-year colleges. Also, older students (ages 19–24) were more likely to stop out during their first year compared with younger students (ages 16–18). The average length of stopout (the first time) was 10 months for four-year college students.

Where do college stopouts and dropouts go once they leave postsecondary education? Slightly less than half of four-year college stopouts later reenroll in the same institution; the remaining students transfer to other institutions. We see the reverse pattern for two-year college stopouts. College dropouts are likely to choose other alternatives to postsecondary education (e.g., civilian employment or military service). Students may choose different paths to a college education that stray from the traditional but most persist and eventually obtain a postsecondary degree.

Demographic Factors and Enrollment

Recent college enrollments are up for all ethnic and socioeconomic groups. The percentage of high school graduates who enroll in college in the fall immediately after high school reflects both the accessibility and marketing of higher education as well as the value high school graduates place on college compared with other alternatives. Although the proportion of minority and low-income students enrolling in college has increased, an analysis of current enrollment data and projections for future enrollment show some barriers to postsecondary access.

Race/ethnicity, parent income, and parents' level of education affect student enrollment and appear to be the best indicators of access to college. In 1999, high school graduates in high-income families were more likely to enroll in college immediately after high school (76 percent) than middle-income (60 percent) or low-income families (49 percent) (U.S. Department of Education, 2001). High school graduates were also more likely to enroll immediately after high school if their parents had at least a bachelor's degree. In 1999, 82 percent of students (ages 16–24) whose parents had bachelor's degrees or higher enrolled in college compared with 54 percent of students whose parents had only high school degrees (U.S. Department of Education, 2001).

In addition to family income and parents' education, race and ethnicity can affect postsecondary aspirations and enrollment. In 1992, of all college-qualified high school seniors who planned to attend college, 88 percent expected to finish school (U.S. Department of Education, 1997). And 82 percent of college-qualified students specifically planned to attend a four-year college. Of those students, Hispanics (77 percent) were less likely to plan to attend a four-year college than blacks (87 percent). Hispanic students were also less likely to enroll in postsecondary education. Moreover, Hispanics who did enroll were more likely to attend a two-year rather than a four-year institution compared with students from other race-ethnic groups. Most students value the importance of a postsecondary education. However, various barriers to access may thwart the intention to apply, enroll, or attend college.

Financing Postsecondary Education

Another potential barrier to attending college is the cost of a postsecondary education, which has risen significantly over the years. To help reduce financial barriers to college enrollment, students may receive state subsidies or federal grants and loans and participate in work-study programs. Financial aid to students has doubled over the past decade, primarily due to an increase in loan aid (College Board, 2001). Grant-based aid has increased as well, but to a lesser extent. Furthermore, more federal awards and college scholarships have been granted to students in recent years in the form of college grants, student loans, or Pell grants (U.S. Bureau of the Census, 2000b).

However, because of increases in nonneed-based borrowing, federal aid on the basis of need has decreased since the mid-1980s, from 80 to 60 percent (College Board, 2001). Since the early 1990s, use of nonneed merit scholarships has grown substantially, although a greater proportion of state financial aid is need-based (75 percent). While the number of need grants grew faster than merit aid, the amount of merit aid was larger (Heller, 2000). In addition, institutional spending on merit-based aid has grown, with a shift away from need-based aid and a greater share of merit-based aid going to middle- and high-income students. Yet, need-based financial assistance has grown significantly in importance to students from lower-income families.

With the help of financial aid, students in low-income families (below $30,000) may pay a lower net price for tuition, board, and fees than students in middle- (between $30,000–$59,999) or high-income families (over $60,000). However, average tuition has increased substantially since the 1980s, and the share of family income required to pay for college tuition and fees has increased for many families, particularly those at low- or middle-income levels (College Board, 2001). The cost of postsecondary education also varies across institutions and the cost not covered by aid (unmet need) remains the responsibility of the students or their families.

Black and Hispanic students, who are overrepresented in low-income families, tend to have the lowest expected family contribution and the highest unmet need. These financial barriers limit low-income students' educational choices. As a result, they either work or get help from family and friends to cover remaining costs. Consequently, these students may have higher loan debt, work more hours, attend less expensive schools, go to school part-time, and live off campus, which all lead to a lower probability of degree completion and negative implications on lifelong earnings.

Over time, the real cost of attending college has risen as a percentage of family income and continues to rise. Financial aid per full-time-equivalent student has increased but has not kept pace with increasing tuition. Such financial barriers to postsecondary education disproportionately affect minorities and lower-income students. These students may rely on various sources of aid to increase their chances of obtaining a postsecondary education. Programs that combine opportunities for access to postsecondary education and service in the military may be one major source of this aid.

Returns to Postsecondary Education

Trends in Earnings for High School and College Graduates

Changes in the labor market (e.g., globalization), the structure and type of occupations (e.g., service and technology), and the strong economy are important factors that contribute to the rise in intentions to obtain a postsecondary education. Nevertheless, the key factor remains the financial incentive for postsecondary education, which appears to have increased substantially over time.

As indicated in Figure 5-4, the earnings of college graduates are significantly greater than the earnings of high school graduates, or even those with some college. Importantly, the returns to investing in a postsecondary education have grown significantly over time, as has the earnings premium for a college degree, relative to a high school diploma. In 1985, for example, college graduates earned about 30 percent more than high school graduates of comparable experience (see Figure 5-9). By 1995, this premium had grown to over 45 percent. Interestingly, the earnings of those with some college (less than a four-year college degree) have not grown relative to the earnings of high school graduates, as illustrated in Figure 5-10.

Average annual earnings for college graduates have kept pace with

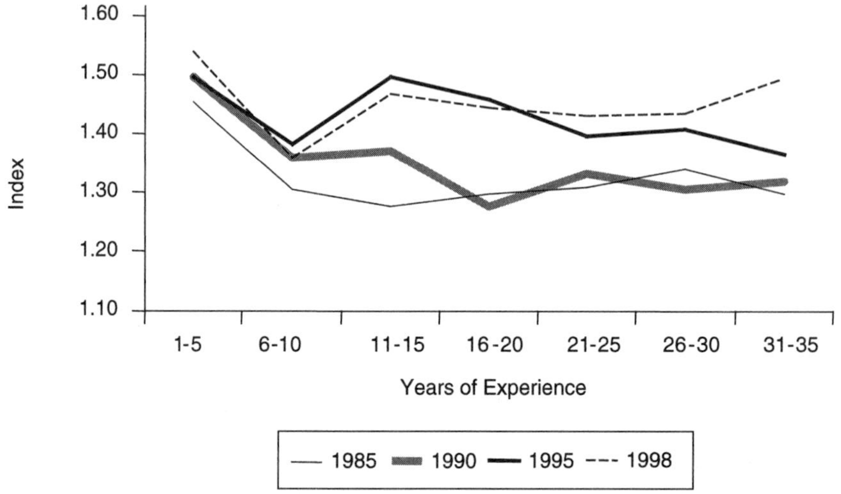

FIGURE 5-9 Earnings of college graduates relative to high school graduates.
SOURCE: Data from the U.S. Bureau of the Census (2002d).

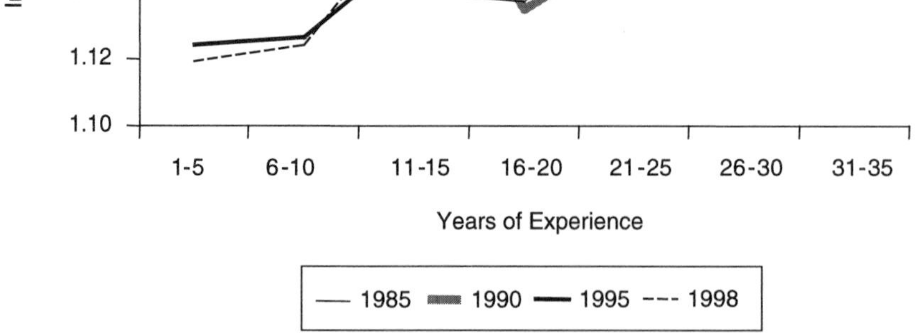

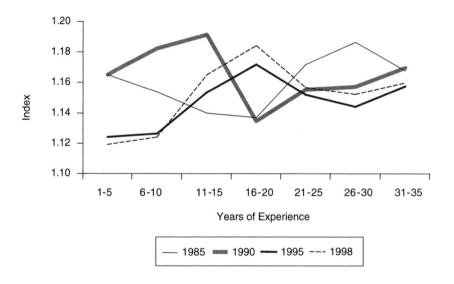

FIGURE 5-10 Earnings of those with "some college" relative to high school graduates.
SOURCE: Data from the U.S. Bureau of the Census (2002d).

or have exceeded the rising costs of attending college. The trend remains even when accounting for earnings lost while attending college, attendance beyond four years, or amount of borrowed funds (e.g., loans). By contrast, the value of a college education is diminished when a degree is not completed. For persons who have completed only some college (one to three years), the result is less favorable. Noncompletion is detrimental to mean earnings such that these earnings have not kept pace with the growth in college costs. Time to recover expenses for a college education is lengthened even more if loan repayment is required.

The increasing earnings premium for college graduates compared with high school graduates is undoubtedly one of the most important factors affecting today's military recruiting market. As noted earlier, military compensation for enlisted members compares favorably with that of high school graduates in the civilian sector but falls significantly short of the compensation of college graduates. Because the Services recruit from a college-eligible population—high school graduates who score in the upper half of the AFQT—this apparent increase in the financial return to college undoubtedly has made recruiting more difficult. One approach to easing recruiting problems is to design a credible program through which the enlistees could serve one or more terms and obtain a college degree at the same time.

Postsecondary Educational Benefits in the Military

Currently the military relies on educational incentives to attract young people to join; thus the choice between enlisting and postsecondary education is not mutually exclusive. There is a broad variety of educational incentive programs offered to recruits serving in the military. The three main options are tuition assistance, the Montgomery GI Bill, and the Veteran's Educational Assistance Program. These programs are described and evaluated in extensive detail in a variety of sources, including DoD and Service web sites (e.g., <http://www.voled.doded.mil>).

Several publications prepared by the National Defense Research Institute at RAND describe opportunities for combining military service with postsecondary education and comprehensively review various educational incentive programs (Asch et al., 1999, 2000; also, see Warner et al., 2001). Here we briefly review several of the postsecondary education programs and educational incentives offered by the military.

Combining Postsecondary Education and Military Service

To aid service members in obtaining a postsecondary education, the Services have various benefits and programs that provide tuition assistance and loan repayment, administer college or graduate degrees in programs such as Community College of the Air Force and Naval Post Graduate School, and offer college course credits for military training and job experience (see Thirtle, 2001). Through the DoD Voluntary Education Program, military services provide funding to military personnel for taking college courses. The military Services also collaborate with educational institutions to offer opportunities to learn at civilian high schools, vocational and technical schools, and undergraduate and graduate programs.

The military Services are increasingly providing assistance for youth to pursue a postsecondary education and serve in the military. Five options currently exist:

1. Officer track: attend four-year college then enter the Service as an officer.
2. College-enlisted track (e.g., College First): attend college or receive some college credit then enter the Service as an enlistee.
3. Enlisted-college track (e.g., Montgomery GI Bill): enter the Service as a high school graduate, complete the service obligation, leave the Service, and go to college as either a veteran or Reserve or National Guard member.
4 Enlisted-officer track: enter as an enlisted member, leave temporarily to attend four-year college, and return to the Service after getting a degree.
5. Concurrent track: obtain college credits while in service.

Each track consists of various programs (some described below) that offer different educational benefits. Most individuals enter the largest tracks—the enlisted-college and concurrent tracks (Asch et al., 1999). Other tracks, such as the college-enlisted track, are small, but the Army and the Navy are considering expanding their programs to better attract college-bound youth (Asch et al., 2000).

All of the Services offer tuition assistance and special academic programs. For instance, participants in the concurrent track receive tuition assistance to take college courses while serving or receive college credits for experience and training in their military jobs (Asch et al., 1999; Thirtle, 2001). Tuition is typically paid in full for members entering high school certification programs and up to 75 percent of tuition for members entering a college program. Loan repayment programs also exist for enlisted Service members with college-related federal loans. Individuals in these programs must enlist in a critical occupational area and be a highly qualified recruit. The Army and the Navy reimburse loans for education obtained prior to enlisting. The Army pays $1,500 of an outstanding eligible loan for each year of service up to $65,000; the Navy's maximum repayment is $10,000 with a requirement of a four-year enlistment.

The Montgomery GI Bill is a benefit program that attracts college-bound youth to join the military. In this program—available to all new recruits in the enlisted-college track— individuals typically enlist first, then, after they leave the service, enroll in college. As mentioned earlier, the Montgomery GI Bill program often serves as a transitional benefit when recruits leave the service and reenter civilian life. (However, enlistees may also use the benefits while on active duty—Asch et al., 1999; Thirtle, 2001.) The maximum monthly benefit is $536 for active-duty Service members and $255 for Reserve and National Guard members. These benefits are available for up to 10 years after leaving the service.

Highly qualified recruits in hard-to-fill occupational areas and recruits who enlist for a specified number of years are eligible to receive even more money through the Army College Fund or the Navy College Fund. These programs are an enlistment incentive to supplement the Montgomery GI Bill; therefore, when combined with the GI Bill, the benefit may reach up to $50,000 (depending on years of service). Warner et al. (2001) evaluated the effects of various college benefits (e.g., the Navy College Fund) and enlistment bonuses on recruiting outcomes and found that these programs were a valued incentive and increased the number of high-quality enlistments.

In contrast to the Montgomery GI Bill program, the College First program allows enlistees to first attend college then enlist at a higher pay grade (college-enlisted track). Participants in this program attend college first for up to two years while in the delayed entry program; in addition, they receive a monthly allowance of $150. After college, the enlistees serve a term of service in the Army or the Navy. The College First program helps expand the enlisted supply and may lead to reduced accession requirements. Furthermore, enlistees with some college acquire skills that make them more productive in their military jobs.

According to Asch et al. (2000), the Montgomery GI Bill program may not be beneficial to the Services in terms of returns yielded if individuals leave the Service to attend college. It is more advantageous to have a skilled Service member return after completing college or one who takes courses concurrently with service. Furthermore, relatively few enlistees obtain a degree after serving or even complete some college courses while serving (Asch et al., 1999). The College First program may therefore be more beneficial for several reasons: individuals enlist at a higher pay grade, earn more while in the Service, and earn more civilian pay after leaving because they accrue more experience and training while in the military. Because enlistees do not leave the Service to receive the benefit, the military realizes a return on its education investment and reenlistment rates increase.

Asch et al. (1999) suggest that further research is needed to evaluate which options are most effective at attracting college-bound youth. They proposed several policy alternatives for attracting youth, including recruiting two-year college students, recruiting college dropouts and paying them at a higher pay grade, and offering some combination of getting an education while serving in the military. According to a recent survey of youth attitudes toward serving in the military, most respondents (high school seniors, college students, and college dropouts) seem to prefer benefit options that combine working with going to school (Asch, 2001). This finding suggests that enlisted personnel may be more likely to obtain a college education while serving in the military and provides further support for a military enlistment strategy that combines military service with postsecondary education concurrently.

Educational incentives are probably one of the most promising policy change options for attracting college-bound youth. However, the growth in military college benefit programs lags behind the costs of a postsecondary education, and legislative proposals for enhancing these benefits are pending (Asch et al., 1999).

Collaborating with Community Colleges

Community colleges provide technical education in skill areas needed by the armed services (e.g., electronics and computers). Compared with high school graduates, community college students have better literacy skills, are more likely to complete initial service obligations, and often possess technical skills they would otherwise have to learn at military expense. Despite overall increasing attendance at community colleges, less than 1 percent of armed services recruits are community college graduates (Golfin and Blake, 2000).

Military programs such as the Community College of the Air Force and Army College First provide recruits with opportunities to earn degrees while on active duty (Golfin and Blake, 2000). However, there are several limitations to obtaining postsecondary education while in the Service. For instance, in the Navy, sea duty, deployments, and training requirements make it difficult for sailors to pursue an education compared with airmen or soldiers.

Recent advances in technology are stimulating the growth of distance education offerings in postsecondary institutions and in the military (e.g., improved means to allow interactivity between students and faculty and improved rapid exchange of information). According to the U.S. Department of Education (2000b), two- and four-year postsecondary institutions are expanding their distance technology and education offerings. Between 1994–1995 and 1997–1998, the number of public and private postsecondary institutions involved in distance learning increased by one-third. Approximately 8 percent of all two- and four-year higher education institutions offered certificates or college-level degrees entirely through distance learning.

Similarly, the DoD education and training community is undertaking initiatives to leverage use of technological capabilities for distance education (e.g., the Advanced Distributed Learning Initiative). One example, the U.S. Army University Online, provides access to education for Service members wherever they are on the globe. Members can choose from 85 different programs at 20 different schools to earn a certificate or an associate's, bachelor's, or master's degree by attending web-based courses. Some benefits of this program include full tuition and fees, personal computer equipment, email and Internet access, and 24-hour technical support.

One-quarter of the 1.3 million Service personnel are involved in some form of training at any one time (Fletcher, 2000). Most of this training involves preparing for and implementing tasks and building occupational skills. A vision for defense training in the future is to deliver instruction and provide assistance to individuals and groups in real time and on demand. Emerging technology will help provide high-quality instruction to individuals anytime and anywhere.

SUMMARY

The primary alternatives to military service are civilian employment and postsecondary education. Each of these alternatives emphasizes different characteristics that affect the choice of highly qualified youth. Currently, the greatest challenge to military recruiting is attracting college-bound youth. The armed forces compete directly for the same portion of the youth market that colleges attempt to attract—high school graduates that score in the upper half of the AFQT.

Both the proportion of high school graduates who continue on to college and the earnings of college graduates compared with high school graduates have increased substantially over the past decade. Both of these related factors suggest that it has become increasingly difficult for the armed forces to compete successfully with postsecondary education for college-eligible youth.

By contrast, the armed forces have been generally successful in offering a compensation and benefits package that is competitive with civilian employment opportunities. Moreover, the resources applied to recruiting including recruiters, bonuses, education benefits, and advertising effectively increase the enlistment rates of high-quality eligible youth. This mix of resources has been reasonably and efficiently applied.

While civilian employers are highly variable in what they offer prospective employees, neither the military nor the civilian sector is clearly superior in job opportunities. For example, compensation and benefits vary but are not necessarily better or worse in the civilian sector. However, in the civilian sector, employees can better respond to employee and market demands and employ a variety of compensation strategies.

In two areas there are substantial differences between the military and civilian employment: working conditions and opportunity to serve a higher purpose. A typical individual might prefer working conditions in the civilian sector to those in the military, where conditions may be onerous or life threatening. However, that individual is more likely to find a transcendent purpose (e.g., duty to country) serving in the military than being employed in the civilian sector.

The notion of a higher purpose, combined with extrinsic rewards, such as competitive compensation and training and educational opportunities, provides a foundation from which the armed forces can compete successfully with civilian employers and postsecondary educational institutions in attracting qualified youth. Therefore, while postservice benefit programs like the Montgomery GI Bill have been successful in the past, innovative approaches that integrate higher education with military service may be necessary in the future to attract youth.

Recruiting has become more difficult in the last several years, but not because of a change in the effectiveness of recruiting resources. The recruiting difficulty was in part due to a booming economy and in part due to the failure of recruiting resources to keep pace with the recruiting mission. However, as previously discussed, a portion of the downturn cannot be explained by either of these factors. Instead, a portion may have been due to a change in tastes for military service in the eligible youth population or other factors that have not been measured. In the next chapter, we examine what unexplained factor may lead to recruiting challenges.

6

Youth Values, Attitudes, Perceptions, and Influencers

Youth analysts are increasingly speaking of a new phase in the life course between adolescence and adulthood, an elongated phase of semiautonomy, variously called "postadolescence," "youth," or "emerging adulthood" (Arnett, 2000). During this time, young people are relatively free from adult responsibilities and able to explore diverse career and life options. There is evidence that "emerging adults" in their 20s feel neither like adults nor like adolescents; instead, they consider themselves in some ways like each. At the same time, given the wide variety of perceived and actual options available to them, the transition to adulthood has become increasingly "destructured" and "individualized" (Shanahan, 2000). Youth may begin to make commitments to work and to significant others, but these are more tentative than they will be later. Jobs are more likely to be part-time than at older ages, particularly while higher education, a priority for a growing number of youth, is pursued. There is increasing employment among young people in jobs limited by contract, denoted as contingent or temporary. Such jobs are often obtained through temporary job service agencies. Young people are also increasingly cohabiting prior to marriage or as an alternative to marriage.

This extended period of youth or postadolescence is filled with experimentation, suggesting that linking career preparation to military service might be attractive to a wider age range of youth than among traditionally targeted 17–18-year-olds who are just leaving high school (especially extending to youth in their early and mid-20s). But what about their values of citizenship and patriotism? Are young Americans motivated to serve? Are their parents and counselors supportive? Is there a link between volunteering in the community and a desire to serve in the military?

This chapter is divided into two sections: the first deals with youth values and the second focuses on the individuals and events that influence or reinforce youth values. Our analysis of youth values includes (1) whether and how values have changed over time, (2) what trends can be anticipated in the future, and (3) changes in youth views of the military. Some primary questions regarding trends in youth values are

- What are the important life goals for young people?
- How do youth think about and act on values related to citizenship, civic participation, and patriotism?
- What are their educational goals beyond high school?
- What are considered the most important or desirable characteristics of a job?
- How do youth feel about work settings?
- What do youth believe about military policy and missions and about the military as a place to work?

The data we draw on to address these questions come from three large national survey databases supplemented by a locally based longitudinal study, several cross-sectional studies, and small observational studies as available.[1] The three national databases are

- **Monitoring the Future**, a nationwide study of youth attitudes and behaviors covering drug use plus a wide range of other subjects, conducted annually by the Survey Research Center at the University of Michigan (Bachman et al., 2001b; Johnston et al., 2001; see also <http://www.monitoringthefuture.org>). In-school questionnaire surveys of high school seniors have been conducted each year since 1975; similar surveys of 8th and 10th grade students have been conducted since 1991. The survey sample sizes range from approximately 14,000 to 19,000. The study includes follow-up surveys of smaller subsamples of graduates from all classes from 1976 onward.
- **Youth Attitude Tracking Study**, a nationwide survey of youth attitudes about various aspects of military service, their propensity to enlist, and the role of those who influence youth attitudes and behavior, conducted by the Department of Defense from 1975 through 1999. In the

[1]The committee is aware that responses to questions designed to elicit attitudinal responses are subject to varying interpretations by respondents and, therefore, must be treated accordingly. This is one of the reasons why our analysis focuses on changes over time rather than the absolute value of the response.

latest survey, approximately 10,000 telephone interviews were conducted. The age range of participants was 16–24.

- **The Alfred P. Sloan Study,** a nationwide longitudinal study of students conducted by the National Opinion Research Center, University of Chicago, from 1992 to 1997. The goal was to gain a holistic picture of adolescents' experiences with the social environments of their schools, families, and peer groups. The methodology included survey, telephone interviews, experience sampling, sociometric reports, and supplemental interviews with those who might influence youth (Schneider and Stevenson, 1999). The sampling was designed to ensure diversity, not to provide population representativeness. Students were drawn from grades 6, 8, 10, and 12. The total sample was 1,211 students.

The locally based study is the Youth Development Study, conducted at the Life Course Center, University of Minnesota (Mortimer and Finch, 1996). The main purpose of this study is to address the consequences of work experience for youth development, mental health, achievement during high school, and the transition to adulthood. One thousand 9th graders were randomly selected in 1987 from the St. Paul, Minnesota, public school district; these youth have been surveyed annually through 2000, from the ages of 14–15 to 27. Selected subsamples of the respondents have been interviewed to develop a better understanding of the subjective transition to adulthood.

The second section of the chapter reviews the scientific literature and data characterizing youth influencers, drawing on (1) the literatures of socialization, attitude formation and change, and youth development as they inform decisions about early career development and (2) information regarding the role of influencers to the extent that it informs early career decisions. The focus of our analysis is on the aspects of the career decision-making process that bear most directly on youth propensity to enlist in the military.

TRENDS IN YOUTH VALUES

Although a primary source of data for this section is the Monitoring the Future survey, we rely more particularly on a report examining these data concerning high school seniors' and young adults' views about work and military service (Bachman et al., 2000a). That report covers trends from 1976 through 1998. Where useful, certain findings have been updated through 2001. The dominant finding from that report was stability rather than change over time in youth views about work and about military service, although there were also important changes that are examined in the following sections. The topics presented include: (1) important goals in life, (2) citizenship, civic participation, and volunteerism, (3) education and work, and (4) views of the military.

Important Goals in Life

What are the life goals of youth, how have they changed over time, and what are the implications for military enlistment? Table 6-1 shows percentages of high school seniors in the Monitoring the Future (MTF) surveys who rated as "extremely important" each of five goals in life (these five were selected from a longer list as being potentially relevant to military service decisions). The table shows young men and women separately and compares recent graduating classes (1994–1998) with classes nearly two decades earlier (1976–1980). Among young men, the percentages shifted only modestly over two decades, and the rank ordering was unchanged. Among young women, the shifts in percentages were small also, and the rank orderings showed only one trivial change (fourth and fifth places reversed). There were, on the other hand, some consistent differences between the male and female ratings, as noted below.

Among the items listed in the table, the highest importance ratings by far were assigned to "finding purpose and meaning in my life." Just over half of the males in each time period rated it as extremely important, and the proportions of females were even higher. This item showed little correlation with military propensity during the last quarter of the 20th century, and from this we might infer that military service during that period was not seen as superior or inferior to other pursuits as a means for finding purpose and meaning in life. But the fact that most young adults still rate this as extremely important suggests that if military service in future years can provide such opportunities—and be perceived as doing so—the appeal is likely to be strong.

"Having lots of money" grew in importance among young men and young women between the classes of 1976–1980 and the classes of 1994–1998, as can be seen in the table. The percentage of young people in the National Center for Education Statistics (NCES) surveys who, two years after high school, thought having lots of money was very important also increased from 1974 to 1994 from slightly over 10 percent to approximately 35 percent (Larson presentation, 4th Committee Meeting—Irvine, December 2000).

According to data from the MTF surveys, "making a contribution to society" was somewhat less likely than money to be rated as extremely important by young men, whereas among young women it was a bit more likely. So are today's youth altruistic? Or materialistic? And have young people been shifting in one direction or the other? The data show only modest changes over time, and the gender differences also remain

TABLE 6-1 Importance Placed on Various Life Goals: Comparison of Rank Orders

How important is each of the following to you in your life?

Rank	% Extremely Important	Males 1976–1980	Males 1994–1998	Males Change
1	Finding purpose and meaning in my life	54.9	51.9	–3.00
2	Having lots of money	23.3	33.4	10.04
3	Making a contribution to society	18.0	21.3	3.24
4	Getting away from this area of the country	11.7	14.4	2.65
5	Living close to parents and relatives	8.2	13.0	4.84
Rank	**% Extremely Important**	**Females 1976–1980**	**Females 1994–1998**	**Females Change**
1	Finding purpose and meaning in my life	72.2	64.8	–7.40
2	Making a contribution to society	17.6	23.7	6.09
3	Having lots of money	12.7	20.2	7.46
4	Living close to parents and relatives	9.5	16.1	6.62
5	Getting away from this area of the country	11.9	14.0	2.15

SOURCE: Data from Monitoring the Future surveys.
NOTE: Rankings were assigned based on respondent ratings from the class years 1994 to 1998. Significance tests were calculated using *t* tests with pooled variance estimates based on percentages and adjusted for design effects, $p < 0.05$.

much the same. "Finding purpose and meaning in one's life" seems more personal and possibly more selfish than "making a contribution to society," but it is perhaps also more realistic and less grandiose. However, data from the Sloan study show that participants considered altruism as the most important value among a host of job values (Csikszentmihalyi and Schneider, 2000:50).

Two items relating to geography were included in Table 6-1 because they were hypothesized to affect willingness to enlist in military service. It was expected that propensity and actual enlistment would be below average among those who placed high priority on living close to parents and relatives, but above average among those who considered it important to "get away from this area of the country." These expectations were correct with respect to the latter dimension, but the findings with respect to living close to parents and relatives were more complex. Specifically, young men who entered military service had been lower on this dimension when they were high school seniors, but after enlistment the importance of this dimension increased significantly, and they no longer were below average in the importance they attached to living close to parents. (This pattern did not appear consistently among the small numbers of military women in the samples.) In any case, the items concerning geography were at the bottom of the importance rankings for both males and females at both times. It thus appears that military service may have some extra appeal for those who want to move to a different area, but for most individuals that is not a matter of great importance.

Citizenship, Civic Participation, and Patriotism

As shown in Table 6-1, a substantial portion of youth feels that it is "extremely important" to make a contribution to society. We need to consider how this finding relates to (or manifests itself in) civic participation, volunteerism, and the propensity to enlist in the services. There are many potential determinants of Americans' national or civic-related attitudes and behaviors and many reasonable indicators of these phenomena. As a result, it is not easy to determine whether change has occurred and, if so, what the sources of such change might be. Some observations, however, are noteworthy. Trust in government, responsiveness to proximal political events, voting in national elections, and many other forces may be pertinent. Some commentators believe that growing materialism and individualism have diminished civic society in America; they provide evidence that political participation and civic engagement in general are declining (Putnam 1995a, 1995b; Bellah et al., 1985; Easterlin and Crimmins, 1991). Survey researchers find that trust in government declined from the 1950s to the 1990s (Alwin, 1998). The attacks on the United States on September 11, 2001, may have altered this picture.

At the same time, youth are described as having relatively little interest in national politics, and they have low rates of voting in national and congressional elections (Tables 6-2a and 6-2b). This has been the case since 1972, when the voting age was lowered to 18 by the 26th amendment to the Constitution. Youth's relative disinterest in traditional formal politics appears to be a trend that extends beyond U.S. horizons (Youniss et al., 2002). Recent surveys in the United States show that adolescents have little accurate knowledge about global issues or national political processes and at least until recently have felt little sense of threat, or potential threat, to their country (Schneider, 2001). This lack of knowledge may also have been altered by the September 11 events. Furthermore, recent studies at the National Opinion Research Center suggest that middle school and high school students (grades 6, 8, 10, and 12) are the most bored and the least engaged when they are attending history classes (Schneider, 2001; Csikszentmihalyi and Schneider, 2000:152).

TABLE 6-2a Presidential Voting by Age

	Percentage Reporting They Voted in Presidential Elections by Age, 1964–2000					
Year	18–20	21–24	25–34	35–44	45–64	65 and older
1964	39.2	51.3	64.7	72.8	75.9	66.3
1968	33.3	51.1	62.5	70.8	74.9	65.8
1972	48.3	50.7	59.7	66.3	70.8	63.5
1976	38.0	45.6	55.4	63.3	68.7	62.2
1980	35.7	43.1	54.6	64.4	69.3	65.1
1984	36.7	43.5	54.5	63.5	69.8	67.7
1988	33.2	38.3	48.0	61.3	67.9	68.8
1992	38.5	45.7	53.2	63.6	70.0	70.1
1996	31.2	35.4	43.1	54.9	N/A	N/A
2000	28.4	24.2	43.7	55.0	64.1	67.6

SOURCE: Statistical Abstract of the United States.
NOTE: The voting age population for 1964 and 1968 covers the civilian noninstitutional population 18 years old and over in Georgia and Kentucky, 19 and over in Alaska, 20 and over in Hawaii, and 21 and over elsewhere. The voting age population for 1972–2000 covers persons 18 years old and over in all states.

Those who are more politically active among today's youth often do not champion causes or goals that could be considered national in focus; instead they tend to direct their energies toward the resolution of global problems or to issues that might be more aptly described as promoting the welfare of humanity at large. Such issues are worldwide, not national, in scope and include human rights, poverty within and between nations, discrimination in all its manifold forms, the eradication of disease, animal rights, and environmental protection.

Table 6-2b Congressional Voting

	Percentage Reporting They Voted in Congressional Elections by Age, 1974–1998					
Year	18–20	21–24	25–34	35–44	45–64	65 and older
1974	20.8	26.4	37.0	49.1	56.9	51.4
1982	19.8	28.4	40.4	52.2	62.2	59.9
1986	18.6	24.2	35.1	49.3	58.7	60.9
1990	18.4	22.0	33.8	48.4	55.8	60.3
1994	16.5	22.3	32.2	46.0	56.0	60.7
1998	13.5	N/A	28.0	40.7	53.6	59.5

SOURCE: Statistical Abstract of the United States.

Although some evidence suggests that civic participation is declining, there also is evidence that volunteerism among youth is on the increase (Wilson, 2000). As Figure 6-1 shows, the MTF surveys provide some confirmation. From 1990 to 2000 the proportion of high school seniors who participated in community affairs or did volunteer work at least a few times a year rose gradually from about 65 to about 75 percent, and the proportion who did so at least once or twice a month also rose by about 10 percentage points—from just over 20 to more than 30 percent. Furthermore, the proportions of MTF seniors who considered it quite or extremely important to be a leader in the community increased from 21 percent in 1976 to 36 percent in 1990 and to 39 percent in 2000. These findings suggest that the avoidance of national concerns has been accompanied by an emphasis on local as well as global problems. While representing a more specific scope, local activities can be a crucial element in the development of civic engagement. Youniss and his colleagues, focus-

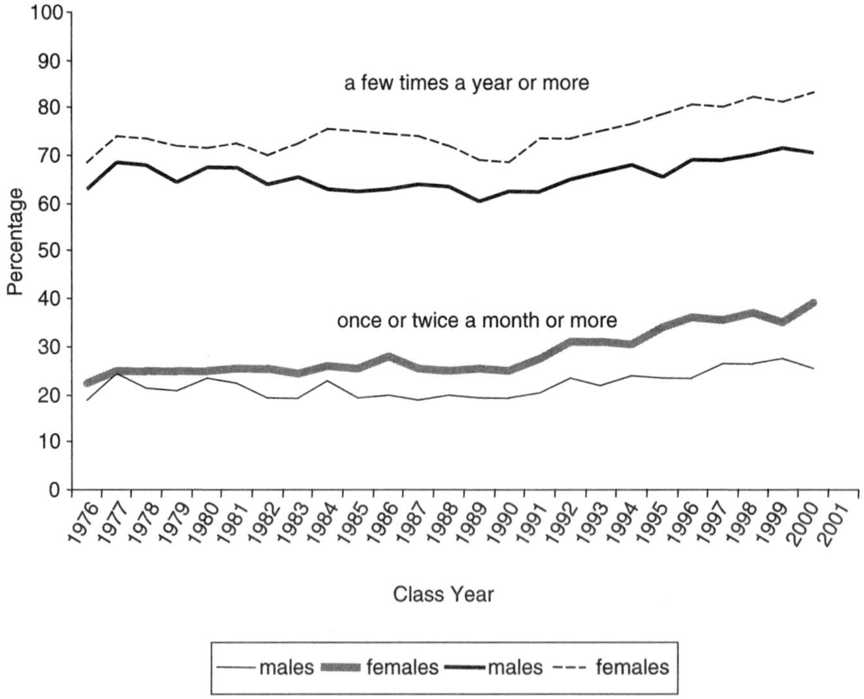

FIGURE 6-1 Percentage of high school seniors who participate in community affairs, class years 1976–2001, by gender and survey form number.
SOURCE: Data from Monitoring the Future surveys.

ing on volunteering in the community, stress that civic activity during adolescence has lasting consequences (Youniss et al., 2002; Youniss and Yates, 1997; Youniss, McLellan, and Yates, 1997; Yates and Youniss, 1996). Their work is based on the theory that behavior drives attitude change. For example, high school volunteers in a soup kitchen, over the course of their service, developed empathy for the homeless as fellow human beings, reflected on their own advantages, and more generally began to consider broader political and moral issues as they thought about the circumstances of their own lives. In so doing, these youth had the opportunity to experience themselves as citizens, to develop a sense of efficacy as effective political agents, and to become more highly motivated to engage in their communities as adults.

Confirming evidence for the benefits of volunteerism has been found in data from the Youth Development Study. The data show that volunteer participation during high school is part of the lives of a substantial minority of Minnesota youth; 37 percent reported at least some volunteer activity while in high school (Johnson et al., 1998). Youth select themselves to volunteer on the basis of previous orientations (e.g., high educational aspirations, higher educational plans, higher grade point averages, higher academic self-esteem, and a higher intrinsic motivation toward school work). However, when the effects of previous attitudes are taken into account, participation in volunteer work was found to foster intrinsic work values, including the importance of service to society as well as enhanced anticipation of future involvement in the community as adults (Johnson et al., 1998). Volunteering also reduced the propensity toward later illegal activity as the respondents began the transition to adulthood (Uggen and Janikula, 1999). In this study, volunteering did not exert an independent effect on educational plans, academic self-esteem, or grade point average.

Furthermore, Verba et al. (1995) found that high school extracurricular activities, particularly participation in school government and clubs (but not sports), predict later political participation. Studies of social movement activists likewise support the conclusion that civic participation during adolescence and young adulthood encourages responsibility in youth, as well as more responsible and active political participation in adulthood (e.g., McAdam, 1988; Fendrich, 1993). Moreover, the effect of volunteering in high school on volunteering during the following four years has been shown to be significant, when numerous relevant background variables as well as prior altruistic and community-oriented values are taken into account (Oesterle et al., 1998; Astin, 1993). The extent to which volunteerism influences activities immediately after high school, such as postsecondary education or military service, is not known.

Education and Work

Educational and Occupational Aspirations

As discussed in Chapter 5, high school graduates have a number of competing opportunities open to them in the worlds of both education and work. During the last quarter of the 20th century, aspirations to complete four-year college programs rose dramatically. As Figure 6-2 shows, fewer than 30 percent of MTF high school seniors in the late 1970s expected "definitely" to complete college, but by the mid- to late 1990s about 60 percent of female seniors and about 50 percent of male seniors expected to do so. If those "probably" expecting to complete college are included, the shift is from about 50 percent in the late 1970s to about 80 percent among women and 75 percent among men in the late 1990s.

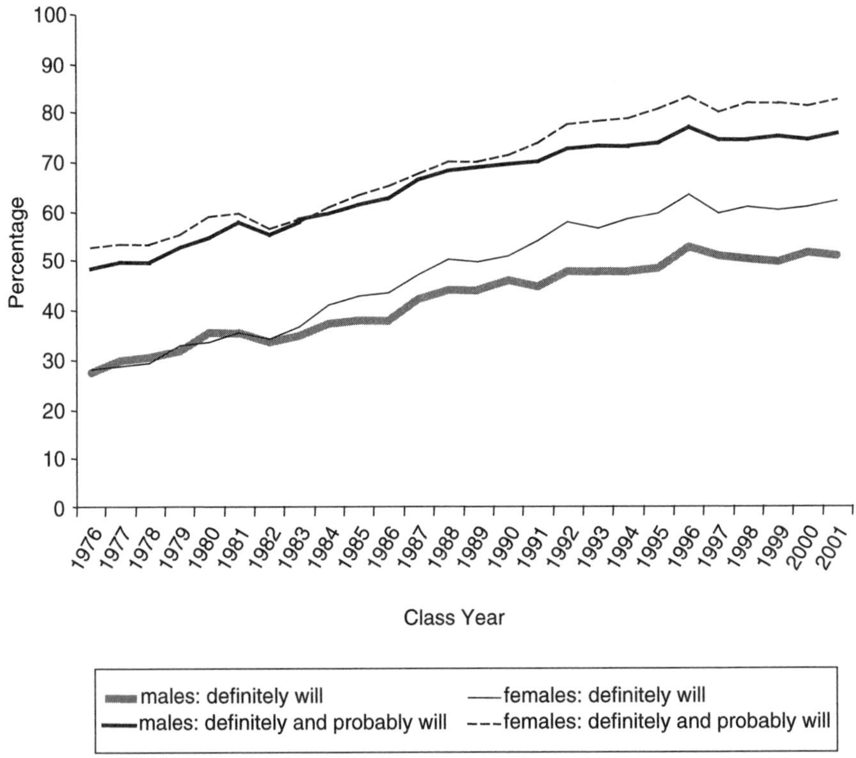

FIGURE 6-2 Trends in plans to graduate from a four-year college program: High school seniors, 1976–2001.
SOURCE: Data from Monitoring the Future surveys.

Looked at another way, it appears that during the space of about two decades the proportions of high school seniors *not* expecting ("probably" or "definitely") to complete college was cut in half—from about 50 to less than 25 percent.

If military recruitment were limited to the noncollege-bound, this great reduction in the target population would be exceedingly problematic. In fact, however, in recent years the majority of high school senior males with high military propensity have also planned to complete four years of college (Bachman et al., 2001a). Nevertheless, it is also the case that average levels of military propensity are lower among the college-bound than among others, so the rise in college aspirations has added to recruiting difficulties.

It is important to note that the proportions expecting to complete college in the MTF reached a peak in 1996 and after that changed little through the latest available data (2001). It may be that in a booming economy, some high school seniors feel less certain that college is the only route to high-quality employment. This may be particularly true of high school students who already possess high levels of computer skills, for example.

Consistent with the increase in college aspirations, there has been a rise in proportions of high school students, especially young women, expecting to obtain high status jobs. For example, between 1976 and 1995 the proportions expecting to become "professionals with a doctoral degree" increased by about half among young men and more than doubled among young women, with some decline thereafter (Figure 6-3). However, young people are generally not well informed about the kinds of educational credentials or other experiences that are required in particular kinds of work. Indeed, the National Survey of Working America (Gallup Organization, 1999) shows that 69 percent of young people ages 18–25 would "try to get more information about jobs or career options than [they] did the first time." Yet adolescents tend to avoid courses or other experiences that could be construed as specific occupational preparation. Apparently, such courses are seen as a diversion from a college degree program and not commensurate with the goal of obtaining high-quality employment. Furthermore, vocational and technical schools and technical certification programs are neither popular nor esteemed (Kerckhoff, 2002).

Youth may realize that the occupational world that they will enter after finishing their education may be quite different than the one that exists while they are in high school (Mortimer et al., 2002). At the same time, the importance of computer literacy in general is widely recognized (Anderson, 2002; Hellenga, 2002). However, instead of regarding technical courses in computer programming or any other technical programs as

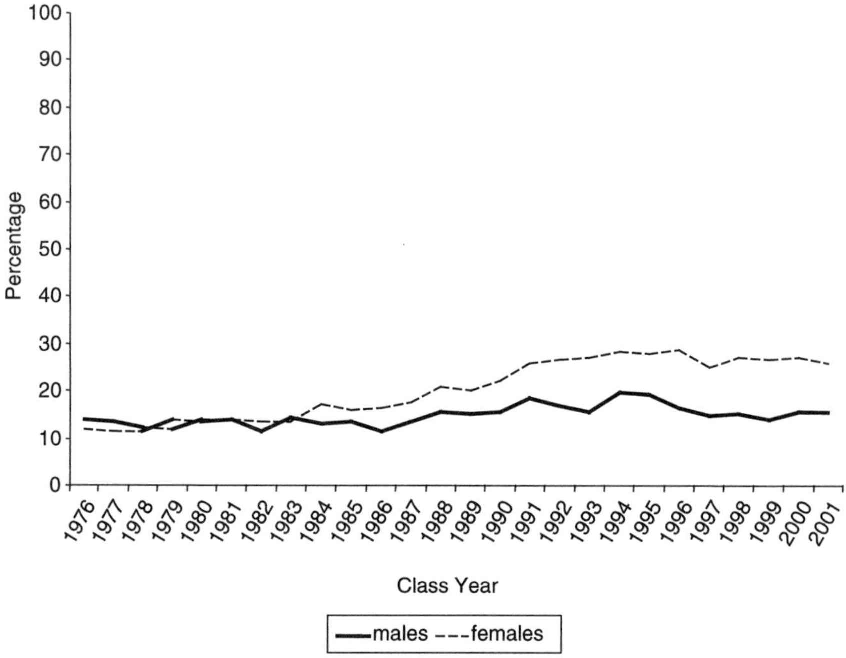

FIGURE 6-3 Trends in expectation for kind of job at age 30 among high school seniors, by gender, 1976–2001 (percentage "professional with doctoral degree").
SOURCE: Data from Monitoring the Future surveys.

vocational insurance, there is a tendency to take an all or nothing approach. Thus, the vast majority of youth plan to obtain a baccalaureate degree and do, in fact, enroll in colleges and universities after high school graduation.

These circumstances present challenges, as well as opportunities, for military recruitment. On one hand, the single-minded focus on getting into college and the lack of attention to the kinds of activities or experiences that would constitute useful preparation for high-quality careers lessen teenagers' serious consideration of opportunities for technical training and related work experience in the military. On the other hand, college dropouts may be more receptive than high school students to these incentives. The decision to leave college may have been prompted by little success in the academic context or by mounting financial pressures. Older youth have had more time to pursue the extended postadolescent moratorium and may have become more attuned to the need for longer-term career planning and preparation. It is noteworthy that among 18–25-

year-old respondents to Gallup's National Survey of Working America (June 2000), 80 percent thought they needed "more training or education to maintain or increase [their] earning power during the next few years."

Preferred Job Characteristics

What do youth see as the most desirable characteristics of a job, and have these values changed over time? The most general observation, based on MTF surveys (Table 6-3), is that job characteristics rated as very important by high school graduates for any job they might hold were quite similar for the classes of 1994–1998 and 1976–1980 (indeed, product-moment correlations between the two sets of mean ratings were 0.97 for males and 0.96 for females). Another general observation is that the ratings for males were similar in most respects to those for females; notably, both genders gave highest importance ratings to a job "which is interesting to do" and lowest ratings to jobs with "an easy pace" and jobs "where most problems are quite difficult and challenging." In other words, it appears that young people prefer interesting jobs that are neither too hard nor too easy.

Participants in the 1999 Youth Attitude Tracking Study (YATS) also gave high marks to "interesting job"—82 percent of the men and 85 percent of the women thought that this feature of work was extremely or very important (figures for the MTF respondents for a job "which is interesting to do" were highly comparable: aggregating data for 1994 to 1998, they were 82 percent for men and 87 percent for women).[2]

The Youth Development Study (YDS), conducted in the city of St. Paul, Minnesota, shows that three years after the 9th grade, when most youth were seniors (N = 930), teenagers of both genders placed high value on work that uses their skills and abilities (among males, 36 percent considered this feature "very important" and 44 percent considered it "extremely important"; comparable figures for females were 32 and 53

[2]It should be noted that many differences between the YATS study and the MTF make responses to the two surveys not directly comparable. For example, although both ask youth to rate 24 items, most of the items are not comparable in wording. Thus, each item is being rated in relation to a different set of issues that provide a comparative referent or framework. Degree of endorsement, as well as the rank order based on importance ratings, could thus differ, depending on the constellation of all other dimensions that are rated. Even referencing the same issues, slight differences in item wording can yield response variation. In addition, the age base is somewhat different. Aggregate data, from which the percentages in the text are derived, are based on 16-21-year-olds in 1999 for the YATS; for seniors in high school (mostly age 17 or 18) in 1994–1998 for the MTF.

TABLE 6-3 Preferences Regarding Job Characteristics: Comparison of Rank Orders

Rank	% Very Important	Males 1976–1980	Males 1994–1998	Males Change
How important is having a job				
1	Which is interesting to do?	86.1	82.1	–3.95*
2	How important is being able to find steady work?[a]	67.0	70.5	3.47*
3	Which uses your skills and abilities—lets you do the things you can do best?	68.0	66.3	–1.67
4	Where the chances for advancement and promotion are good?	64.4	64.9	0.44
5	Where you do not have to pretend to be a type of person that you are not?	64.0	64.2	0.15
6	Which provides you with a chance to earn a good deal of money?	58.1	62.7	4.60*
7	That offers a reasonably predictable, secure future?	63.9	61.2	–2.65*
8	Where the skills you learn will not go out of date?	55.5	50.7	–4.82*
9	Which leaves a lot of time for other things in your life?	45.8	48.8	3.00*
10	Where you can see the results of what you do?	56.6	48.6	–8.00*
11	That most people look up to and respect?	34.3	41.5	7.19*
12	Where you can learn new things, learn new skills?	44.1	41.3	–2.76*
13	Where you have the chance to be creative?	34.6	40.1	5.57*
14	That gives you a chance to make friends?	48.2	39.8	–8.47*
15	Which allows you to establish roots in a community and not have to move from place to place?	40.9	38.7	–2.24
16	Where you get a chance to participate in decision making?	30.3	35.4	5.15*
17	Where you have more than two weeks' vacation?	23.7	34.4	10.71*
18	That is worthwhile to society?	37.8	34.1	–3.70*
19	Which leaves you mostly free of supervision by others?	31.3	33.8	2.55*
20	That gives you an opportunity to be directly helpful to others?	37.0	33.7	–3.27*
21	That has high status and prestige?	24.7	30.5	5.85*
22	That permits contact with a lot of people?	26.3	26.9	0.61
23	With an easy pace that lets you work slowly?	11.5	16.5	4.96*
24	Where most problems are quite difficult and challenging?	14.8	13.8	–0.95

continues

TABLE 6-3 Continued

Rank	% Very Important	Females 1976–1980	Females 1994–1998	Females Change
How important is having a job				
1	Which is interesting to do?	91.2	86.9	–4.25*
2	Where you do not have to pretend to be a type of person that you are not?	81.9	77.4	–4.48*
3	How important is being able to find steady work?[a]	61.3	76.4	15.12*
4	Which uses your skills and abilities—lets you do the things you can do best?	75.4	74.1	–1.37
5	That offers a reasonably predictable, secure future?	62.7	66.6	3.88*
6	Where the chances for advancement and promotion are good?	60.3	61.6	1.27
7	That gives you an opportunity to be directly helpful to others?	60.9	56.9	–4.05*
8	Where the skills you learn will not go out of date?	52.6	53.5	0.95
9	Where you can see the results of what you do?	62.7	53.2	–9.52*
10	Which provides you with a chance to earn a good deal of money?	45.9	51.8	5.81*
11	That is worthwhile to society?	50.2	48.6	–1.60
12	That gives you a chance to make friends?	60.9	46.8	–14.05*
13	That most people look up to and respect?	38.0	46.6	8.56*
14	Where you can learn new things, learn new skills?	51.9	44.9	–6.98*
15	Which allows you to establish roots in a community and not have to move from place to place?	39.2	41.2	2.06
16	Where you have the chance to be creative?	39.9	40.1	0.19
17	Which leaves a lot of time for other things in your life?	35.6	39.2	3.55*
18	That permits contact with a lot of people?	41.3	38.3	–3.06*
19	Where you get a chance to participate in decision making?	26.4	37.2	10.77*
20	That has high status and prestige?	22.4	28.0	5.56*
21	Which leaves you mostly free of supervision by others?	22.3	26.3	4.01*
22	Where you have more than two weeks' vacation?	12.6	21.8	9.18*
23	With an easy pace that lets you work slowly?	8.5	12.6	4.08*
24	Where most problems are quite difficult and challenging?	11.4	11.4	0.01

[a]% extremely important

*Rankings were assigned based on respondent ratings from the class years 1994 to 1998. Significance tests were calculated using *t* tests with pooled variance estimates based on percentages and adjusted for design effects, $p < 0.05$.

SOURCE: Data from Monitoring the Future surveys.

percent, respectively. These youth were much less interested in work that could involve a lot of responsibility. (These jobs might be analogous to the jobs the MTF respondents also considered relatively unattractive, jobs "where most problems are quite difficult and challenging.") Like the MTF seniors, YDS youth of the same age expressed the least interest in jobs that were "easy" (only 16 percent of the total sample attached high importance to easy work).

Against the backdrop of overall stability in work values during the last quarter of the 20th century, as demonstrated by the MTF, there are some changes and other distinctions that may be relevant to military recruiting. Several job characteristics increased in importance, and we note a few of these next.

"Having more than two weeks' vacation" rose from 24 percent of males in 1976–1980 rating it very important to 34 percent of males in 1994–1998; among females the increase was from 13 to 22 percent. Interestingly, although military service vacation allowances are far above the standard two weeks a year associated with most jobs in the civilian sector, this item was not correlated with enlistment propensity; among those who actually did enter military service, the vacation item remained somewhat important, whereas for others it declined in importance. Perhaps because vacations of more than two weeks are one of the perquisites of military service, those serving come to value that as part of their benefits package. (In general, work values tend to change in directions that make them more congruent with the rewards that are available—Mortimer and Lorence, 1979; Mortimer et al., 1986; Johnson, 2001.) In any case, the rising importance of vacation time among high school seniors has some implications for military recruiting efforts: specifically, advertising may need to stress the vacation benefits of military service.

MTF data show that proportions rating a chance to participate in decision making as very important in a job rose from 30 to 35 percent among males and more sharply from 26 to 37 percent among females. Even higher portions of YATS respondents considered decision making important in 1999: 67 percent of men and 62 percent of women. (Differences in endorsement could be linked to variation in question wording: YATS refers to "role in decision-making"; MTF says "Where you get a chance to participate in decision-making.") However, the ranking of this item is quite close in the two surveys: 14th and 16th for men and women, respectively, in the YATS, and 16th and 19th for men and women in the MTF. Both lists contained 24 items.

Ratings of high status and prestige as very important also rose from 25 to 31 percent of males, and from 22 to 28 percent of females. These changes took place during the late 1970s and throughout the 1980s; there was no further increase (actually, a slight decline) during the 1990s. In the

MTF, having a job that most people look up to and respect was rated as important by increasing proportions of males and females; once again, this increase took place prior to the 1990s. However, this indicator of job status was ranked 11th and 13th of 24 items by males and females, respectively. In the YDS, a "job that people regard highly" was considered relatively unimportant, receiving a ranking of 11 out of 12 criteria.

Having a chance to earn a good deal of money assumed considerable importance by youth responding to the MTF surveys, and the level of importance rose over time among both males and females; here again, the increase occurred throughout the late 1970s and the 1980s, with a slight decline thereafter. In addition, higher portions of YATS respondents considered pay to be extremely or very important compared with MTF participants: 90 percent of men (versus 63 percent of male MTF respondents); and 88 percent of women (versus 52 percent of female MTF respondents). However, the particular wording of the YATS item ("job with good pay"), given its less specific and more modest referent compared to the MTF item ("which provides you with a chance to earn a good deal of money"), could have generated the higher level of endorsement.

In 1999, a large portion of YATS men and women (87 and 91 percent, respectively) considered "job security" important. A comparable item in the MTF "How important is being able to find steady work" also yielded relatively high endorsement in 1994–1998 (70.5 percent of men and 76.4 percent of women). In the YDS, a steady job, good chances to get ahead, and good pay were the highest-rated job features (of 12 criteria). Thus, in all three surveys—MTF, YATS, and YDS—extrinsic rewards of work—that is, those that are obtained as a consequence of having a job rather than from the work itself—rank high in importance.

As Table 6-3 shows, several factors declined in importance among both male and female high school MTF seniors, most notably having a job that provides a chance to make friends and having a job "where you can see the results of what you do."

Some of the other factors did not change much but show important gender differences. Jobs providing opportunities to be helpful to others, jobs that are worthwhile to society, and jobs that permit contact with a lot of people are rated as very important by higher percentages of women than men. Women YDS participants also considered opportunities to work with people rather than things and the opportunities to be useful to society more important than did men. Finally, according to MTF data, fewer women than men consider it very important to avoid supervision by others—a dimension that correlates negatively with military propensity. However, these values were not as important for either gender as the extrinsic cluster (steady work, getting ahead, good pay) or the use of skills and ability.

Preferred Work Settings

Table 6-4, taken from the MTF surveys, shows preferences for different work settings among male and female high school seniors, comparing recent classes (1994–1998) with earlier ones (1976–1980). Self-employment was the top choice among males and was high in the ratings of females also. Working for a large corporation was the second most popular setting among both young men and young women; its popularity grew during the late 1970s and early 1980s, with little change thereafter. Schools and universities also gained in popularity, particularly during the late 1980s and early 1990s. Other work settings showed relatively little overall change.

Military service consistently received the lowest ratings among both males and females (although among males social service agencies were about equally unpopular as places to work). Not surprisingly, individuals who considered military service to be a desirable workplace were above average in military propensity and actual enlistment.

Views About the Importance of Work

High school seniors' views about the importance of work in their lives have been mostly stable, although a few modest changes are worth noting. Overwhelming majorities of MTF respondents agree or mostly agree that work will be a central part of their lives. "Being successful in work" was rated most important (of five worthy objectives) among NCES respondents two years following high school in 1974, 1984, and 1994, with some increase being demonstrated over time (to approximately 90 percent at the last year). It is especially noteworthy how much more important success in general in the work sphere is rated by NCES survey participants, as opposed to the extrinsic indicators of such success (e.g., "having lots of money," less than 40 percent in 1994).

However, in the MTF, the proportions considering work as a central part of life declined from about 85 percent for the classes of 1976–1985 to about 75 percent for recent classes (1996–1998). During the same period there was a roughly 10 percent increase in proportions of seniors who viewed work as "only a way to make a living." Also, after the 1970s the proportion of young men who said they would choose not to work if they had enough money to live comfortably rose from about 20 percent to about 30 percent in the late 1990s, whereas among women it remained steady at about 20 percent. Finally, as noted earlier, there has been a modest but steady increase in the proportions of high school seniors who consider it very important to have more than two weeks of vacation per year.

TABLE 6-4 Desirability of Different Working Arrangements and Settings: Comparison of Rank Orders

Rank	% Desirable and Acceptable Combined	Males 1976–1980	Males 1994–1998	Males Change
Apart from the particular kind of work you want to do				
1	How would you rate working on your own (self-employed) as a setting to work in?	80.8	77.7	–3.17*
2	How would you rate a large corporation as a place to work?	64.6	74.2	9.54*
3	How would you rate a small business as a place to work?	69.8	72.3	2.56*
4	How would you rate a small group of partners as a setting to work in?	63.2	64.8	1.66
5	How would you rate a government agency as a place to work?	46.1	52.9	6.85*
6	How would you rate a school or university as a place to work?	34.7	45.5	10.78*
7	How would you rate a police department or police agency as a place to work?	35.8	41.0	5.22*
8	How would you rate a social service agency as a place to work?	28.4	28.6	0.13
9	How would you rate the military service as a place to work?	27.8	28.1	0.23

Rank	% Desirable and Acceptable Combined	Females 1976–1980	Females 1994–1998	Females Change
Apart from the particular kind of work you want to do				
1	How would you rate a small business as a place to work?	73.7	74.6	0.88
2	How would you rate a large corporation as a place to work?	62.0	74.3	12.31*
3	How would you rate working on your own (self-employed) as a setting to work in?	64.4	69.8	5.40*
4	How would you rate a small group of partners as a setting to work in?	58.7	67.5	8.75*
5	How would you rate a school or university as a place to work?	51.7	60.9	9.21*
6	How would you rate a social service agency as a place to work?	64.7	59.9	–4.72*
7	How would you rate a government agency as a place to work?	49.7	49.4	–0.36
8	How would you rate a police department or police agency as a place to work?	41.3	39.6	–1.66
9	How would you rate the military service as a place to work?	23.2	21.9	–1.27

*Rankings were assigned based on respondent ratings from the class years 1994 to 1998. Significance tests were calculated using t tests with pooled variance estimates based on percentages and adjusted for design effects, $p < 0.05$.

SOURCE: Data from Monitoring the Future surveys.

It must be stressed again that the changes noted here in the MTF study were modest and gradual, not at all abrupt. Furthermore, in the NCES study, the aggregate results show that "being successful in work" also became steadily more important. It is abundantly clear that aspirations and ambitions with respect to educational and occupational attainment actually rose substantially among youth during most of the last quarter-century. These shifts in aspirations have had, and will continue to have, important implications for military recruiting efforts.

Trends in Youth Views About the Military

The findings in this section are based mainly on MTF samples of high school students and young adults (Bachman et al., 2000a, 2000b; Segal et al., 1999). We begin by noting recent trends in military propensity among youth, then we turn to their views about the military service as a workplace, and finally we consider their views about the military and its mission more generally.

Trends and Subgroup Differences in Military Propensity

As Chapter 7 and 8 show, military propensity (i.e., planning or expecting to serve in the armed forces) is correlated with actual enlistment, and these correlations can be quite strong (particularly when propensity is measured near the end of high school). Following the introduction of the All-Volunteer Force in the early 1970s, there has been considerable research focused on propensity among American youth. Here, we illustrate the trends only briefly (see Segal et al., 1999, for further details, discussion, and citation of relevant other research).

Figure 6-4 shows the trends in high school seniors' propensity to join the military; the plots are cumulative and the spaces between the plots show the proportions in each propensity category. Specifically, the proportions of male high school seniors expecting "definitely" to serve in the armed forces varied somewhat from year to year during the last quarter of the 20th century, averaging about 10 percent throughout that period, but they have been slightly lower in recent years. Roughly equal or slightly higher proportions of young men expected "probably" to serve. However, follow-up data indicate most of them did not actually serve, whereas most of the "definitely" group did (Bachman et al., 1998). The largest percentage shift among young men was an increase in the proportion who expected that they "definitely" would *not* serve—from fewer than 40 percent in 1983 to more than 60 percent in 1996. This reflects primarily a shift from the "probably won't serve" category to the "definitely won't serve" category.

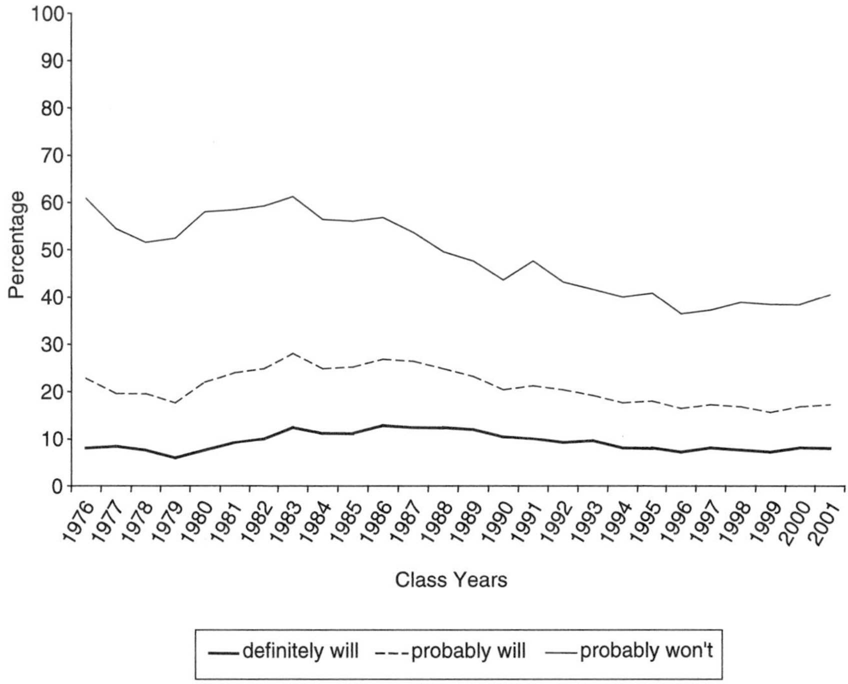

FIGURE 6-4 Trends in high school seniors' propensity to enter the military: Males, 1976–2001.
SOURCE: Data from Monitoring the Future surveys.
NOTE: The spaces between the lines show the percentages in each of the three propensity categories.

One could conjecture that this increased certainty about not serving was simply the result of the growing proportions of young men planning on college. Although that may seem a plausible hypothesis, it is not supported by the data. For example, when the classes of 1976–1983 were compared with the classes of 1992–1998, proportions of male seniors indicating "definitely won't" serve rose from 43 to 60 percent, whereas among the subsamples of male seniors not expecting to complete college (a category that shrank from nearly half of seniors in 1976–1983 to just over one-quarter in 1992–1998) the corresponding shift in "definitely won't" was just about as large—from 37 to 53 percent. So it appears that the shift toward increasing certainty about not serving, shown in Figure 6-4, reflects a fairly pervasive phenomenon: the pool of young men who have not ruled out military service by the end of their senior year of high school has been shrinking—among noncollege-bound as well as college-bound youth.

Figure 6-5 shows a similar but much smaller narrowing in proportions of female seniors who did not rule out military service. More important, the figure shows consistently very low proportions of young women expecting "probably" or "definitely" to serve; moreover, follow-up data show that even among those women "definitely" expecting to serve, most did not (Bachman et al., 1998). It is perhaps worth noting that when asked whether they would *want* to serve "supposing you could do just what you'd like and nothing stood in your way," the proportion of female seniors indicating such a *preference* for serving was consistently higher than the combined proportions "probably" or "definitely" *expecting* to serve; among males the reverse was the case (Segal et al., 1999).

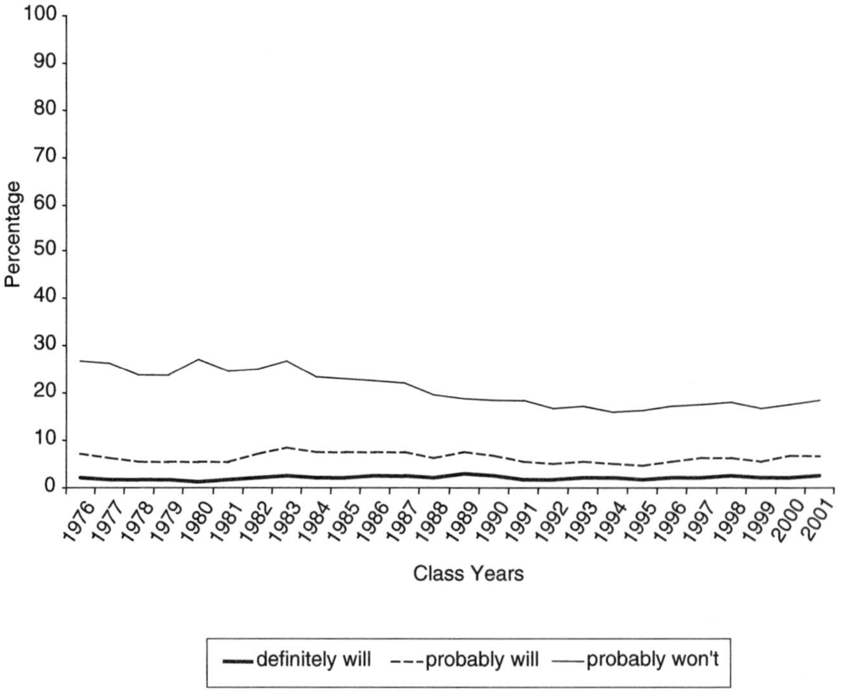

FIGURE 6-5 Trends in high school seniors' propensity to enter the military: Females, 1976–2001.
SOURCE: Data from Monitoring the Future surveys.
NOTE: The spaces between the lines show the percentages in each of the three propensity categories.

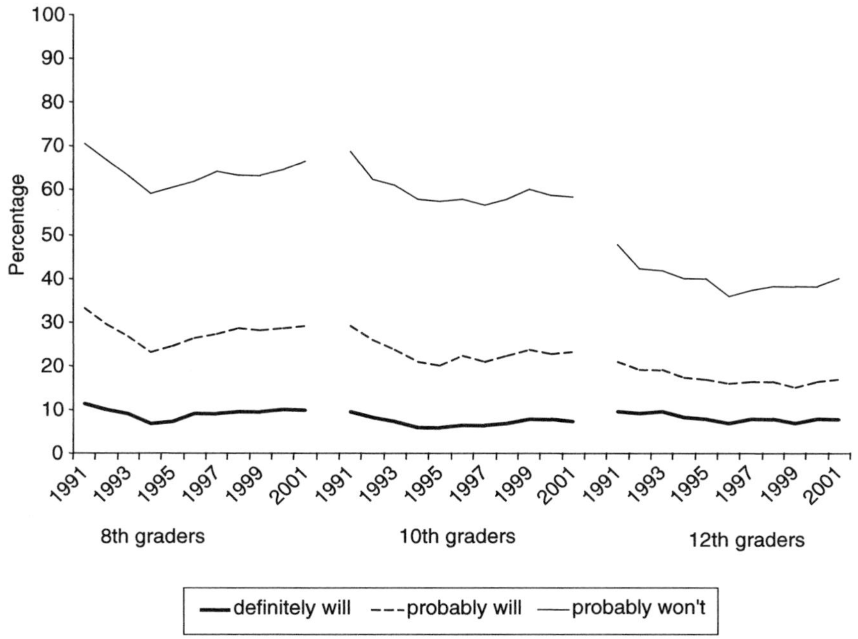

FIGURE 6-6 Comparison of trends in propensity to enlist: 8th, 10th, and 12th grade males, 1991–2001.
SOURCE: Data from Monitoring the Future surveys.
NOTE: The spaces between the lines show the percentages in each of the three propensity categories.

Figures 6-6 and 6-7 make use of MTF data available from 8th and 10th grade students, beginning in 1991, and show how those in lower grades compare with 12th graders. Among males, the 8th and 10th grade data echo the above-mentioned rise in proportions expecting they "definitely won't" serve. More important, these figures show how propensity tends to firm up as students near the end of high school. The findings for 8th and 10th graders differ little from each other, whereas by 12th grade the proportions expecting they "definitely won't" serve are sharply higher, the "probably won't" serve proportions are lower, and there are fewer in the "probably will" category. Among males it appears that at least a few individuals who at lower grades indicated that they "probably will" serve became more "definite" about not serving by the time they reached the end of 12th grade. There is less evidence of that among females.

Arguably, prior to September 11, 2001, World War II was the last large-scale U.S. military effort that was overwhelmingly viewed, by both young people and the population in general, as being "necessary." None of the U.S. military activities during the last quarter of the 20th century

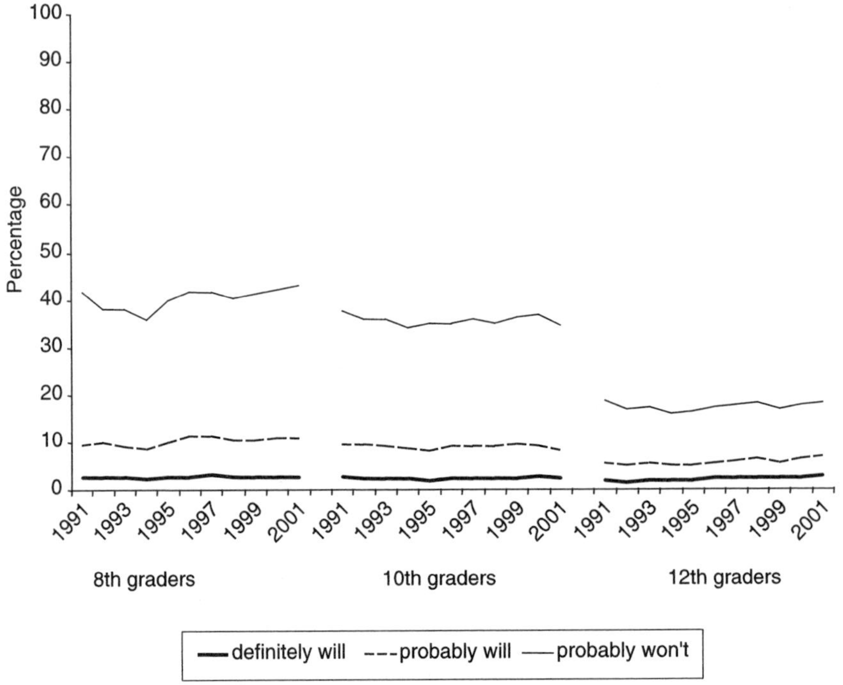

FIGURE 6-7 Comparison of trends in propensity to enlist: 8th, 10th, and 12th grade females, 1991–2001.
SOURCE: Data from Monitoring the Future surveys.
NOTE: The spaces between the lines show the percentages in each of the three propensity categories.

involved that level of support for shared sacrifice. One question included in most of the annual MTF surveys attempts to capture high school seniors' willingness to serve in a "necessary" war. Specifically, the question asks: "If YOU felt that it was necessary for the U.S. to fight in some future war, how likely is it that you would volunteer for military service in that war?" The question is highly hypothetical and thus must be approached with a good deal of caution. Nevertheless, it is of interest to note that the proportions of young men and women who say they would probably or definitely volunteer under such conditions have been far higher than the proportions probably or definitely expecting to enlist under *existing* conditions. Figures 6-8 and 6-9 show that during the 1980s half or more of male seniors said they would volunteer for such a "necessary" war, whereas during recent years just over one-third said so; among female

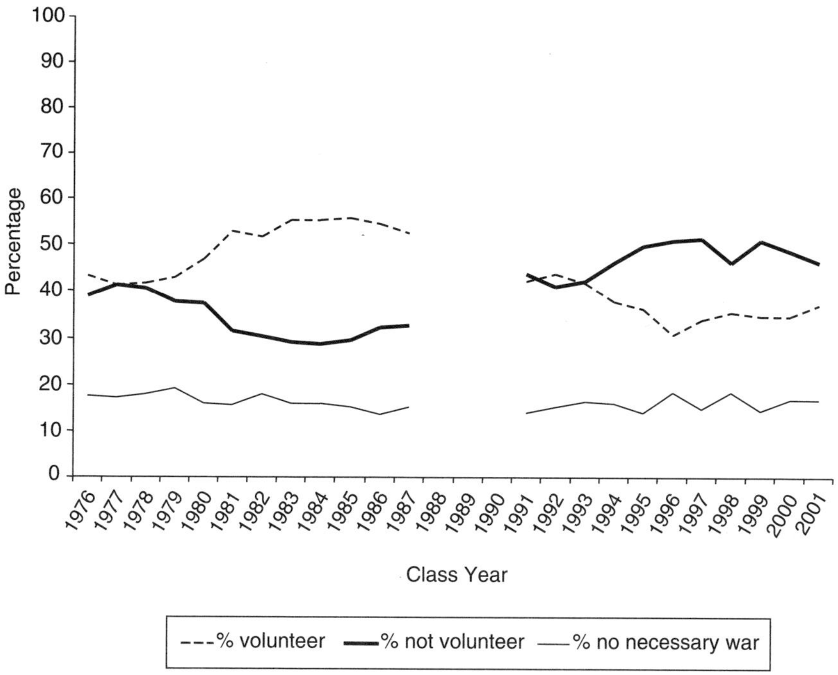

FIGURE 6-8 Trends in high school seniors' willingness to volunteer to fight in a "necessary" war: Males, 1976–2001.
SOURCE: Data from Monitoring the Future surveys.

seniors a similar, albeit smaller, decline was evident. The item includes the response option, "In my opinion, there is no such thing as a 'necessary' war," and about 20 percent or fewer of males and 30 percent or fewer of females chose that option. Probably the most important trend shown by this item is that, among male seniors during the 1980s, substantially more thought they would volunteer than thought they would not—whereas in recent years that pattern was reversed. It is not clear whether this reflects a decline in patriotism, shifting perceptions of how *many* individuals would be needed in a *modern* "necessary" war, or some combination of these and perhaps other factors.

Trends in Perceptions of Military Service as a Workplace

Several survey items focusing on aspects of the military work role have shown relatively little in the way of consistent change during the last quarter of the 20th century. There were, however, consistent gender differences, with higher proportions of female than male high school se-

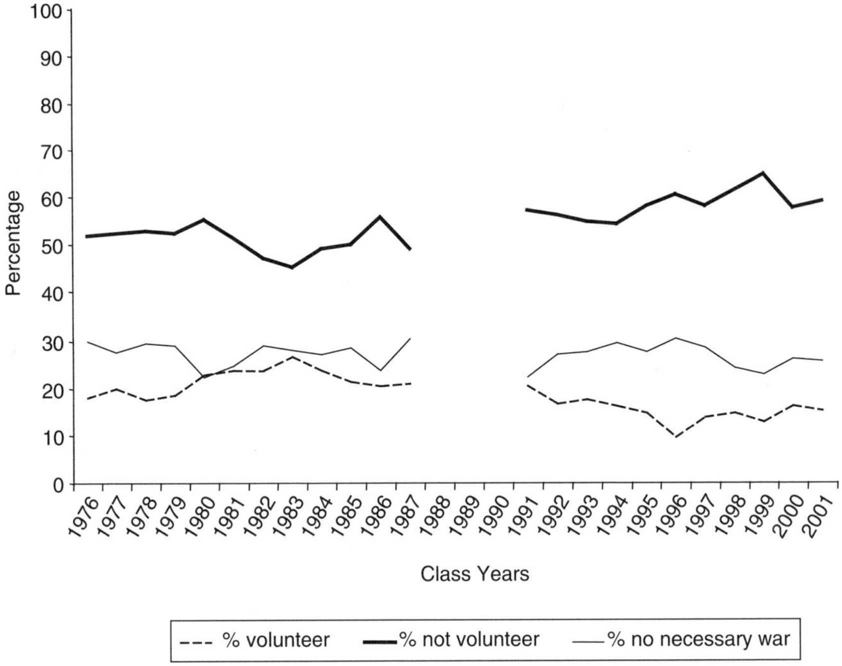

FIGURE 6-9 Trends in high school seniors' willingness to volunteer to fight in a "necessary" war: Females, 1976–2001.
SOURCE: Data from Monitoring the Future surveys.

niors giving high marks to military job opportunities. Thus, about 60 percent of young women, compared with about 50 percent of young men, viewed military service (to "a great extent" or "a very great extent") as "providing a chance to advance to a more responsible position." Roughly similar proportions viewed military service as "providing a chance for more education," although these ratings declined very slightly over the years. Just under half of the women and about 40 percent of the men perceived military service as providing a chance for personally fulfilling work, whereas lower proportions viewed service as providing "a chance to get ahead" or "a chance to get your ideas heard." Fewer than one in five seniors, male or female, viewed military service as a good place for a person to "get things changed and set right if . . . being treated unjustly by a superior."

Two perceptions of military service as a workplace changed in a negative direction in recent years. Figure 6-10 shows a sharp rise in proportions of seniors perceiving that the armed services discriminate to a

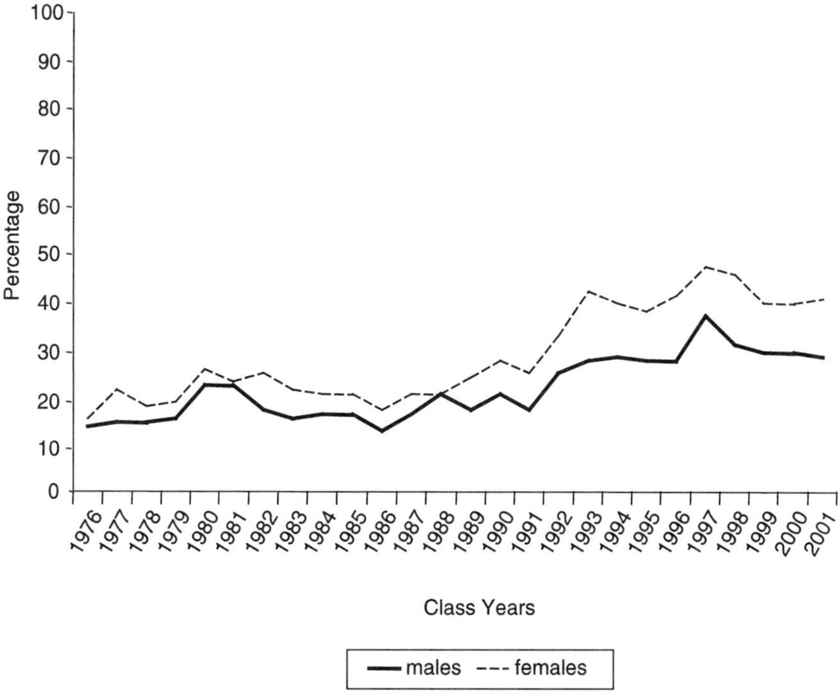

FIGURE 6-10 Trends in perceptions that the military discriminates against women in the armed forces among high school seniors, by gender, 1976–2001. Percentage "to a very great extent" and "to a great extent" combined.
SOURCE: Data from Monitoring the Future surveys.

"great" or "very great" extent against women. Figure 6-11 shows a similar rise in the (generally smaller) proportions of seniors who perceive "great" or "very great" discrimination against blacks in the armed forces. It comes as no surprise that high-propensity seniors generally saw the military workplace in a more favorable light than other seniors. But it does not appear that high-propensity seniors simply view military service through rose-colored glasses; even among the highest-propensity individuals, the perceptions of discrimination against women and against blacks rose in recent years (Bachman et al., 2000a).

Trends and Patterns in Views About the U.S. Military and Its Mission

The preceding section explored views about the military service as a place to work, and surely such views are important in understanding the

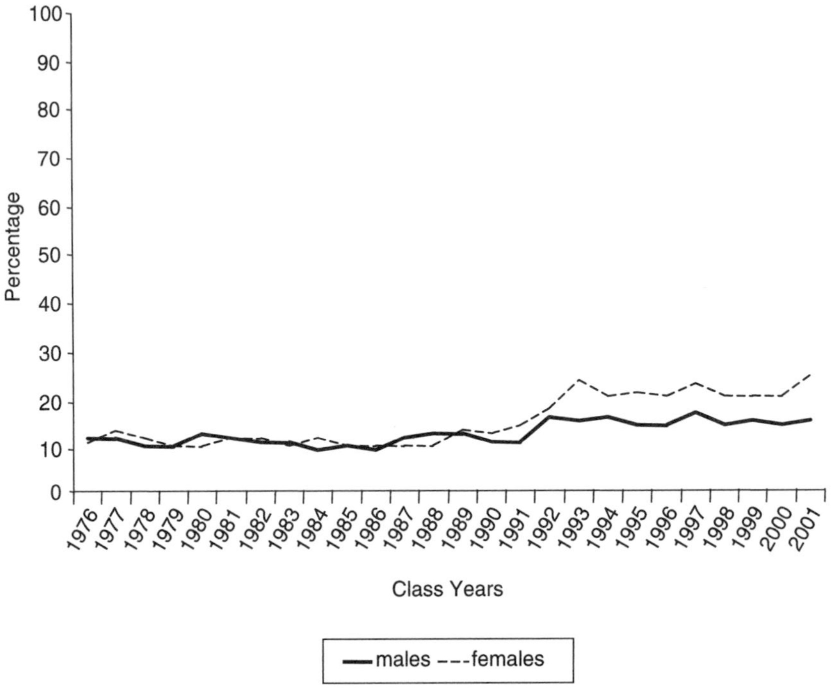

FIGURE 6-11 Trends in perceptions that the military discriminates against blacks in the armed forces among high school seniors, by gender, 1976–2001. Percentage "to a very great extent" and "to a great extent" combined.
SOURCE: Data from Monitoring the Future surveys.

propensity to enlist in the armed forces. But other broader views about the U.S. armed forces and their mission may also be relevant to propensity and enlistment and are thus worth considering here. We take advantage of recently completed analyses of MTF data that compared views of high school seniors with their views in a follow-up survey one or two years later, distinguishing between young men who had entered military service, those full-time in college, and those full-time in civilian employment (the numbers of women in the samples who entered military service were too small to permit reliable estimates, so this analysis was limited to males; see Bachman et al., 2000b, for details).

Figure 6-12 shows, for young men in the high school classes of 1976–1985 and separately for those in the classes of 1986–1995, their senior year and post-high school views about eight aspects of the U.S. military, its mission, and its role in society. All of the attitude dimensions shown in

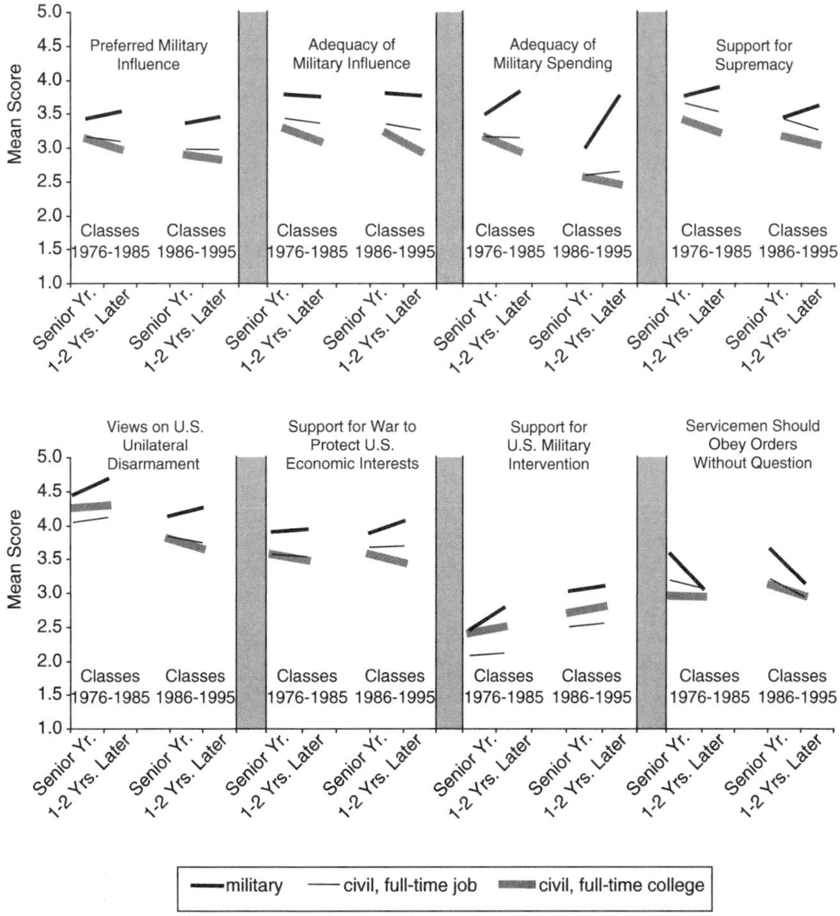

FIGURE 6-12 Young men's military attitudes, high school classes of 1976–1985 and 1986–1995, by post-high school occupational groups.
SOURCE: Data from Monitoring the Future surveys.
NOTE: A score of 5 = highly promilitary, 3 = neutral, and 1 = not promilitary.

the figure are based on 5-point scales with a neutral midpoint (scored 3) and with the most "promilitary" response scored highest (5) and the least "promilitary" scored lowest (1).

Before reviewing the dimensions separately, we offer a few general observations. First, those who entered military service were consistently more promilitary than those who did not; however, in most respects, these differences were not especially large. Second, the differences were evident before the end of high school, with generally little further change after actual enlistment (although two important exceptions are noted below). Third, the comparisons across the two decades (classes of 1976–1985 versus classes of 1986–1995) showed a high degree of replication in subgroup differences, and across most dimensions there were only modest overall changes. We now look at each of the dimensions in the figure, with commentary condensed from the earlier report (Bachman et al., 2000b:568–575).

Views About Military Influence and Spending. The first two dimensions shown in Figure 6-12 have to do with military influence and tell fairly similar stories (although they come from different subsamples of the MTF surveys and thus are independent replications). They show that young men in both decades generally considered military influence to be "about right" and that it ought to remain the "same as now." Among seniors about to enter the armed forces, a slightly higher than average proportion indicated a preference for more military influence; after graduation the differences between subgroups changed only very slightly. The question about adequacy of military spending showed roughly similar differences among high school seniors; however, after enlistment those in the armed forces showed distinct increases in the proportion who viewed current military spending as "too little." Overall preferences for military spending declined from one decade to the next, except among enlistees from the classes of 1986–1995, who enlisted during a period when military salaries rose less than inflation; these respondents showed sharp increases (averaging about three-quarters of a standard deviation) in their preferences for greater military spending.

Views About U.S. Military Supremacy. Overall, young men during the past two decades favored U.S. military supremacy and showed very little enthusiasm for gradual unilateral disarmament, as shown by the fourth and fifth dimensions in Figure 6-12. Again, on average seniors headed for military service showed the most strongly promilitary views; they were significantly higher than the college-bound in support for supremacy and significantly higher than those headed for civilian employment in their rejection of unilateral disarmament. All of the military-civilian differences grew a bit larger after high school, so here again there is some evidence of initial differences (self-selection) that were then enlarged after entrance into different post-high school environments (socialization).

Views About Possible Military Intervention. Young men over the past two decades have been fairly supportive of the proposition that "the U.S. should be willing to go to war to protect its own economic interests"; large proportions of young men in all three subgroups agreed, and few disagreed, with that statement. Support for military intervention was lower when the purpose was to protect "the rights of other countries" rather than U.S. economic interests. Changes between senior year and follow-up (one or two years later) were small and not significant for all groups, indicating no socialization effects attributable to post-high school environments; however, modest selection effects were evident.

Views About Unquestioning Obedience. "Servicemen should obey orders without question: Agree, Disagree, or Neither." The final entry in Figure 6-12 shows responses to this item. This deceptively simple question is actually a bit like a trick question on a test. A correct answer would be "Yes, servicemen should obey orders *provided the orders are lawful.*" Such a response was not offered to respondents. In recent decades, high school seniors split their answers nearly evenly between agreement and disagreement, with many choosing the "neither" midpoint (coded 3). As can be seen in the figure, among young men, the agree responses slightly outnumber the disagree ones (among young women, data not shown, the split is even closer, although there is slightly more disagreement than agreement).

The data in Figure 6-12 show modest selection effects; high school seniors headed for military service were significantly more likely than their classmates to endorse unquestioning obedience. But upon actual entry into military, some of the enlistees changed their views, and most such changes were in the direction of *lessened* support for unquestioning obedience. It thus appears that military socialization largely cancelled the initial differences. The authors of the study commented on this finding as follows (Bachman et al., 2000b:574–575):

> Arguably, the changes found with respect to obedience should meet with the approval of the military—and civilian—leadership, because such socialization has the effect of "correcting" some initial misconceptions about whether obedience in the armed forces should be absolute. Military doctrine maintains that service personnel are responsible to obey only lawful orders and to judge whether orders are lawful before following them. Clearly, such guidelines are quite different from the concept of "unquestioning obedience"; therefore, the changes found among servicemen in response to the obedience item suggest socialization consistent with military doctrine and training.

YOUTH INFLUENCERS

Young people's beliefs, values, and attitudes are learned. They are formed and can be changed in interaction with others. It is therefore useful to inquire about who those influential others might be. In this section the focus shifts from content to agency, from what beliefs, values, and attitudes influence youth propensity to enlist to who influences their propensity and how youth incorporate those influences into their career plans and decisions. While there are many potential influences on the propensity to enlist, the strongest are from a person's social environment, particularly family and friends (Strickland, 2000).

The purpose of this review is not to summarize cumulative theory and research on youth development but, more modestly, to identify those nexuses in the career decision-making process that bear most directly on propensity to enlist in the military. We limit our examination of the literature in three ways. First, we define influencers narrowly, i.e., in interpersonal terms. We do not consider impersonal influences, including TV, film, radio, the Internet, print outlets, or associated media. Second, we maintain a close focus on the dependent variable, propensity to enlist in the military. We draw on the literatures of socialization, attitude formation and change, and youth development only to the extent that it informs early career decision making. We recognize that there are youth exposed to lawless influences that result in them being disqualified from military service (Garbarino et al., 1997), but examination of the causes of deviant behaviors are beyond the scope of this study. Third, we limit inquiry to variables that can be manipulated, by which we mean variables admissible to intervention in military recruitment policy and practice.

Background

The primary domains in which youth function are families, schools, neighborhoods or communities, and, to a lesser extent, the workplace. Briefly stated, people influence one another in three basic ways. A person can exert influence on another through the provision of reward and punishment, through teaching or explicit guidance, or by modeling what is perceived as appropriated or desirable behavior. We accept these processes as given.

Over the past 40 years, theory and research on youth influencers has evolved from efforts to understand youth attitudes and behaviors in terms of largely undifferentiated reference groups to more fine-grained significant-other influences. Taking Coleman's seminal work, *The Adolescent Society* (1961), as a point of departure, the youth cohort was described as a society unto itself, a "world apart" that differs radically from adult society. Coleman's adolescent society was a counterculture, even a contraculture. His youth were rebellious, at odds with their parents and the rest of adult society. He memorably characterized the relationship between adult and youth cultures as a "generation gap," a caricature that dominated popular conceptions of the nation's youth through the remainder of the century.

Challenges to the generation gap thesis surfaced in both the popular and technical literature over the decades that followed. For example, in the mid-1970s DeFleur (1978) reported that male and female Air Force cadets experiencing periods of career indecision turned more to their parents (45 percent) and other adults (48 percent) than to siblings (2 percent) or peers (5 percent) for career advice. When asked who had the most influence on their futures, jobs, and careers, the cadets identified their fathers as the biggest influence and their mothers as second in importance.

Soon thereafter, Stanford University issued a news release reporting that 4 out of 5 Stanford juniors sought advice on career planning from their parents, and 9 out of 10 sought parental advice on personal problems (Stanford University News Service, 1980). Rutter (1980) published an extensive meta-analysis of research on parent-child relationships in the United Kingdom and the United States, concluding that "young people share their parents' values on the major issues of life and turn to their parents on most major concerns." Taking dead aim at the generation gap thesis, Rutter asserted that "the concept that parent-child alienation is a usual feature of adolescence is a myth."

MTF findings from 1976 through 1998 consistently show that a bit fewer than half of high school seniors thought their ideas agreed with their parents' ideas when it comes to how the students spent their money and their leisure time, whereas just over half perceived agreement about what is okay to do on a date. About two-thirds reported that their ideas about what they should wear were mostly or very similar to their parents' ideas. Although roughly two-thirds reported having a fight or argument with parents three or more times during the past year, about two-thirds also indicated that overall they were more satisfied than dissatisfied with how they got along with their parents. Agreement was stronger on more fundamental issues. About three-quarters thought their ideas and their parents' ideas were very similar or mostly similar with respect to religion, politics, what values are important in life, and what they, the students, should do with their lives. Finally, views about the value of education showed high and growing agreement with parents between 1976 and the early 1990s; the proportions perceiving mostly or very different views declined from 14 to about 8 percent, whereas proportions perceiving very similar views rose from about 50 to about 65 percent. Notably, this increase in perceived close agreement about the value of education coincided with the increase in proportions of high school seniors reporting that they definitely expected to complete a four-year college program.

Empirical evidence to the contrary, the generation gap definition of parent-youth relations persisted, which may have had the positive effect of intensifying inquiry into generational differences and parent-child relations in both the popular and technical literatures. For example, a Reader's Digest study of four generations (Ladd, 1995) reported that Americans in every age group share basic values, concluding that the finding "explodes the generation-gap myth—for good." A Sylvan Learning Centers study (International Communications Research, 1998) asked identical questions of teenagers and their parents about perceived and actual career aspirations and reported remarkable similarities in parent-child responses across a variety of issues.

The persistence of the generation gap definition of parent-youth relations illustrates a noteworthy feature of much literature that characterizes youth. There is a penchant not limited to the popular press to dramatize, sensationalize, and otherwise mythologize youth behaviors and attitudes—to the detriment of youth (Youniss and Ruth, 2002). Adelson illustrated the point some years ago in an article in *Psychology Today* in which he cited the 1972 national election as an example. There had been talk that Democrat George McGovern would get the youth vote. After all, young people were supposed to be doves. Young people were supposed to be liberals. And young people would turn out to vote. The popular notion was that the youth vote would carry the election for McGovern. At the same time, youth studies were reporting that young people were as divided in their views on Vietnam as were their elders; that young people were probably more hawkish on war issues than were their elders; and that young people were less likely to vote, not more likely to vote, than were their elders. What happened? McGovern lost, convincingly. The conventional wisdom was wrong.

Against this backdrop, it is instructive to examine what is known about youth influencers and how they affect youth career decision making in terms of accumulated theory and empirical research.

The Achievement Process

The social psychological model of the achievement process provides a useful initial framework within which to identify major influencers and key processes that affect youth career decisions. In *The American Occupational Structure*, Blau and Duncan (1967) documented the fact that sons' educational and occupational achievements depend largely on their parents' educational and occupational achievement levels. To be sure, Blau and Duncan were interested in such issues as how much mobility occurred in occupational careers and how sons' career outcomes compared to their fathers—issues of intra- and intergenerational mobility and stratification. They were not interested in youth influencers and career decision-making processes per se. They asked: What are the ways that family education and occupation advantages or disadvantages transfer from one generation to the next? How is it that family levels of achievement remain relatively stable across generations? Why is there an intergenerational correlation (typically $r = 0.3$) between fathers' and sons' levels of occupational prestige?

Their research and that of others bear directly on the question of who influences youth career decisions. Following Sorokin and Parsons, Blau and Duncan reasoned and provided empirical confirmation for the theses that level of education is the key means by which society selects and distributes youth into occupational roles, and that education serves the critical socialization function of instilling achievement values and orientations in youth. Sewell and colleagues (Sewell et al., 1969, 1970) criticized the Blau and Duncan model for being too structural and too simplistic, as well as for its failure to explain the interpersonal processes that influence youth career decisions. Building on earlier work by Haller and Miller (1971), Sewell and colleagues (Sewell et al., 1969, 1970; Sewell and Hauser, 1975) expanded the model into a social psychological explanation of who influences youth aspirations and achievements and how that process works over time.

Simply stated, the model indicates that parental levels of education, occupational prestige, and income predispose youth in career directions in three ways that sequentially involve a young person's academic ability and performance, the expectations others have for her or him, and the youth's own career aspirations. What a young person aspires to is the critical link in the process. Formed and modified in interaction with other people, young people assess their own educational and occupational prospects in terms of their understanding of their abilities and past performance. Independently, influential others also evaluate the young person's potential and communicate their career expectations to her or him. Because individuals and families live in social networks with similar levels of education, occupational prestige and income, those with influence in a young person's life—including teachers and peers—tend to have levels of education, occupational prestige, and income similar to the youth's parents and therefore provide career encouragement, role models, and expectations that complement parental values. Thus formed, career aspirations set a young person on a career trajectory. A young person's self-reflection (Haller and Portes, 1973) is complemented by the independent evaluations of significant others (Woelfel and Haller, 1971) who communicate their expectations to the young person thus influencing his or her career aspirations, which is the strongest predictor of eventual career achievements.

Sewell and Hauser documented the predictive and explanatory power of the social psychological model of the achievement process at length (1972). Alexander et al. (1975) offered strong independent support based on a national sample. Otto and Haller (1975) provided conceptual replications based on four datasets and, assessing the convergence of theory and research across datasets, concluded that there is strong support for the social psychological explanation of the achievement process (see also Featherman, 1981; Hotchkiss and Borrow, 1996).

Significant-Other Influences

Significant others are people who are important and influential in the lives of others. The social sciences distinguish two kinds of significant others, role-incumbent and person-specific significant others. Role-incumbent significant others have influence because they have power and authority over a young person, who, for that reason, is beholden to them. Examples include parents, teachers, and police officers. Person-specific significant others, by comparison, are chosen by the individual. Examples include best friend, role model, and confidant—relationships based on understanding and trust.

Person-specific significant others have influence not because they have power associated with their role, but because individuals choose to follow them as models and exemplars. Young people follow them not because they have to, but because they want to, which positions person-specific significant others to make a difference in young people's lives.

The empirical literature on the development of aspirations and achievements sketches the achievement process with broad-brushed strokes, and that literature is largely limited to estimating the effects of parents, teachers, and peers on respondents' aspirations and achievements. That literature concludes that parents have the critical influence (Sewell and Hauser, 1975) on sons' career aspirations and achievements. However, the large-scale longitudinal datasets on which achievement research tends to be based do not lend themselves to sharply focused inquiry into youth influencers. It is therefore instructive to intensify inquiry into youth career influencers in two ways: by broadening the inquiry to include more potential influencers and by shifting the focus from selected role-incumbent significant others to person-specific significant others.

Two studies inquired directly of youth about who influenced their occupational choice. Rather than assuming that persons in particular roles influenced youth and then measuring that influence, both studies asked the youth to specify who influenced specific aspects of their career plans.

In the Youth Development Study (Mortimer, 2001) students were surveyed about their experiences in the family, school, peer group, and workplace each year during high school (1988–1991). Following a question about occupational choice, Mortimer and colleagues asked the high school seniors: "Have any of the following people influenced your choice of this kind of work?" Respondents were instructed to circle all that apply from a list of 15 possibilities.

Several observations follow from the results. Two-thirds of seniors identified a friend as the person who influenced their choice of occupation "very much, much, or somewhat." Mothers followed closely. An adult working in the same field was chosen third most often, followed by father and teacher or coach at school. Fewer than half of the seniors identified anyone else as having influenced their choice of work—e.g., siblings, aunts and uncles, grandparents, another relative, a guidance counselor, a work-study coordinator, a neighbor or adult friend, or a priest, minister, or rabbi. It is noteworthy how prominently person-specific significant others were identified as influencers—particularly friends and adults working in the same field. Very few youth mention no influencers on their career choices. Clearly, young people seek and find guidance from others in their vocational decision making.

There were no differences in the number of influencers identified by adolescent gender, race, or nativity, but youth from two-parent families reported more influencers than those from other family structures. Their advantage comes from having two parents to draw on for help in thinking through their vocational goals. Other YDS data indicate that sons were particularly influenced by their fathers' s occupational values. The transmission of values from parents to children was mediated by close, communicative family relationships (Ryu and Mortimer, 1996).

In the Youth and Careers Study, Otto (2000) also examined parent-youth relationships in the context of career development and, like Mortimer, gathered information from the perspective of youth themselves. He asked 362 juniors from six high schools in North Carolina about specific career development attitudes and behaviors that bear on their relations with their parents. A total of 80 to 90 percent of participants reported that they discussed their occupational career plans with their parent or guardian often or sometimes during the past year, and that their ideas regarding the occupation they should enter, the value of a college education, and how they should prepare for a career were all similar to their parents' ideas. More than three-fourths discussed plans for college, two-thirds discussed career preparation possibilities other than college, and half discussed plans for vocational or trade school often or sometimes with their parents during the past year.

Respondents named their mothers most often as the person who offered the greatest help when discussing career plans and was most aware of their career interests and abilities. That is, respondents talked to their mothers most seriously about the occupation they wanted to enter and about the training or education required to enter the selected occupation. Friends or fathers were typically ranked second or third. The findings generally held for young women as well as young men, although young women tended to rank friends higher than fathers, whereas young men ranked fathers higher than friends. The findings also are consistent across race, although more blacks than whites reported career assistance from their parents and from their mothers, while fewer blacks reported assistance from their fathers.

A third set of studies is less rigorous but queried Army and Air Force enlistees at selected time points from 1990 to 1999 about who influenced their decisions to join the military (Strickland, 2000). In 1992, 1994, 1996, and 1999, Army enlistees were asked who was their most supportive influencer. Mothers were cited most often and fathers second, followed by significant others, friends, siblings, and educators. The major influencers were the same for Air Force enlistees in 1992, 1996, and 1999. Mothers ranked highest, followed by fathers, then Air Force veterans and spouse or fiancé, siblings, and counselors. Air Force enlistees ranked Air Force veterans highest in support of their enlistment decisions, followed by father, then mother, spouse/fiancé, sibling, and counselor. When asked who was their most supportive influencer, Air Force enlistees ranked parents highest, followed by spouse or fiancé and other family, siblings, recruiters, and counselors.

The Army and Air Force enlistee reports are not directly comparable to the civilian population studies, and they are not representative of all Service branches. Nonetheless, the influence patterns reported by enlistees largely parallel those reported by civilian youth and underscore the influence of parents, mothers as well as fathers, in the youth career decision making process.

Summary of Youth Influencers

Two questions framed our inquiry: Who influences youth propensity to enlist in the military, and how do young people incorporate those influences into their career plans and decisions? The marked trend in theory development and empirical studies is understanding parent-youth relations as compatible and supportive across a range of substantive issues. We reviewed key studies over four decades in three genres of research: intergenerational mobility in the occupational structure, social-psychological processes affecting achievement outcomes, and current cross-sectional research on youth reports of significant-other influences on their careers. Across studies the single most compelling observation is that parents have a critical influence on their sons' and daughters' career aspirations and achievements. The most recent and most carefully conceptualized and executed studies point to the important role mothers play in affecting youth career plans and decisions. Others to whom youth turn include peers, fathers, other adults, and counselors. These results are generally replicated in Army and Air Force enlistee reports.

Close examination of the empirical data suggests that parents, peers, counselors, and recruiters exert different kinds of influence on youth career decisions. Attitude theories traditionally differentiate between two attitude dimensions, direction and intensity. Directionality indicates whether an individual is pro or con on an issue, disposed to act favorably or unfavorably. Intensity indicates whether an individual is likely to act, whether he or she is sufficiently motivated to behave in the direction of the attitude. Lack of clarity on either dimension, uncertainty of direction or lack of commitment to the position, disposes the individual to passivity.

Career aspirations are attitudes (Woelfel and Haller, 1971) that require both direction and intensity if they are to translate into behavior. Formulating direction for a defined occupational career objective or trajectory requires cognitively processing information, but acting in the direction of a particular career requires motivation. Information and reason may persuade, but behavior is driven by commitment and emotion.

There is some suggestion in the recent empirical research that this attitude-behavior paradigm applies to youth career decision making. When youth are asked about such aspects of career decision making as who is "most aware of their abilities and interests," who is "most helpful to them when talking about career plans," and who "most often influenced" their choice of work, young people clearly look to their parents, especially their mothers. Army and Air Force enlistees also named their mothers, then their fathers, as their most supportive influencers. Only Air Force veterans were ranked slightly higher than mothers and fathers in support of Air Force enlistment decisions. However, when questions sharpen and require more factually based information—e.g., with whom do young people talk most seriously about the training or education they need to enter the occupation they want, with whom would they like to talk more about their career plans, or who was the most important information source for Army enlistees—then they also attribute importance to information providers, to counselors, and to recruiters. The implication is that recruitment effectiveness may be improved by complementing information with messages designed to activate social support for enlistee decisions.

Parents are uniquely positioned to provide encouragement, affirmation, and legitimization of a young person's aspirations and career decisions. One notable observation is that when it comes to making career plans, youth consider parents as more than role-incumbent significant others, but also as person-specific significant others. Youth choose to follow their parents' career advice not because they have to but because they want to. The implication for military recruitment is that enlisting parental support may yield enlistment dividends.

The important role mothers play in the youth career decision-making process should not be lost. Not only do youth credit their mothers with being especially influential in their career decisions but also, traditionally, mothers have been the family voice on social and expressive issues. Mothers are positioned to support youth career decisions, but they are likely to do so only if their sons and daughters career decisions square with their own beliefs and values.

Mother's ideological perspective on military service cannot be ignored in designing effective recruitment programs. Traditionally, women have been more pacifist than men. They have been less experienced in military matters, and they may have less knowledge of military life and careers on which to advise their sons and daughters responsibly. Moreover, data from the General Social Survey, by the National Opinion Research Center, indicate that contemporary women have less confidence in military leaders than do men: only 31 percent of women compared with 43 percent of men express a great deal of confidence in the military (Mitchell, 2000:173). It is tenuous to assume, therefore, that mothers are predisposed to support youth propensity and enlistment. In fact, mothers may be more likely to be lukewarm or even oppose military careers—perceived as action-packed carnage—for their own daughters and sons. However, the same mothers may be more favorably disposed toward the nonwarrior definitions of military service—military as peacemaker, peacekeeper, and administrator of humanitarian aid—which suggests that promoting these forms of military service might energize mothers' support. Mothers are important influencers in the youth career decision-making process, and accurately ascertaining their military ideological stance together with fashioning recruitment messages that appeal to rather than counter their traditional attitudes may hold promise for more effective recruiting.

Young people's propensity to enlist is based not only on their own evaluations of their past performance and potential, but also on the assessments and expectations that influential others make and communicate to them. Their career aspirations portend their career behaviors. Parents are well positioned to provide support for both their aspirations and their career decisions. This raises the possibility that military recruitment effectiveness may be improved by taking into account major youth influencers and the critical processes that affect youth career decisions.

SUMMARY

In this chapter we reviewed three large databases as well as a locally based longitudinal study and a few cross-sectional studies. We also examined the professional literature on socialization, attitude formation, and youth development. Our focus was on (1) youth values and attitudes toward work, education, and the military and (2) the degree of influence that parents, counselors, and peers have on the choices made by youth regarding these options.

Our analysis suggests that, over the past two decades or more, there has been little change in youth ratings of the importance of various goals in life, preferred job characteristics, and work settings. The primary value that has been changing is college aspirations, and this has led to increases in rates of college attendance through the mid-1990s. One useful finding regarding education and military service is that in recent years the majority of high school seniors (both male and female) who reported highest military propensity also expected "probably" or "definitely" to complete a four-year college program. This was also found among young males who actually entered military service (within 5–6 years after high school). Although overall levels of propensity have been shown to be lower among the college-bound, those college aspirants who also plan on military service are just as likely to enter the Service as those without college aspirations. These findings have important implications for military recruiting policy in terms of recruiting college-bound youth and in offering to these youth higher education opportunities in conjunction with military service.

Other aspects of youth attitudes and behavior that provide potential guidance for the design of military recruiting and adverting messages include: (1) the time in which youth make decisions about education and careers has extended well into their 20s; (2) there has been little or no change in youths' views about the military service as a workplace or the value and appropriateness of military missions; (3) there has been some increase in the desire of youth to have two or more weeks vacation—a benefit of military service over the private sector; (4) there is a possible link between youth attitudes toward civic duty and volunteerism and military service (the potential of this link requires further study); and (5) parents (particularly mothers) and counselors have a strong influence on youth decision making with regard to career and educational choices.

7

Determinants of Intention or Propensity

The preceding chapters have demonstrated that demographic and economic variables have not been able to account for all of the important variation in recruiting. In particular, after accounting for changes in the economy, recruiting resources, and demographic factors, there remains an unexplained negative trend in enlistments throughout the 1990s. Similarly, considerations of traditional attitudes and values have also provided little insight into variations in recruitment.

At the same time, however, one variable has consistently been shown to be a very good predictor of recruitment—the propensity to enlist. For example, as reported earlier, Bachman et al. (1998) found that among male high school seniors in class years 1984–1991, enlistment within the next six years was strongly related to propensity ($R = 0.50$, $Eta = 0.55$, $p < 0.001$). Moreover, while 70 percent of those male seniors strongly inclined join the military within six years, only 30 percent of those moderately inclined, and only 6 percent of those who said they "definitely will not join" actually did so.[1] Similarly, Wilson and Lehnus (1999), using data from the Youth Attitude Tracking Survey (YATS), also found a strong association between propensity and enlistment. Although no correlation was reported, these investigators concluded that "positively propensed high school seniors are applying to enter the military about five times as

[1]It should be noted that the Monitoring the Future propensity measures asking for expectations of serving in the military service occurs near the end of the senior year of high school. By that time, many seniors already have made a formal commitment to serve, and others have taken at least some steps in that direction. For some individuals, therefore, the measure captures actual behaviors as well as intentions.

frequently as those who indicated negative propensity. Propensity is an imperfect but very significant predictor of enlistment behavior."

Given that propensity, as it is measured in both the YATS and Monitoring the Future (MTF), is extremely similar to the psychosocial construct of intention, and that intention has played a central role in many behavioral theories,[2] in this chapter we use behavioral theory to help explain why some young adults are, and some are not, inclined to join the military. More specifically, we briefly review some behavioral theories and describe an integrated model of behavioral prediction. We then consider the application of this model to the prediction of propensity (for both males and females) and enlistment, presenting available data. Finally we point out the implications of the model for military recruitment.

THEORIES OF BEHAVIOR

There are many theories of behavioral prediction, such as the theory of planned behavior (Ajzen, 1985, 1991; Ajzen and Madden, 1986), the theory of subjective culture and interpersonal relations (Triandis, 1972), the transtheoretical model of behavior change (Prochaska and DiClemente, 1983, 1986, 1992; Prochaska et al., 1992, 1994), the information/motivation/behavioral-skills model (Fisher and Fisher, 1992), the health belief model (Becker, 1974, 1988; Rosenstock, 1974; Rosenstock et al., 1994), social cognitive theory (Bandura, 1977, 1986, 1991, 1994), and the theory of reasoned action (Fishbein and Ajzen, 1975; Ajzen and Fishbein, 1980; Fishbein et al., 1991). However, there is a growing academic consensus that only a limited number of variables need to be considered in predicting and understanding any given behavior (Fishbein, 2000; Fishbein et al., 2001). Three theories have most strongly influenced intervention research, outlined below.

Social Cognitive Theory: According to social cognitive theory, there are two factors that serve as primary determinants underlying the initiation and persistence of any behavior. First, the person must have self-efficacy with respect to the behavior. That is, the person must believe that he or she has the capability to perform the behavior in the face of various circumstances or barriers that make it difficult to do so. Second, he or she must have some incentive to perform the behavior. More specifically, the expected positive outcomes of performing the behavior must outweigh the expected negative outcomes.

Social cognitive theory has focused on three types of expected outcomes: physical outcomes (e.g., performing the behavior will make me healthy); social outcomes (e.g., performing the behavior will please my

---

[2]See caution in footnote 1 about MTF "propensity" measure.

parents); and self-standards (e.g., performing the behavior will make me feel proud). Note that the expected outcomes of performing one behavior (e.g., enlisting in the Army) may be very different from those associated with performing another behavior (e.g., enlisting in the Marines). That is, one might believe that enlisting in the Army will lead to very different costs and benefits than those that will follow from enlisting in the Marines.

The Theory of Reasoned Action: According to the theory of reasoned action, performance or nonperformance of a given behavior is primarily determined by the strength of one's intention to perform or not perform that behavior; intention is defined as the subjective likelihood that one will perform or try to perform the behavior. The intention to perform a given behavior is, in turn, viewed as a function of two basic factors: one's attitude toward performing the behavior and one's subjective norm concerning the behavior, that is, the perception that one's important others think that one should or should not perform the behavior in question.

The theory of reasoned action also considers the determinants of attitudes and subjective norms. Attitudes are a function of beliefs that performing the behavior will lead to certain outcomes and the evaluation of those outcomes (i.e., the evaluation aspects of those beliefs). Subjective norms are viewed as a function of normative beliefs and motivations to comply with what the referent wants to do. Generally speaking, the more one believes that performing the behavior will lead to positive outcomes or will prevent negative outcomes, the more favorable will be one's attitude toward performing the behavior. Similarly, the more one believes that specific individuals or groups think that one should perform the behavior and the more one is motivated to comply with those referents, the stronger will be the perceived pressure (i.e., the subjective norm) to perform that behavior.

The Health Belief Model: Although developed to help explain health-related behaviors, the health belief model can be adapted to a wide range of behaviors. According to this model, the likelihood that someone will adopt or continue to engage in a given behavior is primarily a function of two factors. First, the person must feel personally threatened by some perceived outcome, such as poor job prospects in the civilian labor market. That is, he or she must feel personally susceptible to or at risk for a condition that is perceived to have serious negative consequences. For example, in difficult economic circumstances, one might perceive a high likelihood of being unemployed and unable to support one's family. Second, the person must believe that, in addition to preventing or alleviating this threat, the benefits of taking a particular action outweigh the perceived barriers to or costs of taking that action.

On the basis of these three theories, one can identify four factors that may influence an individual's intentions and behaviors:

1. The individual's attitude toward performing the behavior, which is based on his or her beliefs that performing the behavior will lead to various positive and negative consequences;
2. Perceived norms, which include the perception that those with whom the individual interacts most closely support his or her attempt to change and that others in the community are also changing;
3. Self-efficacy, which involves the individual's perception that he or she can perform the behavior under a variety of difficult or challenging circumstances; and
4. The individual's perception that he or she is personally threatened by or is susceptible to some negative outcome.

While there is considerable empirical evidence to support the role of attitude perceived norms, and self-efficacy as determinants of intention and behavior (Holden, 1991; Kraus, 1995; Sheppard et al., 1988; Strecher et al., 1986; van den Putte, 1991), this is not always the case for perceived susceptibility (or perceived risk—see Gerrard et al., 1996; Fishbein et al., 1996). Indeed, as we discuss below, it appears that perceived risk may best be viewed as having an indirect effect on intention. On the basis of these and other considerations, Fishbein (2000) has recently proposed an integrative model of behavioral prediction (see Figure 7-1).

An Integrated Theoretical Model

Looking at Figure 7-1, it can be seen that any given behavior, such as enlisting in the military, is most likely to occur if one has a strong intention to perform that behavior, if one has the necessary skills and abilities (i.e., meets military enlistment standards), and if there are no environmental constraints preventing behavioral performance. Indeed, with a strong commitment, the necessary skills and abilities, and no environmental constraints, the probability is very high that the behavior will be performed (Fishbein, 2000; Fishbein et al., 2001).

One immediate implication of this model is that very different types of interventions will be necessary if one has formed an intention but is unable to act on it than if one has little or no intention to perform the behavior in question. In some populations or cultures, a behavior may not be performed because people have not yet formed intentions to perform it, while in others, the problem may be a lack of skills or the presence of environmental constraints. In still other populations, more than one of these factors may be relevant. Clearly, if people have formed the desired

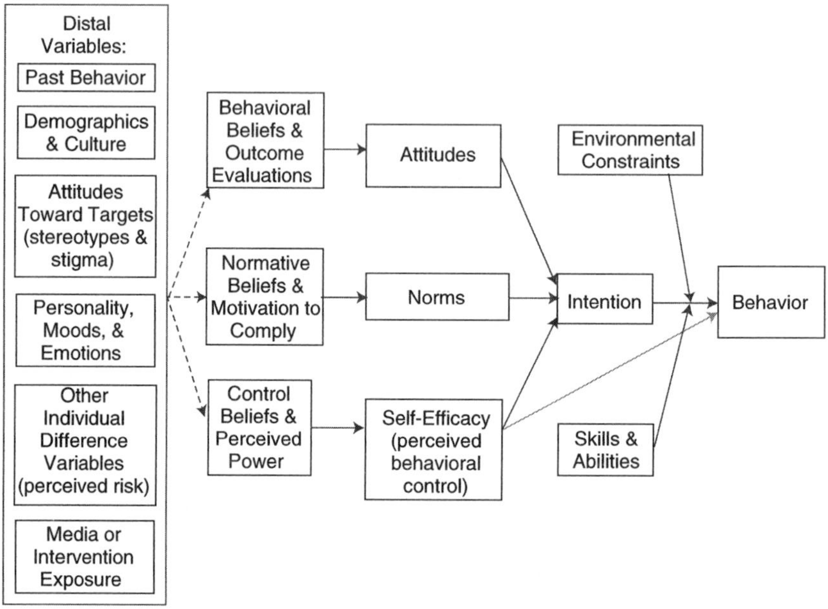

FIGURE 7-1 Determinants of behavior.

intention but are not acting on it, a successful intervention will be directed either at skill building or at removing environmental constraints.

If strong intentions to perform the behavior in question have not been formed, the model suggests three primary determinants of intention to consider: the attitude toward performing the behavior, perceived norms concerning performance of the behavior, and one's self-efficacy with respect to performing the behavior. It is important to recognize that the relative importance of these three psychosocial variables as determinants of intention depend on both the behavior and the population being considered. Thus, for example, one behavior may be primarily determined by attitudinal considerations, while another may be primarily influenced by feelings of self-efficacy. Similarly, a behavior that is attitudinally driven in one population or culture may be normatively driven in another. Thus, before developing interventions to change intentions, it is important to first determine the degree to which that intention is under attitudinal, normative, or self-efficacy control in the population in question. Once again, it should be clear that very different interventions are needed for attitudinally controlled behaviors than for behaviors that are under normative influence or are strongly related to feelings of self-efficacy. Clearly, one size does not fit all, and interventions that are successful at changing a given behavior in one culture or population may be a complete failure in another.

The model in Figure 7-1 also recognizes that attitudes, perceived norms, and self-efficacy are all themselves functions of underlying beliefs—about the outcomes of performing the behavior in question, about the normative proscriptions or behaviors of specific referents, and about specific barriers to behavioral performance. Thus, for example, as was described above, the more one believes that performing the behavior in question will lead to "good" outcomes and prevent "bad" outcomes, the more favorable should be one's attitude toward performing the behavior. Similarly, the more one believes that specific others think one should (or should not) perform the behavior in question, and the more one is motivated to comply with those specific others, the more social pressure one will feel (or the stronger the subjective norm) with respect to performing (or not performing) the behavior. Finally, the more one perceives that one has the necessary skills and abilities to perform the behavior, even in the face of specific barriers or obstacles, the stronger will be one's self-efficacy with respect to performing the behavior.

It is at this level that the substantive uniqueness of each behavior comes into play. For example, the barriers to and consequences of enlisting in the military may be very different from those associated with reenlisting in the military. Yet it is these specific beliefs that must be addressed in an intervention if one wishes to change intentions and behavior. And although an investigator can sit in the office and develop measures of attitudes, perceived norms, and self-efficacy, she or he cannot tell you what a given population (or a given person) believes about performing a given behavior. It is necessary to go to members of that population to identify salient outcome, normative, and efficacy beliefs. To put this somewhat differently, one must understand the behavior from the perspective of the population one is considering.

The above discussion focuses attention on a limited set of variables that have consistently been found to be among the strongest predictors of any given behavior (see, e.g., Petraitis et al., 1995). Although focusing on behavior-specific variables may provide the best prediction of any given behavior, this approach does little to explain the genesis of the beliefs that underlie attitudes, norms, and self-efficacy. Clearly, the beliefs that one holds concerning performance of a given behavior are likely to be influenced by a large number of other variables. For example, one's life experience will influence what one believes about performing a given behavior, and thus one often finds relations between such demographic variables as gender, ethnicity, age, education, socioeconomic status, and behavioral performance. Similarly, young adults who are unemployed may have very different beliefs about joining the military from those who are employed. Other factors that may influence one's beliefs include perceived risk, moods and emotions, culture, knowledge, attitudes toward objects or institutions, types of social networks, and media exposure.

Although these are all important variables that may help us to understand why people hold a given belief, they are perhaps best seen as "distal" or background variables that exert their influence over specific behaviors by affecting the more "proximal" determinants of those behaviors. This can also be seen in Figure 7-1.

According to the model, distal or background factors may or may not influence the behavioral, normative, or self-efficacy beliefs underlying attitudes, norms, and self-efficacy. Thus for example, while men and women may hold different beliefs about performing some behaviors, they may hold very similar beliefs with respect to others. Similarly, rich and poor, old and young, those who do and do not plan to go to college, those with favorable and unfavorable attitudes toward law enforcement, and those who have or have not used drugs may hold different attitudinal, normative, or self-efficacy beliefs with respect to one behavior but may hold similar beliefs with respect to another. Thus, there is no necessary relation between these distal or background variables and any given behavior. Nevertheless, distal variables, such as cultural and personality differences and differences in a wide range of values, should be reflected in the underlying belief structure.

It is probably worth noting that when properly applied, theoretical models such as the one presented in Figure 7-1 recognize, and are sensitive to, cultural and population differences. For example, the relative importance of each of the variables in the model is expected to vary as a function of both the behavior and the population being investigated. Moreover, these types of models require one to identify the behavioral, normative and self-efficacy beliefs that are salient in a given population. Thus, these types of models should be both population- and behavior-specific.

Applying the Model

The first step in using a behavioral prediction or behavioral change model is identifying the behavior that one wishes to understand or change. And unfortunately, this is not nearly as simple or straightforward as is often assumed. First, it is important to distinguish between behaviors, behavioral categories, and goals. Research has shown that the most effective interventions are those directed at changing specific behaviors (e.g., walk for 20 minutes three times a week) rather than behavioral categories (e.g., exercise) or goals (e.g., lose weight; see Fishbein, 1995, 2000).

Note that the definition of a behavior involves several elements: the action (enlisting), the target (the Army), and the context (after graduating from high school). Clearly, a change in any one of the elements changes the behavior under consideration. Thus, for example, enlisting in the Army is a different behavior from enlisting in the Navy (a change in target). Similarly, enlisting in the Army after graduating from high school is a different behavior from enlisting in the Army after completing college (a change in context). Moreover, it is also important to include an additional element—time. For example, enlisting in the Army in the next three months is a different behavior from enlisting in the Army in the next two years. Clearly, a change in any one element in the behavioral definition will usually lead to very different beliefs about the consequences of performing that behavior, about the expectations of relevant others, and about the barriers that may impede behavioral performance.

The second step in applying a behavioral theory is to identify a specific target population. For any given behavior, both the relative importance of attitudes, norms, and self-efficacy as determinants of intention or behavior and the substantive content of the behavioral, normative, and control beliefs underlying these determinants may vary as a function of the population under consideration. Thus, it is imperative to define the population to be considered. Target populations may be defined in very different ways. Often, populations are divided into groups according to characteristics of social diversity, such as age, income, gender, and education. Even a single group characterized by one demographic variable will comprise multiple diverse audiences. For example, the population of men between ages 18 and 25 will include men varying in ethnicity, income, education, age, family history, and beliefs.

Information about intended audiences can be obtained through national or local survey data or through formative research, which involves data collection on audiences' sociodemographic characteristics, behavioral predictors or antecedents such as audience beliefs, values, skills, attitudes, and current behaviors. There is rarely only one way to segment a diverse population. Given limited resources, it is critical to prioritize and select a particular population segment in lieu of others.

Once one or more behaviors and target populations have been identified, theory can be used to understand why some members of a target population are performing the behavior and others are not. That is, by obtaining measures of each of the "proximal" variables in the model (i.e., beliefs, attitudes, norms, self-efficacy, intentions, and behavior), one can determine whether a given behavior such as enlisting in the military is not being performed because people have not formed intentions to enlist or because they are unable to act on their intentions. Similarly, one can determine, for the population under consideration, whether intention is influenced primarily by attitudes, norms, or self-efficacy. And finally, one can identify the specific beliefs that discriminate between those who do or do not intend to perform the behavior. These discriminating beliefs need to be addressed if one wishes to change or reinforce a given intention or behavior. When the beliefs are appropriately selected, changes in these beliefs should, in turn, influence attitudes, perceived norms, or self-efficacy—the proximal determinants of one's intentions to engage in and often the actual performance of that behavior.

Identification of these beliefs requires understanding the behavior from the perspective of the population in question. That is, proper implementation of theory requires one to go to a sample of that population to identify the relevant outcomes, referents, and barriers (Middlestadt et al., 1996). Once this has been done, survey data can be used to identify the beliefs that discriminate between those who do or do not intend to perform the behavior in question (Hornik and Woolf, 1999).

APPLICATION OF BEHAVIORAL THEORY TO MILITARY RECRUITMENT AND RETENTION

Despite the fact that propensity, as measured by most investigators, has repeatedly been found to be the best single predictor of enlistment, and despite the fact that propensity can best be viewed as a measure of intention to join the military, there have been very few attempts to use behavioral theories, such as those presented above, as a framework for predicting and understanding propensity. Although we were able to find no studies that have applied behavioral theory to the prediction of propensity to enlist or actual enlistment, we did find two studies that were concerned with predicting retention.

Predicting Retention in the British Women's Royal Army Corps

In 1976 there were many more applicants for the British Women's Royal Army Corps (WRAC) than could be accepted and a set of selection procedures was adopted. Despite this, a certain number of applicants (about 16 percent) left the Corps after the first six weeks of basic training. Because this was quite expensive, the Army Personnel Establishment wanted to see if they could predict who would leave and who would stay. In an attempt to answer this question, Keenan (1976, reported in Tuck, 1976) conducted a study using the theory of reasoned action. Consistent with expectations, intentions to "stay in the WRAC after my training at Guilford" as measured in the first three days of joining the Army, were very good predictors of retention ($R = 0.68$, $df = 98$, $p < 0.001$).

More important, this intention was significantly related to both their attitudes toward "joining the WRAC" and their subjective norms concerning "my joining the WRAC" (multiple R = 0.59, beta weights not reported), and these two measures were in turn predicted from their underlying behavioral and normative beliefs. More specifically, their behavioral beliefs that "My joining the WRAC" would lead to 12 outcomes (identified through formative research: gives me opportunities to travel, gives me independence, means good pay, means being away from home, etc.) weighted by their evaluations of these outcomes was correlated 0.47 with their attitudes. Their normative beliefs (also identified through formative research: about mother, father, friends in the service, boyfriend, etc.) weighted by their motivation to comply with these referents correlated 0.58 with their subjective norm.

In a further analysis, Keenen found that the main factors influencing retention were the women's beliefs that "My joining the WRAC is a chance to get a career" and that their parents would support this decision. In addition, she found that the more favorable the women were toward "making new friends," "independence," "discipline," and "being away from home," the more likely they were to stay in the WRAC.

Reenlistment in the National Guard

As part of a larger study, Hom et al. (1979) used the theory of reasoned action to predict the propensity to reenlist of Army National Guard members as well as the actual reenlistment behavior of 228 men whose current term of enlistment expired in the next six months. Because the paper was concerned with comparing various models, the researchers measured only attitude, subjective norm, intention, and behavior. Consistent with expectations, the men's intention to "reenlist in the National Guard when my present enrollment expires" was highly correlated with their actual reenlistment ($R = 0.67$, $p < 0.001$). Also consistent with expectations, their propensity to reenlist was highly correlated with both their attitude toward "reenlisting in the National Guard at the next opportunity" ($R = 0.79$) and their subjective norm with respect to "reenlisting in the guard at the next opportunity" ($R = 0.69$). The multiple correlation was 0.81, with both attitude (beta = 0.42) and norms (beta = 0.17) contributing to the prediction.

Other than these two studies, most information concerning factors influencing measured propensity have come from two surveys: Monitoring the Future (MTF) and the Youth Attitude Tracking Survey (YATS). In MTF, high school seniors (surveyed during the spring, just prior to graduation) are asked "How likely is it that you will do each of the following things after high school? The activities listed include "serve in the armed forces." Respondents are asked to choose from the following alternatives: definitely won't, probably won't, probably will, and definitely will.

As we discussed in Chapter 6 (see Figure 6-4, this volume), there has been a modest decline in the proportion of male high school seniors who say they definitely will or probably will serve in the armed forces from the mid-1980s onward. In contrast, during the same period, there has been a steady and significant increase in the proportion who say they "definitely will not serve" and a corresponding decrease in those who say they "probably will not serve." Figure 6-5 shows responses by female high school seniors to the same question. The percentage of females who say they definitely will or probably will serve has remained relatively constant (at about 8 percent) since 1976. However, as with their male counterparts. the percentage who say they definitely will not serve has increased, with a corresponding decrease in those who say they probably will not serve.

The YATS asks respondents "How likely is it you will be serving on active duty in the [Army, Navy, Marine Corps, Air Force]; would you say definitely, probably, probably not, or definitely not?" Those who say "definitely" or "probably" for at least one of the four services are said to have expressed "positive propensity" for military service; those who say they will "probably not" or "definitely not" join, along with those who say they "don't know" or decline to answer, are said to have "negative propensity." In 1991, 34 percent of males ages 16–21 were "positively propensed." This dropped to a low of 26 percent in 1997 and 1998 and then showed a slight increase to 29 percent in 1999. More recently, respondents have also been asked about the likelihood that they would be serving on active duty "in the military."

Predicting Propensity to Enlist from MTF Data

One major attempt to look at the psychosocial determinants of propensity and enlistment has been done by Bachman et al. (2000b). For example, Bachman et al. (2000b) predicted propensity from eight demographic variables including: race/ethnicity; number of parents in household, parents' average education, post/current residence (e.g., farm, city/large metropolitan region), region, intentions to attend college, high school curriculum, and high school grades. While all except high school grades contributed significantly to the prediction of propensity, the two most important predictors were race/ethnicity and college plans. Consistent with previous data, blacks were more likely than other race or ethnic groups to intend to join while those with college plans were least likely to indicate a propensity to join the military. Generally speaking, however, these demographic variables accounted for relatively little of the variance in propensity; R^2 was only 0.086 for men and 0.070 for women.

The researchers then analyzed a wide range of attitudes, values, and behaviors that they thought might be correlated with propensity. They examined roughly 140 possible predictors. They first examined the bivariate correlation of each of these 140 variables with propensity and then, for each variable that had a significant bivariate correlation with propensity, they controlled for the previously described background variables. It is interesting to note that, of the 140 predictors considered, only 14 were significantly related to propensity for men, and only 10 were significant for women. Perhaps not surprisingly, and consistent with the integrated model, the strongest single predictor of propensity for both men and women was an item that Bachman et al. viewed as an indicator of "attitude toward the military as a workplace." More specifically, respondents were asked, "Regardless of your specific job, would you find the military an acceptable place to work?" It seems reasonable to conclude that this variable would be highly correlated with respondents' attitudes toward joining the military. Even after controlling for the demographic background variables, this single item was correlated 0.73 with propensity for men and 0.61 for women. Only one other variable had a correlation over 0.28, and this too was considered as an indicator of the same attitude. Rather than being a single item, this latter variable was an index the authors constructed to assess "opportunities and treatment in the military." This index was correlated 0.38 with propensity for men and 0.16 for women.

While such findings may initially appear discouraging from a recruitment perspective, they are entirely consistent with theories of behavioral prediction and change. That is, from a behavior theory perspective, demographic and economic variables, as well as more traditional measures of attitude and values, have primarily an indirect influence on a person's decision to perform (or not to perform) any given behavior. Indeed, although one can identify hundreds of demographic, economic, and psychosocial variables that may or may not influence any behavioral decision, as we saw above, behavioral theories suggest that there is only a limited number of variables that serve as critical determinants of any given behavior. Consistent with this, MTF data suggest that young adults' attitudes toward the military as a workplace are highly correlated with propensity. However, MTF was not designed to explore insight into why people hold such positive or negative attitudes. In order to gain some insight into this, we now consider the YATS data.

Predicting Propensity from YATS Data

As part of YATS, respondents were asked to indicate the "importance" of 5 randomly selected job attributes from a set of 26 such as "job security," "getting money for education," "preparation for future career or job," "doing something for your country," "personal freedom," and "a job with good pay." In addition, they were asked to indicate whether each of these five randomly selected attributes was more likely to be found in "the military," "a civilian job," or "equally in both."

According to behavioral theory, the more people that believe they can attain valued outcomes from the military rather than from a civilian job, the more favorable should be their attitude toward joining the military and the higher their propensity to join the military. Because all attributes are phrased in a positive manner, ratings of importance are highly correlated with ratings of the value (i.e., extremely good, quite good, slightly good, neither good nor bad) that respondents place on these attributes. In addition, the second set of judgments can be viewed as measures of belief that the attribute would be obtained by "joining the military." Thus, it's possible to compute an expectancy-value score for each attribute. According to behavioral theory, this product should be most highly correlated with propensity. In order to explore this, we conducted a secondary analysis of the data. More specifically, we regressed propensity on (a) the 26 importance ratings, (b) the 26 belief ratings, and (c) the 26 belief × importance.

Because each respondent was only asked about five randomly selected attributes, the amount of missing data makes the interpretation of these regression analyses somewhat problematic. The following should therefore be considered suggestive rather than definitive. Table 7-1 shows the results of these three regressions.

It can be seen that consistent with expectations, the expectancy-value scores provided the best prediction of propensity. Moreover, it can be seen that beliefs were more important determinants of propensity than were values.

TABLE 7-1 Prediction of Propensity from Job Attributes

	R^2	Significance
Importance	0.154	NS
Belief	0.270	$p < 0.02$
Belief × Importance	0.304	$p < 0.005$

SOURCE: Data from the Youth Attitude Tracking Study.

In looking at the particular job attributes that predict propensity, only one importance rating was significant at less than the 0.05 level. More specifically, the more these young adults felt that "doing something for my country" was important, the stronger their propensity to join the military ($p < 0.01$). Because of missing data and the likelihood of multicolinearity, we also consider those attributes that approach significance (i.e., that have p-values at less than 0.20). By relaxing this standard, there are two additional values that appear to influence propensity. The more people felt that a "mental challenge" was important, the lower their propensity ($p < 0.15$), and the more they felt that "opportunity for adventure" was important, the higher their propensity to join the military ($p < 0.15$).

In contrast, several beliefs appear to influence propensity. The more they believed that the military (rather than a civilian job) would allow them to be near family ($p < 0.01$), give them personal freedom ($p < 0.05$), prepare them for future jobs ($p < 0.10$), provide U.S. travel ($p < 0.10$), teach leadership skills ($p < 0.15$), give them something to be proud of ($p < 0.15$), would be doing something for their country ($p < 0.15$), and provide equal opportunity for minorities ($p < 0.20$), the stronger was their propensity to join the military.

Finally when the expectancy-value (i.e., belief x importance) scores are considered, propensity is related to being near family ($p < 0.001$), personal freedom ($p < 0.02$), getting leadership skills ($p < 0.05$); getting parents approval ($p < 0.05$); having an interesting job ($p < 0.10$); job preparation ($p < 0.10$), doing something to be proud of ($p < 0.15$), working with people you respect ($p < 0.15$), doing something for your country ($p < 0.20$), U.S. travel ($p < 0.20$), and equal opportunity for minorities ($p < 0.20$).

What is perhaps most surprising about these findings is the apparent lack of relationship between propensity and attributes that have often been promoted by the military—money for education, pay, job security, and working in a high-tech environment. Since it is possible that the failure of these attributes to emerge as factors influencing propensity could be due to issues of multicolinearity, and since it is probably not appropriate to consider each job attribute as independent, the committee decided it was important to examine the extent to which these 26 attributes represent more general, underlying dimensions. Given the amount of missing data, we were unable to arrive at a meaningful factor solution using traditional factor analysis. Thus we conceptually grouped attributes and used a confirmatory factor analysis procedure in AMOS (which uses all available data and does not impute missing values; see Arbuckle, 1996) to investigate whether various sets of selected attributes could be used as indicators of a more global construct. This analysis suggested that five basic dimensions provided good fits to the data for both men and women:

1. **Learning Opportunities:** develop self-discipline, develop leadership skills, job preparation, learn new skills.
2. **Working Conditions:** work autonomy, work as a team, mental challenge, interesting job, work with people you respect, near family, personal freedom.
3. **External Incentives:** high-tech environment, money for education, good pay, job security, parental approval, earn respect.
4. **Patriotic Adventure:** opportunity for adventure, physical challenge, doing something for your country, foreign travel, domestic travel, doing something to be proud of.
5. **Equal Opportunity:** equal employment opportunity for women, equal employment opportunity for minorities, prevent harassment.

In order to explore the relations of each of these dimensions to propensity, we again used AMOS. More specifically, for each of the five dimensions, three models predicting propensity were tested: one predicting propensity from importance judgments, one predicting propensity from beliefs, and one predicting propensity from the belief x importance product. Based on the above regression analyses, we would expect that "importance" would be a significant predictor of propensity only when attributes concerning "patriotic adventure" are considered, while beliefs and the expectancy-value products would be significant predictors in all five domains. It is important to note that the prediction of propensity from beliefs is also an expectancy-value prediction. That is, since all attributes are phrased positively and represent valued outcomes, the higher the estimated belief score, the more likely it is that one will obtain a valued outcome. Thus one would expect the estimated belief x importance product to improve prediction over and above the estimated belief score only when importance is also a significant predictor of propensity.

Each analysis simultaneously considered both males and females. Figure 7-2a shows how well the estimated importance of "patriotic adventure" predicted propensity; Figure 7-2b shows the relations between propensity and the estimated belief that the military (rather than civilian jobs) would lead to "patriotic adventure"; Figure 7-2c shows the role of the estimated expectancy-value score in influencing young men and women's intentions to join the military.

Looking at Figure 7-2a, it appears that the model fits the data quite well (rmses = 0.024, TLI = 0.991) and, consistent with expectations, the estimated importance of attributes related to patriotic adventure accounts for 12 percent of the variance in young men's propensity, and 5 percent of young women's propensity, to enlist in the military. In Figures 7-2b

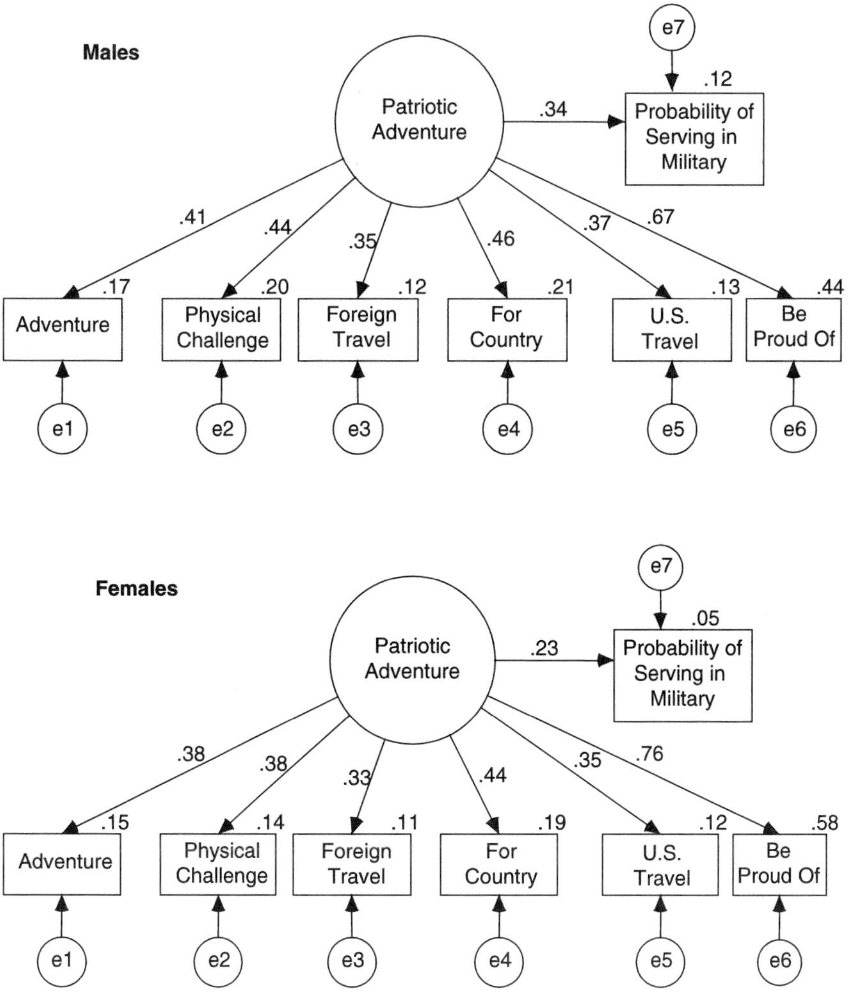

FIGURE 7-2a Predicting propensity from the importance of patriotic adventure. chi = 258.720, df = 38, p = 0.000, rmsea = 0.024, TLI = 0.991
SOURCE: Data from the Youth Attitude Tracking Study.

and 7-2c, it appears that the model holds equally well for beliefs (rmses = 0.025, TLI = 0.990) and the expectancy-value products (rmses = 0.018, TLI = 0.993).

As expected, the more the men believed that they would be more likely to get "patriotic adventure" by joining the military (rather than by taking a civilian job), the higher their propensity to enlist (R^2 = 0.11).

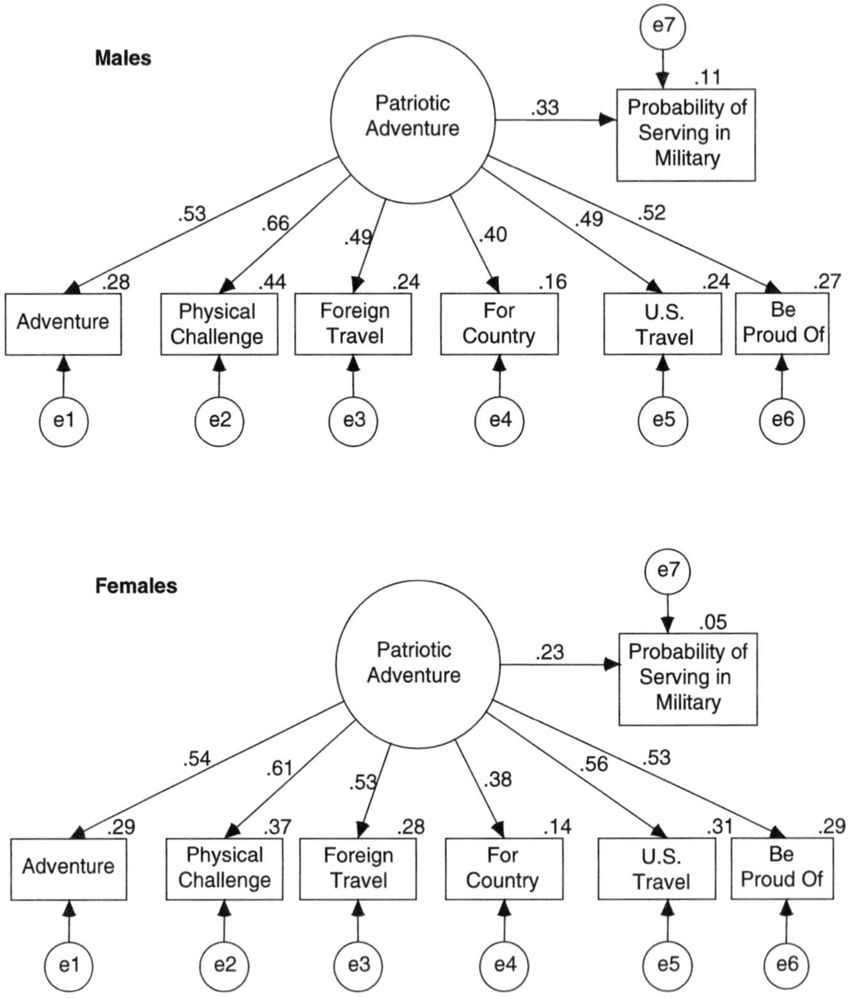

FIGURE 7-2b Predicting propensity from the belief that joining the military will lead to patriotic adventure. chi = 275.385, df = 38, p = 0.000, rmsea = 0.025, TLI = 0.990
SOURCE: Data from the Youth Attitude Tracking Study.

Similarly, the more the women believed that they would be more likely to get "patriotic adventure" by joining the military, the higher their propensity to enlist ($R^2 = 0.05$). Note, too, that taking both belief and importance into account leads to slightly better prediction than considering either one alone. That is, for men, the estimated expectancy value for patriotic adventure accounts for 21 percent of the variance in their propensity to join

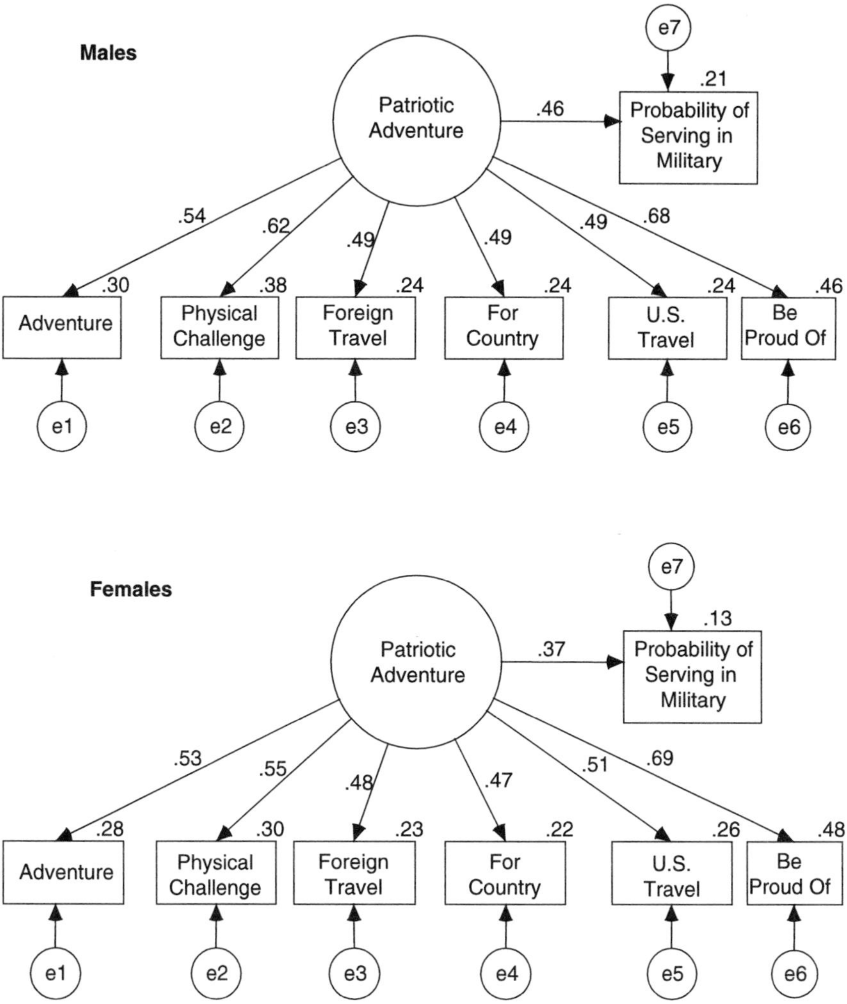

FIGURE 7-2c Predicting propensity from the expectancy-value score associated with patriotic adventure. chi = 165.376, df = 38, p = 0.000, rmsea = 0.018, TLI = 0.993
SOURCE: Data from the Youth Attitude Tracking Study.

the military and, for women, the estimated expectancy value for patriotic adventure accounts for 13 percent of the variance in their propensity to join.

These findings clearly suggest that in order to increase propensity, one should try to increase the importance of attributes such as "doing something for my country" and "having an opportunity for adventure." In addition, one should try to increase beliefs that these "patriotic adventure" attributes are more likely to be obtained in the military than in civilian life.

Figures 7-3a, 7-3b and 7-3c provide similar data for the role of "external incentives." Once again it can be seen that all three models (i.e., for importance alone, for beliefs alone, and for the expectancy values) provide a good fit to the data. Here, however, it can be seen that the importance of this dimension contributes very little to the prediction of propensity ($R^2 = 0.02$ for both men and women). In marked contrast, both beliefs and the belief x importance products are significantly related to propensity. More specifically, beliefs account for 22 percent of the variance in propensity for men and 10 percent of the variance for women's and the belief x importance products account for 19 percent and 11 percent for men and women, respectively.

Similar analyses were done with respect to the remaining three dimensions. Table 7-2 summarizes the findings by showing the regression weights and R^2 between propensity and (a) the estimated importance of each dimension, (b) the estimated beliefs that the attribute dimension would be most likely to be obtained in the military, and (c) the estimated expectancy-value score for each attribute dimension, for both men and women. There it can be seen that only the importance of "patriotic adventure" significantly contributes to propensity. In contrast, when beliefs are considered and when importance ratings are weighted by the belief that the attributes defining a given dimension are more likely to be obtained in the military than in civilian life, four of the five dimensions are found to significantly contribute to propensity. Somewhat surprisingly, these young adults appear to pay little attention to equal opportunity considerations. Note that, consistent with the earlier findings of Bachman, Freedman-Doan, and O'Malley (2000) men's propensity is more predictable than women's propensity. More important, considerations of external incentives and patriotic adventure are the most important determinants of propensity for both men and women.

Generally speaking, then, these analyses suggest that considerations of patriotism, adventure, external incentives, opportunities for learning, and working conditions all influence young adults' intentions to join the military. These findings strongly suggest that, in order to increase the propensity to enlist, it will be useful to increase (a) the value young adults place on "doing something for my country" and on "opportunities for adventure," as well as (b) their beliefs that they are more likely to be doing something for the country and that they will have a greater opportunity for adventure in the military than in civilian life. In addition, it might be useful to strengthen their beliefs that they are more likely to get

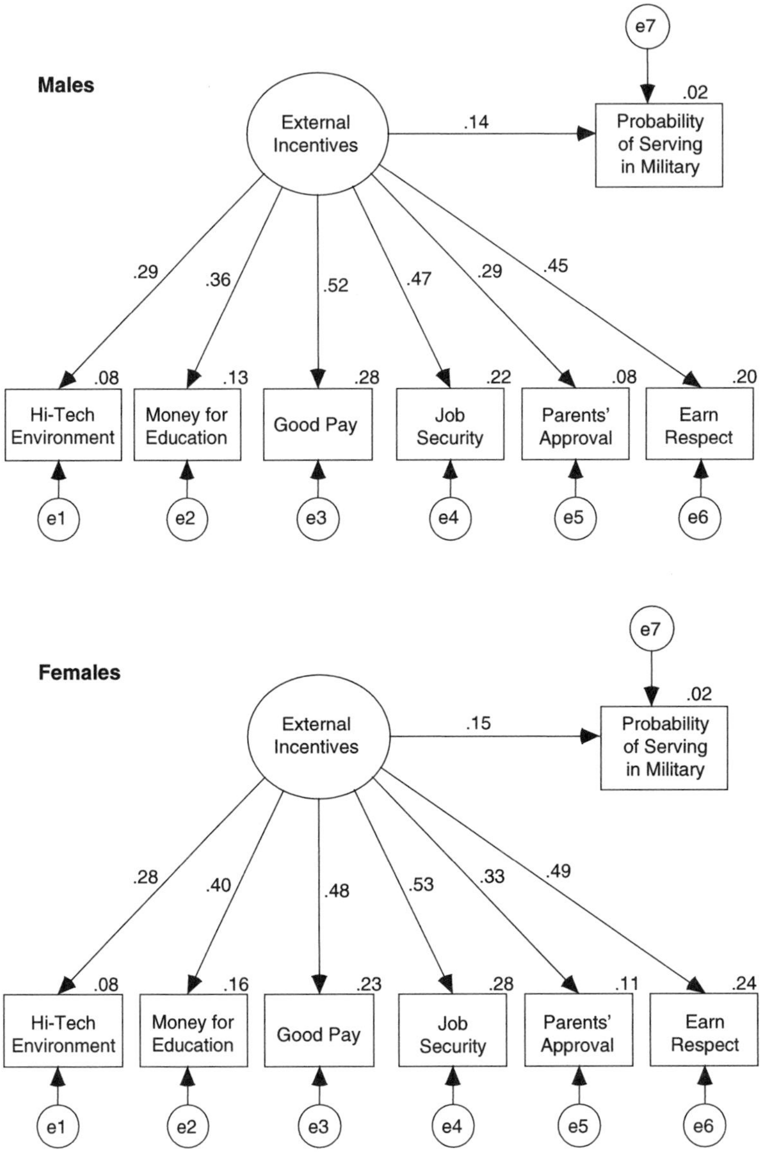

FIGURE 7-3a Predicting propensity from the importance of external incentives.
chi = 92.874, p = 0.000, df = 38, rmsea = 0.012, TLI = 0.998
SOURCE: Data from the Youth Attitude Tracking Study.

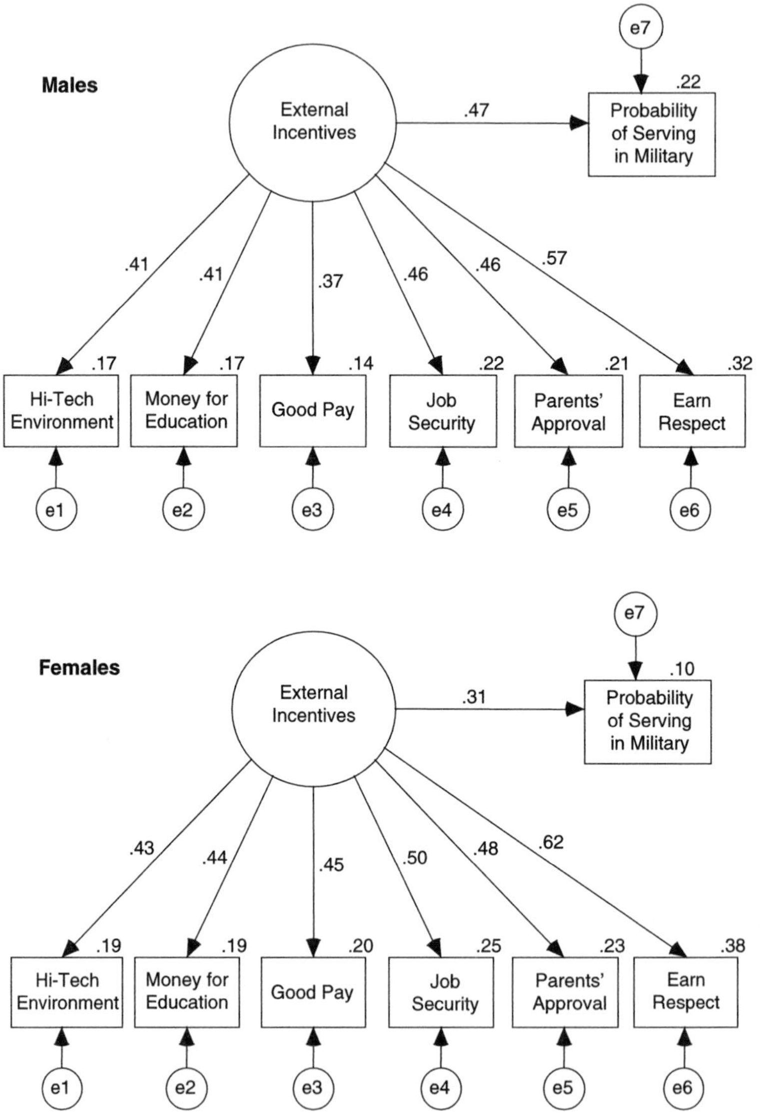

FIGURE 7-3b Predicting propensity from the belief that joining the military will lead to external incentives. chi = 84.003, p = 0.000, df = 38, rmsea = 0.011, TLI = 0.998

SOURCE: Data from the Youth Attitude Tracking Study.

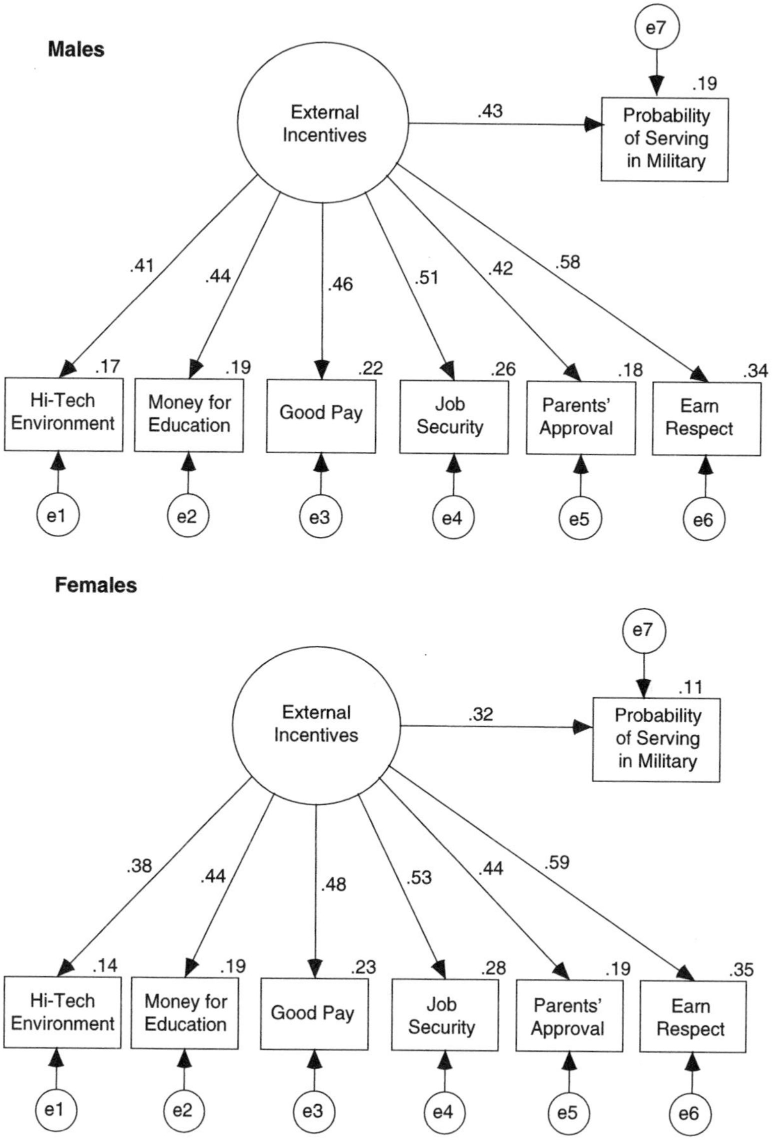

FIGURE 7-3c Predicting propensity from the expectancy-value score associated with external incentives. chi = 101.661, p = 0.000, df = 38, rmsea = 0.013, TLI = 0.997

SOURCE: Data from the Youth Attitude Tracking Study.

TABLE 7-2 Predicting Men and Women's Propensity from Importance Estimates, Belief Estimates, and Expectancy-Value (Importance x Belief) Estimates

	Importance		Belief		Importance x Belief	
	Beta	R^2	Beta	R^2	Beta	R^2
Men						
Working conditions	0.02	0.00	0.48	0.23	0.34	0.11
Learning opportunities	0.15	0.02	0.42	0.17	0.37	0.14
External incentives	0.14	0.02	0.47	0.22	0.43	0.19
Patriotic adventure	0.34	0.12	0.33	0.11	0.46	0.21
Equal opportunity	0.03	0.00	0.07	0.00	0.09	0.01
Women						
Working conditions	–0.04	0.00	0.30	0.09	0.20	0.04
Learning opportunities	0.14	0.02	0.23	0.05	0.29	0.09
External incentives	0.15	0.02	0.31	0.10	0.32	0.11
Patriotic adventure	0.23	0.05	0.23	0.05	0.37	0.13
Equal opportunity	0.00	0.00	0.06	0.00	0.05	0.00

SOURCE: Data from the Youth Attitude Tracking Study.

a number of external incentives (e.g., money for education, good pay, job security, parents' approval) from the military than from civilian life.

In order to illustrate this, Figure 7-4 shows how beliefs and values associated with "doing something for my country" are related to propensity. More specifically, the figure shows the percentage of young adults with positive propensity to join the military as a function of their beliefs that they are most likely to do something for the country in the military or in a civilian job (or equally in both), as well as of the degree to which they think that "doing something for the country" is important. Consistent with the above analyses, it can be seen that both importance and belief are highly related to propensity.

Looking at the three bars on the left of the figure, it can be seen that those who believe that they are more likely to be doing something for the country in the military are almost 10 times more likely to have a positive propensity to join the military (23.8 percent) than are those who believe that they are more likely to do something for the country in a civilian job (2.4 percent). Equally important, those who think that doing something for the country is "extremely" important are more than three times likely to have a positive propensity (27 percent) as are those who think that "duty to country" is "not at all important" (8.4 percent).

A somewhat different picture is presented in Figure 7-5. Recall that in the above analyses, we found that although the importance one placed on

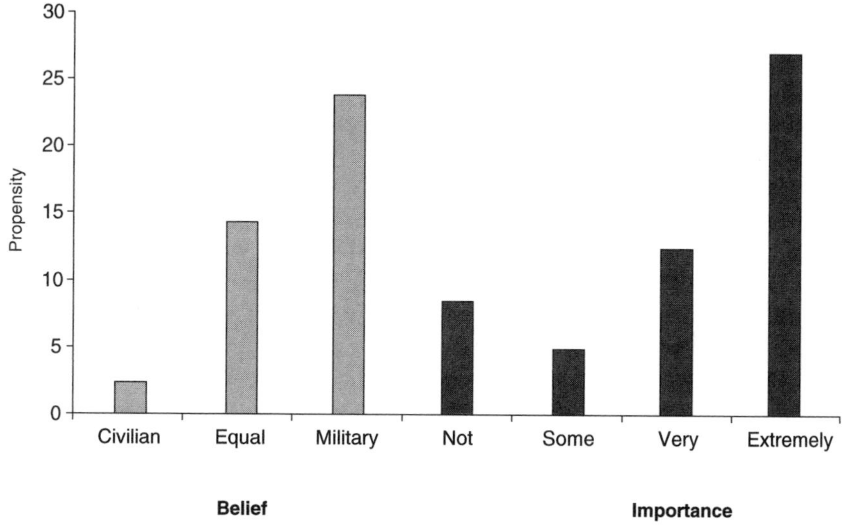

FIGURE 7-4 Do something for country.
SOURCE: Data from the Youth Attitude Tracking Study.
NOTE: Survey responses on "do something for my country." See text for explanation.

external incentives had little to do with propensity, the more one believes that he or she is more likely to obtain these incentives in the military (compared with a civilian job), the more likely one is to have a positive propensity to join the military. Figure 7-5 considers the external incentive of "good pay." Note that while those who believe that they are most likely to get good pay in the military are six times as likely to have a positive propensity (31.4 percent) as those who believe they are most likely to get good pay in a civilian job (5.2 percent). Those who think "good pay" is "extremely important" are only slightly more likely to have a positive propensity (14.1 percent) than are those who think that "good pay" is "not at all" important (10 percent).

One additional individual attribute is worth considering. In the previous chapter, we saw the important role that social influencers, and in particular one's parents, play in young adults' decisions to enter the military. Figure 7-6 shows that the more young adults believe that they are likely to gain parental approval in the military (rather than in a civilian job), the more likely they are to have a propensity to join. More specifically, 40 percent of those who believe that they are more likely to get parental approval by being in the military intend to join, while only 5.7

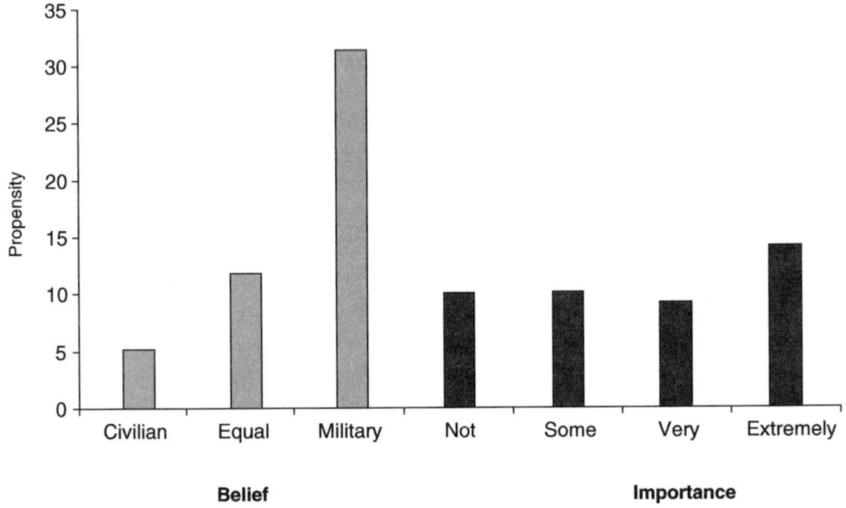

FIGURE 7-5 Good pay.
SOURCE: Data from the Youth Attitude Tracking Study.
NOTE: Survey responses on "good pay." See text for explanation.

percent of those who believe they will gain more approval in a civilian job have a propensity to join. In marked contrast, we can again see that the value these young adults place on parental approval plays little or no role in influencing their propensity to enlist.

These findings indicate that, at any given point in time, the greater the number of young people who value "patriotism" and who think that the "opportunity for adventure" is important, the greater should be the number of young people who have a positive propensity to join the military. Similarly, the greater the number of young people who believe that they are more likely to "do something for the country" or to have an "opportunity for adventure" in the military and who believe that they are more likely to obtain external incentives, such as job security, good pay, and parental approval, from the military than from a civilian job, the greater should be the pool of young adults with a propensity to join the military.

In Chapter 6, we saw that there have been relatively few changes in youth values over time. The two changes that have occurred—an increase in the value placed on educational attainment and a decrease in the value placed on doing something for the country—have undoubtedly contributed to reduced propensity and recruiting difficulties. Perhaps even more

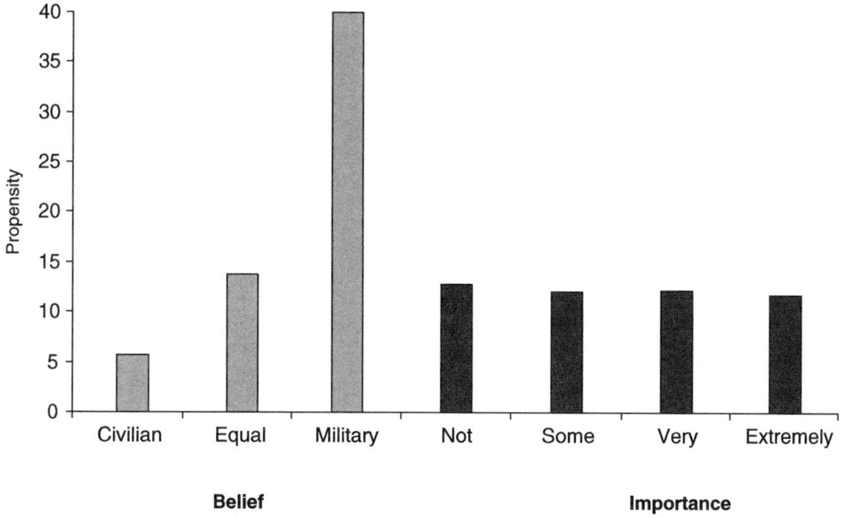

FIGURE 7-6 Get parents' approval.
SOURCE: Data from the Youth Attitude Tracking Study.
NOTE: Survey responses on "get parents' approval." See text for explanation.

problematic, however, there have been a number of substantial changes in young adults' beliefs that valued attributes would be obtained from the military. For example, as recently as the beginning of the 1990s there was considerable agreement that one would be most likely to be doing something for the country by serving in the military. Since that time, however, there has a been a gradual erosion of this belief to the point at which more people now feel that they will be doing something for their country in a civilian job than in the military. More specifically, the net percentage attributing "doing something for the country" to the military rather than civilian jobs (i.e., the percentage attribution to the military minus the percentage attribution to civilian jobs) declined from 37 percent in 1992 to –5 percent in 1999 for men and from 39 percent in 1992 to –17 percent in 1999 for women.

Indeed, what is most startling about our analysis of the YATS data is that they clearly show a steady loss of a number of valued attributes to civilian occupations. That is, there has been a consistent erosion in the likelihood that many of these values are more likely to be obtained from the military and a corresponding increase in the likelihood that they will be obtained from a civilian job or equally in both. For example, among men, the net percentage attributing "working in a high-tech environment" to the military over civilian jobs declined from 20 percent in 1992 to –1 percent in 1999 for men and from 15 to –8 percent for women. What this means is that more young adults now believe that they will be working in a high-tech environment in a civilian job than in the military.

As another example, consider the attribute of "getting money for education." Although the value of educational attainment has significantly increased over the past decade, the value young adults place on "getting money for education" has remained very high and has not changed over time. Once again, however, there has been an erosion in the belief that one is more likely to get money for education from the military than from civilian life and a corresponding increase in the belief that one is equally or more likely to get money for education from a civilian job.

If these trends continue, the pool of young adults with a propensity to join the military will begin to dry up. It is incumbent upon the military to develop strategies to increase the number of young adults with a propensity to join. The strategy for the past decade seems to have been to fish where the fish are (let's go after those with a propensity and try to close the deal). At this point in time, however, it would appear that the pool needs to be restocked and enlarged. In order to do this, it will be necessary to renew a sense of patriotism and adventure and make it clear that these important values are best attained through the military. In addition, the military must continue to make young people and particularly women aware that it provides a number of extrinsic incentives, such as good pay, job security, health benefits, and vacations. Finally, the military must develop strategies to increase young adults' beliefs that their parents (and other influencers) would approve of their going into military service. As pointed out in Chapter 6, one possible way of doing this is to direct messages toward influencers as well as to the youth per se.

The above analyses suggest that a change in advertising strategy might significantly increase the number of young adults with a positive propensity to join the military. As we discuss more fully in the next chapter, much of the military advertising to date has focused on external incentives and, except in the Marine Corps, little attention has been paid to issues of patriotism or duty to country.

At the very least, advertising should try to increase the number of young adults who have a propensity to join the military. And here it may be necessary to consider two potential audiences. On one hand, advertising could be directed at trying to move people who say they "probably will not" join the military to saying "they probably or definitely will join." On the other hand, advertising could also be used to move people from "moderate" to "strong" propensity (i.e., from "probably will join" to "definitely will join." If advertising were used to increase the pool of young adults with a propensity to join, it would seem quite reasonable for recruiters to concentrate their efforts on signing up those who already have a propensity to join. However, this may often include attempts to strengthen the propensity of those with a moderate one. Issues of advertising and recruitment are addressed more fully in Chapter 8.

SUMMARY

In this chapter we have tried to show that proper application of behavioral theory can help to increase understanding of the factors influencing propensity and enlistment. However, the data necessary for a more comprehensive analysis at the individual level are currently not available. Clearly, it would be reasonable for the military to begin systematically obtaining data on the behavioral, normative, and efficacy beliefs that underlie young adults' attitudes, perceived norms, and feelings of self-efficacy with respect to joining the military. If data of this type are obtained, it is important to recognize that all of the salient beliefs about joining the military should be considered. For whatever reason, most research to date has focused only on beliefs concerning the benefits (or positive attributes) of joining the military. There is, however, one very important potential negative consequence of joining the military: one could be injured or killed. To fully understand propensity it will be necessary to consider all of a person's beliefs about joining the military. Clearly, if fear of death or injury is a critical determinant of propensity, then very different strategies and messages will be necessary to increase it. The committee could find few data to address this question.

8

Military Advertising and Recruiting

This chapter examines military recruiting practices in the context of recent trends in the interests of the youth population. It focuses on the role that advertising by the military Services' plays in providing information to the youth population and promoting youth interest in military service, as well as the role that recruiters play in identifying prospective applicants for military service. The chapter begins with an overview of the current situation, which identifies a 15-year decline in overall level of youth interest in military service as a significant factor affecting the success of military recruiting. Message strategies employed in advertising and recruiting techniques are examined in terms of their likely effects on the level of youth interest in military service. The military recruiting process is briefly described. The chapter concludes with a brief look at recruiting practices in private industry to determine if there are lessons to be applied to the military situation.

CURRENT SITUATION IN MILITARY RECRUITING

Scale of Military Recruiting

The U.S. Department of Defense (DoD) is the nation's largest employer. There are 1.2 million men and women on active duty, who are supported by 672,000 civilian employees. By way of comparison, *Fortune Magazine* reports the number of employees for each company in the Fortune 500. The largest employer is Wal-Mart Stores with 1.3 million employees. McDonald's is next with 395,000 employees. United Parcel Service is third with 370,000 employees.

With the possible exceptions of extraordinarily large employers with very high turnover (e.g., Wal-Mart Stores), most businesses and organizations hire far fewer than the 200,000 people the military Services enlist on an annual basis. Military recruiting is among the most challenging human resources staffing operations conducted by any large-scale organization. In addition to the size of the recruiting effort, the military Services must also meet high standards unlike those of many civilian employers. Successful applicants for military service must demonstrate the physical and mental capabilities to master complex military systems and operations as well as meet high moral standards, age limits, and citizenship requirements. In addition, successful applicants for military service usually place high priorities on service to the nation, teamwork, dedication to duty, and readiness to face personally challenging circumstances—priorities that are often not shared by the entire youth population.

A DoD briefing presentation reported that the total mission for U.S. military recruiting mission was 203,522 for FY 2002. This total recruiting mission included 79,000 for the Army, 53,000 for the Navy, 37,283 for the Air Force, and 34,239 for the Marine Corps. The recruiting mission for the enlisted Reserves totaled 74,950.

To meet its manpower needs, the DoD engages in an array of recruiting activities to inform and interest members of the youth population concerning military service. The recruiting process involves national and local advertising to efficiently supply information on a widespread basis; informational visits by recruiters to schools and student groups; traveling military exhibits to provide information to schools and the public; direct mail advertising and telephone solicitation to identify interested youth; web sites to provide information on military services; and contacts and visits with recruiters to qualify leads and to assist youth in gaining needed information about the decision to enlist and the selection of a particular Service.

Environment for Recruiting

Military recruiting is challenging not only because of the scale of the recruiting mission, but also because of the current environment in which recruiting takes place. Alternatives to military service are more attractive to many young people. As shown in Chapter 3, an increasing proportion of high school youth is attending college, and the recent economic conditions have provided ample employment opportunities for students who opt not to pursue college education or delay post secondary education. Although surveys indicate public confidence in military leadership and the military as an institution, military service is not seen as one of the more attractive choices for young people following high school. However, since September 11, 2001, there has been a greater interest in military service (Wirthlin, 2002).

Trend in Propensity to Enlist

Figure 6-4 shows that the propensity to enlist among high school males has been declining since the mid-1980s. In this key group for recruiting, the proportion saying "definitely will" enlist has declined from 12 to 8 percent during that time period. Although there is a decline in those saying "probably won't," the percentage of those saying "definitely won't" has increased. Thus, the proportion least interested in military service has increased during the past two decades from about 40 to about 60 percent. A similar trend is noted for young women in Figure 6-5.

The decline in the proportion of "propensed youth"—those saying they definitely or probably will enlist—means that the Services increasingly find themselves in competition with each other to meet their recruiting goals. If the recruiting focus remains on this declining group of propensed youth, the message strategies for military service advertising and military recruiters may become increasingly competitive, focusing on claims that differentiate one Service from another. In such a recruiting environment, enlistment bonuses would become an important means of attracting (or buying) applicants.

Importance of the Negative Propensity Groups

The vast majority of youth now say they do not plan on enlisting. Current events such as the Gulf War of 1990–1991 show only short-term effects in terms of arresting or altering the long-term trend of declining youth interest in military service.

The continuing transition between "probably won't" and "definitely won't" responses is of particular concern because it has been shown that an important proportion of those who do eventually enlist come from these two groups. A RAND Corporation study examined enlistment behavior following the Youth Attitude Tracking Survey (YATS) from 1976 to 1980 and found that 46 percent of those who enlisted came from the two negative intention groups (Orvis et al., 1992:17).

A subsequent study from the Monitoring the Future Project examined accession into the armed forces during the first five or six years after high school. This study based on data from graduates of the high school classes of 1984 through 1991, found that, among males, 32 percent came from the negative intention groups. For females, 62 percent of those who enlisted came from the negative intention groups (Bachman et al., 1998: 64).

Indeed, the proportion of propensed youth has declined to the extent that the Army (the service branch with the largest annual recruiting mission) must enlist youth from the negative propensity group to fulfill its manpower goals, because there are not enough propensed youth to meet the Army's goals. This circumstance is particularly challenging because the least interested "definitely won't" group now comprises over 60 percent of high school seniors.

Another indication of the increasing difficulty of meeting recruiting goals in the current environment is the rapid increase in the costs of military recruiting over the past decade. For example, the annual investment in advertising expenditures and recruiting bonuses has increased by a more than threefold factor in the years from 1993 to 2000 (see Figure 8-1).

Figure 8-1 places the recent increase in military advertising in the context of yearly advertising expenditures since 1976. This figure also shows that advertising expenditures are largely allocated to meet the recruiting goals of the individual Services. A relatively small proportion of military advertising expenditures has been allocated to joint Services advertising. Other than a continuous period from 1982 to 1992, there is an irregular pattern to joint Services advertising, with frequent gaps showing little or no investment in this advertising approach.

The decreasing size of the group of youth with a propensity to enlist coupled with the increasing amounts invested in advertising and enlistment incentives points to a new marketplace dynamic. While military recruiting once relied on communication strategies focusing on "selective demand" as the dominant approach, the new market situation calls for greater emphasis on the stimulation of "primary demand." Strategies to encourage selective demand portray the unique or differentiating characteristics of competing brands (such as the specific military Services), while strategies for primary demand focus on information that promotes favorable reactions to the entire category of products (such as general interest in joining any military Service) (Borden, 1942:167).

Message strategies that stimulate primary demand convey information that highlights the benefits of the promoted product or service and are designed to increase the interests of a larger group of potential consumers. For example, communication strategies to stimulate the primary demand for military service would convey information designed to increase the salience of values or goals that can be uniquely satisfied by military service, such as duty to protect the country, self-sacrifice to maintain a free society, and the noble virtues of effective military service. These factors can differentiate and promote military service as an attractive alternative to other pursuits available to youth. This approach could

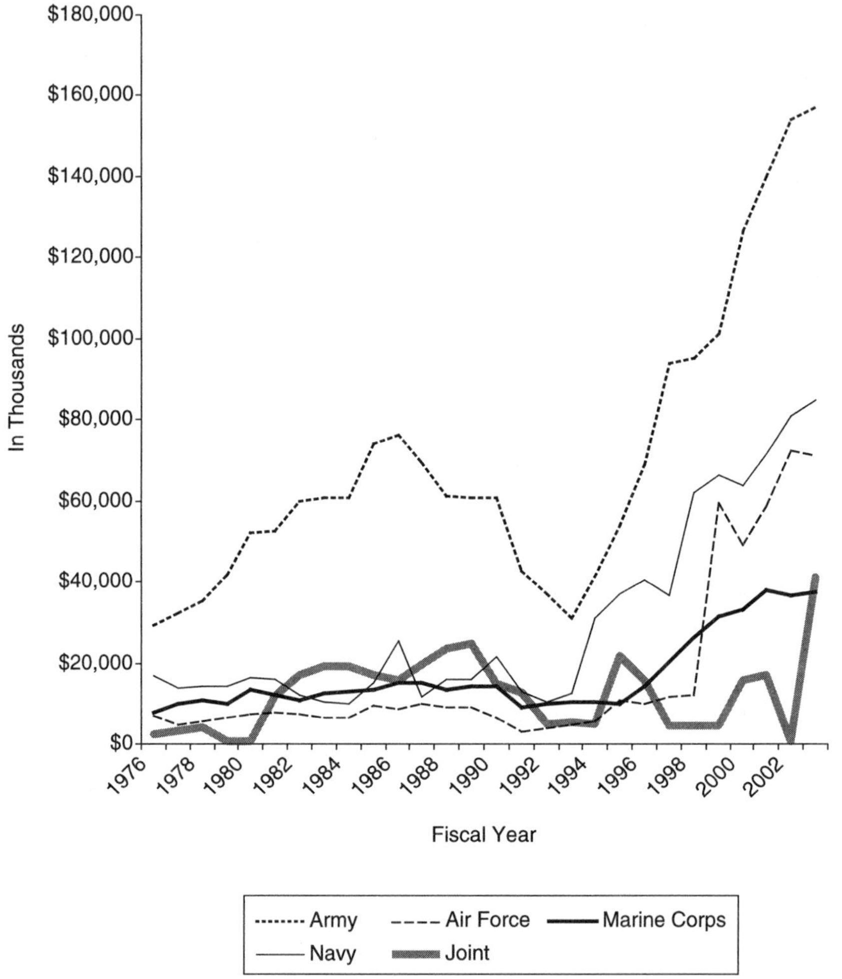

FIGURE 8-1 Enlisted advertising resources FY 1976–2003 in current dollars.
SOURCE: U.S. Department of Defense (2002).

be an important role for joint Services advertising, although the message approach employed in previous joint Service campaigns has not focused on the unique attributes of military service.

In some cases, the absence of information to support primary demand in a product market can be associated with contraction of the demand for the product. The American public's transition from coffee to soft drinks is a notable example of this process. Thirty years ago, when soft drink consumption was accelerating, coffee marketers continued to concentrate on "selective demand" message strategies, such as "less bitter than other brands." The coffee brands competed with each other rather than recognizing the need to promote public interest in coffee as an attractive and satisfying beverage.

Moreover, in addition to information provided by military advertising and recruiting, youth may encounter information concerning the military from family and friends, teachers and advisers, news media, books, movies, music, and other information found in popular culture. The visibility of the military in the day-to-day information environment is largely influenced by world events and the work of writers and producers in the entertainment media. Certainly, the extensiveness of this information and the nature of its content (favorable or unfavorable concerning military service) is an important factor in the degree of youth interest in military service. The decline in the proportion of the youth population that has a family member (parent or other) with military service experience is a example of decline in the general information environment of the youth population.

MODEL OF THE MILITARY RECRUITING PROCESS

The model in the figure presents a schematic of the primary factors involved in military recruiting (Penney et al., 2000:1). The model indicates that recruiter production—that is, a recruiter's level of success in meeting recruiting goals—is a function of recruiter performance in the context of current youth interests in military service, as measured by the propensity to enlist.

The model in Figure 8-2 indicates that propensity to enlist is primarily influenced by military advertising and by a range of environmental factors, such as the youth employment rate, the size of the youth population, youth attitudes and knowledge concerning the military, attitudes relating to occupational prestige and career choice, the relative attractiveness of military compensation and benefits, and the interests of youth in pursuing further education. Recruiter contacts may also influence the propensity of members of the youth population, but that linkage was not featured in the diagram as originally proposed by its authors (Penney et al., 2000).

Recruiter performance is seen as a function of three factors: (1) personal characteristics such as level of interest in the recruiting mission, communication skills, and problem-solving skills; (2) training and development programs to prepare them for the recruiting mission; and (3) technical and organizational support in the form of information and resources (e.g., presentation materials, promotional materials, leads on po-

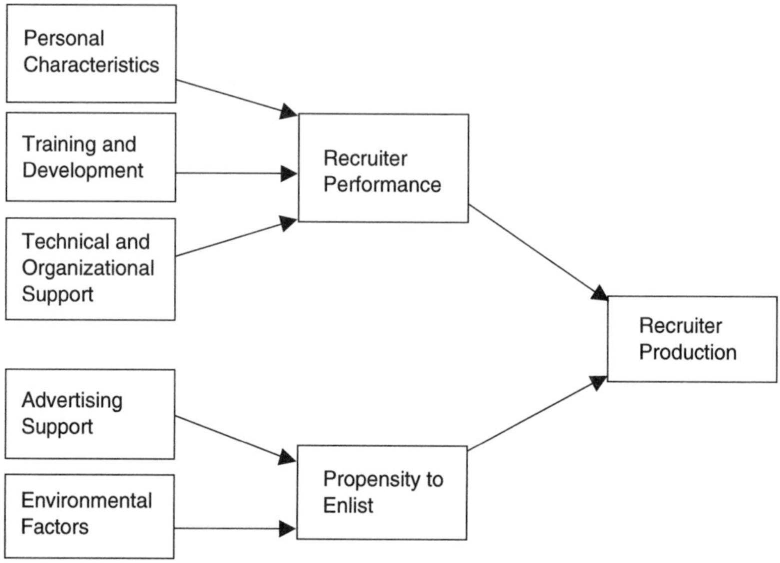

FIGURE 8-2 Model of military recruitment.

tential applicants, and procedures for keeping and organizing information on applicants).

The challenge faced by recruiters is demonstrated by the data in Table 8-1. This table makes apparent how many contacts are required to enlist one person. For example, the Army reports that, on average, recruiters

TABLE 8-1 Stages in the Recruiting Process from Lead Generation (Contacts) to Accessions (Contracts)

Recruiting Stage	Army	Navy	Marine Corps[a]
Leads/contacts	120	80	90
Appointments scheduled	17	21	*b*
Appointments conducted	10	7	15
ASVAB tests	2.3	2	*b*
Qualified applicants	1.5	*b*	*b*
Contracts	1.2	*b*	1
Accessions	1	1	*b*

[a]Marine Corps information shown represents command expectations; actual experience is available only at the local level.

[b]Data not tracked or not made available. Comparable data for the Air Force are not available.

make 120 contacts with specific individuals for every non-prior service soldier accessed into active duty. From those 120 contacts, 17 appointments are scheduled, and 10 appointments are actually conducted. From those 10 appointments, 2.3 applicants take the enlistment test (the Armed Services Vocational Aptitude Battery [ASVAB]), and 1.5 applicants pass all the enlistment qualification standards. Of those 1.5 qualified applicants, 1.2 will sign an enlistment contract, and 1 new recruit will actually enter the Army. Each of these stages present opportunities to examine and possibly improve recruiter performance.

ROLE OF ADVERTISING IN MILITARY RECRUITING

Informing Interests and Tastes

The purpose of advertising is to distribute information designed to influence consumer activity in the marketplace. Advertising messages convey information to support and promote consumer choices between or among competing brands (Hallmark greeting cards versus American Greetings) and between or among alternative product categories (greeting cards versus phone calls). This information may be in such forms as appeals or claims, product demonstrations, and imagery. Often the messages employ creative approaches to expression in order to attract the attention of consumers and to engage them in the content of the advertising. Effectively designed advertising messages, placed in appropriate mass media, provide an efficient means of reaching large numbers of people.

Some economists view advertising as a wasteful expenditure designed to interfere with consumer sovereignty. Others view advertising as an economic force that sustains individual brands, product categories, and society's overall propensity to consume. In his critical assessment of the "new industrial state," economist John Kenneth Galbraith described the role of advertising as "the conditioning of attitudes necessary for the operation and prestige of the industrial system" (Galbraith, 1967:210). Galbraith also spoke to the importance of advertising as a means of supporting primary demand for a product category. "Were there but one manufacturer of automobiles in the United States, it would still be essential that it enter extensively on the management of its demand. Otherwise consumers, exercising the sovereignty that would be inconsistent with the company's planning, might resort to other forms of transportation and other ways of spending their income" (1967:207). Galbraith's assessment of the "management of specific demand" points to the importance of advertising message strategies that can provide information about products and about the value or importance to consumers of the benefits delivered by products. The absence of consumer access to sufficient information of this kind inevitably leads to declining consumer interest. This is the current situation for U.S. military recruiting.

Fairfax Cone, founder of a leading advertising agency, further emphasized the role of advertising with the observation "I don't think anyone would use advertising if he could see all of his prospects face-to-face. It is our belief that advertising is good in ratio to its approximation of personal selling" (Wright and Warner, 1963:468). Cone's observation leads directly to the role of advertising in military recruiting. It is to support military recruiters as they identify prospective applicants and as they inform and assist those individuals who show an interest in military service. Advertising messages can help stimulate the interest of the youth population in military service and provide information bearing on the particular Service a youth might select.

The model in Figure 8-2 indicates that the primary role of advertising in military recruiting is to support recruiting by influencing youth attitudes about military service (as reflected by measures of the propensity to enlist). Indeed, it seems likely that military advertising may be the most audible voice in society conveying information about careers and life in the military. The concept of propensity to enlist applies to military service in general and to the individual military Services. Hence, propensity can be seen, in advertising terms, as a measure of the primary demand for military service when applied to the youth population's overall interest in military service, and selective demand when specifically applied to youth interests in the specific Services.

Amount of Military Advertising

Turning to the amount of military advertising, Figure 8-1 shows the advertising expenditures for four military services and joint services advertising since 1976. The pattern of advertising expenditures shown in the figure suggests three phases: a period of growth leading to a general leveling in the years 1986 to 1990, a decrease of 44 percent from 1990 to 1993, and an increase of 318 percent from 1993 to 2000.

Figure 6-4 showed that propensity to enlist by began a pattern of continuing decline beginning in about 1984. The noticeable decline in advertising expenditures between 1990 and 1993 suggests that, at that time, the budgeting of military advertising may not have been based on considerations relating to the direction of youth propensity to serve in the military.

The reduction in advertising expenditures coincided with a period of drawdown in the scale of the military. Thus, planners may have merely questioned the need to continue to advertise.

The propensity to enlist has not increased despite a 318 percent increase in advertising over the seven-year period from 1993 to 2000. This suggests that the elasticity of advertising may be low; however, a more material question to ask is about the message strategies employed in military advertising and their effectiveness in promoting propensity.[1] One might hypothesize that the recent increases in military advertising reflect increasing competition among the Services to attract youth in the declining positive propensity groups. Rather than focusing on market expansion, or primary demand, it appears military advertising has continued to concentrate on selective demand (individual Service advertising) within the declining positive propensity groups.

Message Strategies in Current Military Advertising[2]

The traditional approach to the selection of a message strategy is to identify a customer need that is important and widely held and then to stress a product or brand attribute that is responsive to that customer need. It is desirable that the selected attribute be unique to the advertised brand (Overholser and Kline, 1975: 82). The message strategies employed in military advertising have substantially focused on what recruiters call the "package" or "offer" made to applicants for military service. Army advertising is particularly noteworthy in this regard because it is the most likely of the military Service advertising to be seen by the youth population. For example, during the five-year period from 1996 to 2000, advertising for the Army constituted 45 percent of all military advertising in the United States. The message strategies in Army advertising have generally focused on how military service can provide youth with job skills, college credits, and money to pursue education following military service (Eighmey, 2000), material considerations for youth making comparisons among specific military Services (selective demand) as well as for youth making choices among military service, civilian employment, and further education (primary demand).

These package-oriented considerations do not involve attributes that are unique to military service or to primary demand. That is, few messages focus on the aspects of military service that are unique, such as the higher level of public service associated with duty to country.

A review of the content of current advertising for the Army indicates the apparent audience to be youth who are dissatisfied with their current circumstances and who would welcome a challenge to become something

[1]Econometric models make no distinctions between dollars and impressions.

[2]Data from the DoD on the effectiveness of individual or joint Service advertising campaigns was not available.

more. Recently produced television and web commercials present basic training as a kind of a quest with stages of accomplishment, such as marksmanship, teamwork, and live fire situations. Although the copywriting and art direction have changed recently, the basic message strategy remains "Be all you can be." "An Army of One" was recently adopted as an advertising campaign idea; however, most recently this idea has become secondary to such claims as "212 ways to become a soldier."

The inclusion of references to web site goarmy.com in Army advertising contributes to the evaluation of this advertising through measures of visits to the Army web site, profiling of the visitors, and number of inquiries turned into leads for recruiters. The strength of the Army advertising appears to be in its appeal to restless youth who want something more. However, that may not be sufficient. Within the message content of Army advertising, there is little that addresses either selective demand, the aspects of the Army that differentiate it from the other Services, or primary demand, the broader issues of the virtues of duty to country, service to others, heroism, and self-sacrifice for freedom and the benefit of others. A review of the current content of advertising for the Air Force indicates the apparent audience to be intrigued with engines, technology, and speed. These are not restless youth; they have self-assurance. Success in basic training is a given. Within this area of interest, the appeal is "Cross into the blue" where "Nothing you do is ordinary" and "Everything is different and important, especially you." The visuals focus on sophisticated planes and equipment.

Like Army advertising messages, the Air Force offers youth a transforming experience. However, the offer of a transformation is pitched at a higher level. Also, the advertising copy "When you cross into the blue, everything is different and important, especially you" indicates the potential for this advertising to attract the interests of youth with no propensity to enlist. An "out of the blue" idea is generally unexpected yet welcome.

A review of the content of current advertising for the Navy indicates the apparent audience to be youth who may perceive their present circumstances to be ordinary and are seeking to escape. Navy advertising responds to this group of youth by showing dramatic action visuals of Navy ships, equipment, and people. The advertising promises that life in the Navy will be anything but dull. The message challenges youth with the question, "If someone wrote a book about your life, would anyone want to read it?" The promise of the advertising is that the Navy will "accelerate your life" and that "you will do more in a few years than most people do in a lifetime, if you are ready." There is an implied promise that a youth's life will rocket forward and that the educational experience in the Navy will substitute for college or graduate school.

A review of copies of television advertising provided to the committee by the DoD in the spring of 2002 for the Marine Corps indicates a specific focus on an audience of youth possessing high propensity for military service. Marine advertising stresses the importance of noble virtues and the value of people who never fail to defend those virtues. The television commercials speak of honor, courage, commitment, and pride. The message emphasizes that there will always be a need for people who can live up to these values. The implied message is "we know you want to serve your country; are you good enough to be with the very best?" Visual elements include the sword, the globe and anchor symbol, and the American flag. The potential for emotional response to this advertising is strong. For example, one current television commercial opens with a view of the earth at its most innocent, from outer space. Views of the earth transform themselves into the globe and anchor symbol in which the anchor can be appreciated as protecting the earth. At the completion of the commercial, the Marine symbol stands alone against the blue background of the uniform. This commercial continues a tradition of exceedingly well-crafted and powerful advertising for the Marines.

The advertising strategies of the Army, the Air Force, and the Navy appear to be somewhat coordinated in their appeal to youth with differing personal circumstances. Army advertising promises the acquisition of a variety (212) of professional directions and provides reassurance about making it through basic training. Air Force advertising promises the opportunity to work with sophisticated aircraft. Navy advertising promises adventure and accelerated education and experience. However, the advertising for none of these Services focus on what could be called the primary demand for service in the military. Only the advertising for the Marines consistently addresses the noble virtues that can be associated with all military service. The capability of Marine advertising to generate interest in military service is reflected in the comment of an Army recruiter who was interviewed as a part of this project. The recruiter spoke of the impact of Marine advertising on the youth he approached and stated that his recruiting problems might be lessened if DoD invested more resources in advertising for the Marines.

Other Advertising Directed to the Youth Population

The youth population is a primary audience for many sellers of products and public information campaigns concerning such issues as the decision to use cigarettes or drugs and sexual activity. The content of product advertising focuses on positive benefits, while public information campaigns sometimes employ "fear appeals" concerning personal risks.

The lessons learned from these areas generally demonstrate the importance of well-selected message strategies, rather than communication styles (such as testimonials or demonstrations, or the use of attention-getting video techniques, popular music, etc.). It is important that advertising have a contemporary look or approach in the eyes of the target market, but the underlying strength of the advertising comes from the use of message strategies based on the differentiating and audience-relevant qualities of the product itself.

For example, the soft drink Mountain Dew continues to develop its market acceptance with imaginative advertising directed to the youth population. But whatever the fresh approach to storytelling employed in the advertising, the underlying message strategy remains focused on a product that delivers a highly energizing combination of product ingredients. An effective message strategy is the key ingredient in successful advertising.

In this volume, our focus is on the selection of effective message strategies rather than on an evaluation of various communication styles.

Youth Perceptions of the Benefits of Military Service

Until very recently the YATS provided a means of examining the relationships between propensity to enlist and youth perceptions of the benefits of military service. Question 526 in the 1999 YATS survey was an open-ended item concerning the main reasons to join the military. Tabulation of the first responses to this question given by each respondent can reveal the top-of-mind reasons for enlisting in the military. This is a common means of detecting whether there is correspondence between the content of advertising campaigns and the product-related perceptions reported by the audience for the advertising.

About 60 percent of the survey respondents gave answers related to the career preparation benefits of the kinds depicted in recent advertising. Pay for education led the responses, with 27.1 percent of the respondents giving that answer. Development of work skills followed at 12.6 percent, pay at 10.9 percent, job security at 2.9 percent, and retirement benefits at 1.9 percent. These responses suggest a means-end or "cognitive outlook" on the benefits of military service. Military recruiters describe this as the "What's in it for me?" viewpoint exhibited by many applicants.

About 25 percent of the respondents gave answers associated with the more immediate benefits realized by military service. Of these, duty to country led the responses with 8.4 percent, travel followed at 5.9 percent, self-esteem and self discipline at 5.6 percent, family tradition at 2.5 percent, and national defense at 1.9 percent. These responses reflect social-normative or "value outlook" on the benefits of military service.

Moreover, duty to country and national defense are benefits that are unique to military service and therefore distinguish military service from other career alternatives.

These survey results suggest the career advancement, or means-end, message that typifies military advertising is top-of-mind for about 60 percent of the youth ages 16 to 24. Nevertheless, despite a relative lack of reinforcement in mass communication, about 25 percent of youth respond with value-oriented answers that are presumably uniquely associated with military service.

Youth Values and Propensity

About 25 different areas of youth values were examined in the 1999 YATS survey. These included such considerations as the importance of money for education, preparation for a career, learning trade-related skills, learning about information technology, duty to country, and personal freedom.

Three of these values sample the areas associated with (1) the package orientation, (2) the product attributes that can be unique to military service, and (3) the barriers to enlistment. The value "career preparation" reflects a primary theme in the content of the military advertising, "duty to country" reflects the particular public service role that defines military service, and "personal freedom" reflects a principal barrier to enlistment.

Figure 8-3 shows the relative importance, among youth in grades 10 to 12, of these three benefits for each of the four levels of propensity commonly used in YATS and other propensity surveys.

Among these three critical benefits, the importance of duty to country shows the greatest difference across the four levels of propensity (0.0001 level of significance). Youth in the four propensity groups see career preparation and personal freedom as similarly important, and the lower importance of duty to country differentiates the youth in the negative propensity groups.

Youth Perceptions of the Benefits of Military Service

For each of the previously mentioned 25 values, YATS survey participants were also asked whether joining the military would provide them with the specific benefit. Figure 8-4 shows how youth in grades 10 to 12 evaluated the extent to which they thought service in the military would provide the benefits of "duty to country," career preparation, and "personal freedom."

Of the three benefits shown in the figure, personal freedom is least associated with military service. Moreover, the negative propensity groups show the lowest endorsement levels for all three of the benefits

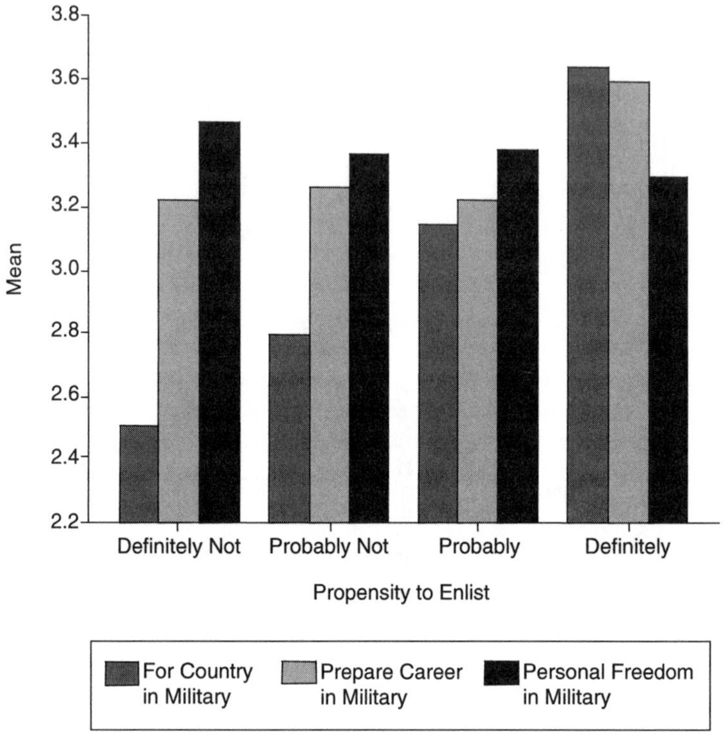

FIGURE 8-3 Importance of three value items by propensity group for youth in grades 10 to 12. The four items were rated as "not important," "somewhat important," "very important," and "extremely important." The range is 1 to 4 with "extremely important" scaled as 4.
SOURCE: Data from the Youth Attitude Tracking Study.

(0.01 for "duty to country," and 0.0001 level of significance for "career preparation" and "personal freedom").

Advertising Message Strategies and the Propensity to Enlist

The patterns shown in Figures 8-3 and 8-4 suggest the roles that the three key values may play as youth consider the decision to enlist in the military. As this report makes clear, there is a need to reinforce the importance of such values as well as their association with military service. Indeed, the discussion in Chapter 7 pointed to a consistent erosion in the extent to which values such as doing something for the country are associated with the military.

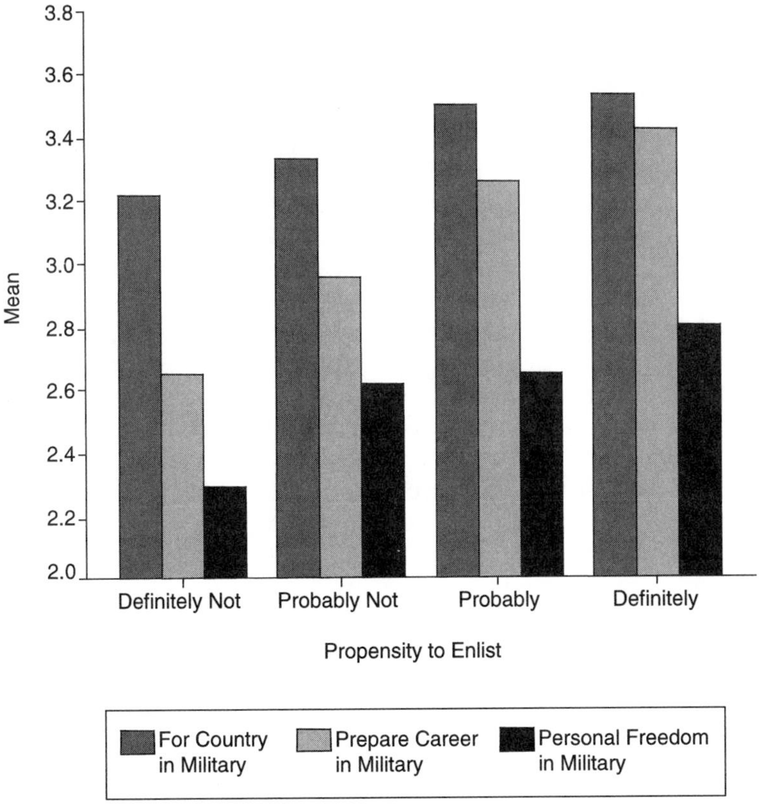

FIGURE 8-4 Likelihood of military service providing three value items by propensity group for youth in grades 10 to 12. The four items were rated as "definitely not," "probably not," "probably," and "definitely." The range is 1 to 4 with "definitely" scaled as 4.
SOURCE: Data from the Youth Attitude Tracking Study.

a. Duty to country is the value seen by high school youth as most associated with military service. But duty to country is also the least important of the three values to those in the negative propensity groups. The lower level of youth support for the value of duty to country among youth with negative propensity suggests there is a need to provide American youth with more information on the importance of duty to country. As mentioned earlier, the noble virtues of military service are emphasized only in Marine advertising, and there are questions about the extent to which these virtues are taught or reinforced elsewhere in society. This situation suggests declining trends in propensity to enlist might be abated by the deployment of advertising messages designed to teach the role of the military and the noble virtues of service to one's country.

b. Personal freedom is a strongly held value among youth in all four propensity groups, and it is least associated with military service. This suggests the possibility of exploring advertising strategies designed to focus on issues related to personal freedom. For example, perhaps a connection can be made between freedoms enjoyed by youth and the willingness of youth to protect the society that makes them free.

c. Career preparation is a strongly held value for all four of the propensity groups, and it is in what might be called the "middle range" in terms of its association with military service. This suggests that the recent military advertising message strategies focused on the career advantages gained through military service may not have been as effective as might be desired. For example, the review of current military advertising indicated that Army, Air Force, and Navy advertising generates leads by focusing on three different types of interests held by youth. This advertising, or portions of these advertising campaigns, might be strengthened by including more specific information on how military experience translates to career success following military service.

THE MILITARY RECRUITING PROCESS

Overall Description

Currently, information concerning military service is largely conveyed to youth in a Service-specific manner. As discussed previously, military advertising has largely utilized Service-specific message strategies that promote the individual Services. The practice of focusing informational programs on the advantages of individual Services, rather than the shared virtues of military service (stimulation of primary demand) originated prior to the initiation of the All-Volunteer Force. This approach ignores current issues regarding the size of the youth population and its interest in military service.

Each Service has well-structured, formal processes in place for recruiting that rely on a sales force of active-duty, uniformed, enlisted personnel to recruit youth who have not previously served in the military.[3]

[3]For purposes of this chapter, we consider only the process leading to an active component, non-prior-service enlistment contract. We are therefore not addressing the process for Reserve or National Guard recruiting. Neither are we concerned with the process for bringing back onto active duty those veterans who have previously left a Service, but who now wish to again join that Service or enlist in another Service.

This recruiting sales force is supported by a management structure that provides policy guidance, advertising and promotion support, training, and other administrative support. Recruiters are distributed across the country according to Service-specific needs. Within a Service, each recruiter has an exclusive geographic zone usually defined in terms of specific high schools and the areas those schools serve. For example, a specific active-component, nonprior-service Army recruiter is assigned responsibility for all youth attending a specific high school and for the geographic area where those youth live. That recruiter receives credit for any student who attends that high school who enlists as a non-prior-service active Army recruit. It is important to note, however, that there is a similar recruiter from every other Service who also has "exclusive" rights to students at that same high school on behalf of their Service, and there may be Reserve, National Guard, or Reserve Officer Training Corps recruiters who also recruit among that high school's students.

With some exceptions, each recruiter has a personal, monthly goal for new enlistment contracts. Goals are based on such factors as market size (e.g., number of high school seniors in the recruiter's zone), the propensity of youth in that zone to enlist in any Service or in the recruiter's Service specifically, and the number of new contracts required by the command for a given month. In some cases, a goal for new enlistment contracts is shared jointly among recruiters assigned to a given office location (or station), and all the recruiters in that office are either successful or not as a team.

Recruiters—with management assistance—develop plans for generating leads. They then follow up leads, with the intention of setting up appointments to discuss service opportunities. Those appointments may be in the recruiting office, in the high school, or in the applicant's home. When meeting potential applicants, recruiters use an interview process to ensure that the applicant is at least potentially qualified (age, education, medical history, history of involvement with law enforcement authorities, etc.) They may also use a short cognitive ability test to estimate whether the applicant will score sufficiently high on the ASVAB to justify further testing. Recruiters schedule the applicant to take the ASVAB at a convenient testing site and to take a physical examination at the nearest Military Entrance Processing Station (MEPS). After successfully passing all the enlistment standards described in Chapter 4, the applicant meets with a counselor at the MEPS and signs the enlistment contract. The enlistment contract specifies when the applicant will enter active-duty service and what guarantees the Service is making regarding job training and assignment, enlistment bonuses, educational benefits, and so forth. Once the contract is signed, the recruiter receives credit for one new contract toward his or her monthly goal.

The applicant spends the time between signing a contract and actually entering on active duty in the Delayed Entry Program (DEP). That time can be a matter of days, or—as in the case of a typical high school senior—until the end of the school year. The recruiter regularly meets with all of his or her applicants who are in the DEP. During this time, the recruiter has the job of ensuring that applicants are committed to entering active duty. The recruiter loses credit for a new contract if an applicant drops out of the DEP. In that case, the recruiter must replace the applicant in order to keep making satisfactory progress toward his or her assigned goals.

There are many variations on this general process, of course, too many to discuss in any detail. For example, recruiters get reassigned; there are procedures in place to ensure that each applicant in the DEP always has an assigned recruiter, even if it is not always the recruiter who first met with the applicant. Some applicants who do not meet all enlistment standards may be eligible for waivers of some standards, and recruiters must be knowledgeable in procedures for handling those waiver requests.

Sources for Leads

Recruiters obtain leads from many sources. Table 8-2 lists some common types of leads and their sources. Some leads are generated without much involvement by the recruiter. For example, if a potential applicant sees a Service magazine ad and calls the appropriate toll-free number (or mails back the postage-paid card), the recruiter responsible for the geographic zone where the respondent lives would be given whatever contact information the respondent made available. The recruiter is then held

TABLE 8-2 Common Types of Leads and Their Sources

Recruiter Generated	National, Service-Specific	National, Joint
Referrals from other applicants	Postage-paid mailback cards from magazines	Career Exploration Program participants
Posters set up in businesses	Calls to 1-800 numbers from magazines, radio, TV, newspaper ads	Purchased high school directories
Displays at high schools	Inquiries from recruiting web sites	National drivers' license lists
Direct responses from local newspaper or radio ads		

accountable for following up with that person to provide further information and to complete the sale, i.e., enlist the individual.

Other leads result directly from the recruiter's own efforts. For example, a recruiter may set up a display in a high school or place an ad in the window of a cooperative local business. Often, recruiters use their current applicants—especially those already in the DEP—to provide leads to their friends and acquaintances.

One crucial source of leads—the names, addresses, and telephone numbers of students in high schools—is often made available as a result of cooperative efforts among recruiters for the various Services who have been assigned responsibility for that school. Similarly, leads result from a school's participation in the DoD Career Exploration Program, a program that provides schools with a free multiple aptitude battery, interest inventory, and interpretive materials.

In recent years, each Service has established a recruiting presence on the Internet. These sites (e.g., <http://www.goarmy.com>, <http://www.airforce.com>, <http://www.navy.com>, <http://www.marines.com>) provide information about Service missions and opportunities and vignettes about specific enlistees. Some sites offer real-time "chats" with knowledgeable recruiting representatives and interactive games or simulations designed to appeal to high school (or younger) youth. The Army reports that visits to the recruiter chat area of its site have increased from 3,600 in January 2000 to 27,500 in January 2002 (Browning, 2002).

Effectiveness of Various Leads Sources

The ratio of conversion of leads (or contacts) to accessions can vary dramatically across sources of leads, across recruiters, and at every step along the way. That is, leads obtained from some sources (e.g., referrals from people already in the DEP) are more likely to result in accessions than leads obtained from other sources (e.g., a list of new applicants for a driver's license). Similarly, some recruiters demonstrate better performance in developing leads, obtaining appointments, convincing applicants to take the ASVAB, and closing the sale.

The Services have been very pleased with the effectiveness of their web sites in generating leads from qualified and interested young people. Because of the interactive nature of the web, the Services can use their web sites to provide information that is appropriate to both the age and the interests of those people who reach the site. By the time a "prospect" reaches the point of asking specifically to talk to a recruiter or to have further information sent directly to him or her, the Service is reasonably sure he or she is at least minimally interested in enlisting. The task then falls to the assigned recruiter to convert that prospect into an applicant and then into a new contract. The Army reports that visitors to their web site have increased by 126 percent, and—more important—leads have increased 76 percent in recent years.

Of the leads generated from the Army's online chat, 10 percent end up being converted into enlistment contracts—far better than the less than 1 percent conversion rate for all sources of leads combined (Browning, 2002).

Process for Determining Individual Goals

As noted above, individual recruiters have specific monthly (and annual) goals for new contracts. In addition, recruiters must replace any applicant who signed a contract and then leaves the DEP. (These "DEP losses" can be a significant factor in a recruiter's achieving his or her goal. Although rates have been declining somewhat over time, they have averaged between 15 and 20 percent in the Army.) The recruiter's success in meeting or exceeding these goals largely defines whether the recruiter is perceived as successful.

Because of large differences in the number of new contracts required and the number of assigned recruiters, goals can vary significantly across services. As shown in Table 8-3, the Army had 6,054 recruiters assigned in FY 2001, with a requirement for 75,800 new recruits. For that same year, the Air Force had 1,650 recruiters authorized, with a requirement for 34,600 new recruits. Thus, at a gross level, the Army required 13 recruits per recruiter while the Air Force required 21 recruits per recruiter. Because not all authorized recruiters are actually assigned at any one time, and not all those who are assigned recruit for non-prior-service new contracts, the actual goal for each recruiter is somewhat higher than this gross number. For example, in FY 2001, the actual average goal per Air Force recruiter was 28 new contracts for the year. Annual goals in the Air Force for new contracts have varied from 45.2 in 1984 to 22.6 in 1990.

Each Service has access to a significant amount of market data to

TABLE 8-3 Recruiting Staff and Service Goals for the Army and the Air Force

FY2001	Army	Air Force
Non-prior service goal	75,800	34,600
Authorized recruiters	6,054	1,650
Gross goal per recruiter	13	21
Average goal per non-prior service production recruiter	*	28
Net contracts per non-prior service production recruiter	*	28.2

assist in the process of assigning goals to individual recruiters (or, in some cases, teams of recruiters). The increasing sophistication of available market data, down to the zip code and block level, allows recruiting management to set goals that, by design, are equally difficult for each subordinate recruiting organization to attain. In addition to the numbers of potentially qualified youth in each recruiting region, recruiting management considers economic and employment factors and past productivity of each region in establishing goals. Each level of recruiting management receives its goals from the next higher level, assigning goals to the next lower level in turn. Eventually, this process results in a recruiter's being given a goal for new contracts for the month.

Incentives for Production

As with most sales forces, military recruiters generally participate in an incentive system to recognize high performers. Unlike most sales forces, however, military recruiting incentive systems do not include financial rewards such as commissions for meeting or exceeding goals. (Recruiting duty itself does bring with it increased compensation in the form of special duty assignment pay. Such pay is not contingent on meeting or exceeding goals; rather, it is paid at a standard rate to all recruiters and based only on the length of time they have been assigned as recruiters.) Instead of monetary incentives, explicit rewards for exceeding goals take the form of plaques, trophies, watches, and other items of nominal value. In addition, military decorations may be awarded based on performance. Military promotions, which determine the recruiter's salary, may be contingent on successful performance.

Office Locations

Military recruiting organizations are national in scope, with recruiters located in communities in every state and territory (and on military installations overseas). Each Service uses market information to allocate its assigned recruiters to office locations, with the intention that every recruiter has sufficient market to be successful. The Army Corps of Engineers generally leases space on behalf of all Services' recruiting efforts. Use of a common lessor and the same market information often leads to recruiting offices in adjoining space at any given location. Because the Services differ dramatically in the numbers of their recruiters, some Services have only one or two recruiters in an office, while another Service might have four or more. Until recently, the costs of leasing office space took precedence over appearance of that space when the leasing agent was finalizing the specific building in which to locate an office. DoD supports the idea of offices that will provide a "contemporary, inviting atmosphere," in high-traffic locations (Sellman, 2001).

Staffing Issues

Recruiter Selection Process

As in many other respects, the Services differ in the manner in which they select and assign personnel to recruiting duty. Some of these differences may be minor administrative issues. For example, Service members in certain career fields may be ineligible for recruiting duty because their Service needs their specific skills elsewhere. Other differences represent more fundamental issues. For example, until very recently, the Air Force relied solely on volunteers to fill its recruiting ranks, whereas the Marine Corps identified all candidates centrally. The Army and the Navy relied on a mix of volunteers and centrally identified candidates for recruiting positions. More recently, the Navy has been moving toward an all-volunteer recruiting force, while the Air Force has had to rely on some non-volunteers to fill its recruiting ranks.

Regardless of whether the Service is using volunteers or not, the recruiter selection process typically involves a review of the candidate's military record and interviews by the candidate's current supervisors and commanders. Health screens and family financial checks may also be part of the process. Some Services are gathering data on various measures purported to provide good predictions of sales success in other contexts. It is not always the case that major weight is given to an evaluation of a candidate's potential for success in a sales occupation when the candidate is finally approved for recruiting duty. Rather, the Services typically rely on the recruiter training process to identify those approved candidates who are not likely to be successful as recruiters.

In response to recruiting difficulties in recent years, the Services are currently investigating alternative methods for selecting recruiters. Borman and colleagues (2001) describe a number of initiatives are being tested that involve the use of ability or interest measures as a predictor of recruiting success. They note promising developments using in-basket (a work sample test in which the candidates respond to the contents of a job incumbent's in-basket) and situational judgment (scenarios used to test the candidate's decisions among multiple courses of action) tests and behavioral job analysis to improve the recruiter selection process. At the same time, they note research initiated in the 1970s—but never implemented operationally—that found reasonable validity predicting military recruiting success for personality constructs (including "working hard," "spontaneity," "leading and showing off," and "leading and influencing") and interest constructs (including "extraverted interests" and "sports interests").

As noted above, studies of military recruiter effectiveness and predictors of recruiter effectiveness have been undertaken many times over the years. McCloy et al. (2001) reviewed literature specifically addressing military recruiter selection and noted encouraging results for to the prediction of recruiting success. Borman et al. (2000), Penney et al. (2000), and Bearden and Fedak (2000) all reviewed the links between sales success and the military recruiting job. All noted past research establishing the validity of numerous measures in the prediction of sales success in general and success as a military recruiter specifically. However, current practices of recruiter selection generally do not incorporate any formal predictor of sales success.

Recruiter Training

Each Service operates an extensive in-residence training program for new recruiters. In the Army, this course lasts for seven weeks and is designed to teach the skills, knowledge, and techniques to succeed as a U.S. Army or Army Reserve recruiter. The course has six major segments:

1. Introduction to recruiting
2. Management
3. Eligibility
4. Prospecting and sales techniques
5. Communication skills
6. Computer skills

The largest segment is devoted to the study of sales techniques and communication skills. Although there is some practice with fellow trainees, there are no simulated exercises involving real-life applicants. In addition to these six segments, the course also covers a number of special topics, including recruiter expectations, computer program awareness, extremist group awareness, physical fitness/stress management, recruiting support command, public speaking, uniform inspection, quality of life/family support programs, MEPS orientation, knowing the competition, ethics, enlistment standards, advertising, and the Internet. After graduating from the in-residence training program, new recruiters then enter an on-the-job training program with their peers, supervisors, or designated training personnel.

Summary of Military Recruiting Issues

The military Services have been recruiting in an all-volunteer force environment for 30 years. In that time, they have developed and refined their recruiting processes into systems that depend upon a network of active-duty, uniformed recruiters in local offices as the sales force for the recruiting effort. The Services have significant market analysis data available to guide their efforts when they select office locations and assign goals to individual recruiters or recruiting teams. They support their recruiters with substantial advertising resources. The combination of local presence plus advertising support provides the recruiting sales force with their leads.

The process for converting a lead into an active-duty accession involves losses at every step along the way. Some of those losses are unavoidable and result from the recruits' inability to meet the physical, mental, and moral enlistment standards required by law or policy. At the same time, recruiters vary in their ability to convert fully qualified leads into accessions. Several factors may affect sales performance:

- Are people chosen for sales positions based on the knowledge, skills, and abilities necessary for success in sales?
- Are people adequately trained on those tasks required for success in sales?
- Do adequate incentives exist to motivate success in sales?

Klopp and Hemenway (2001) note that military recruiting organizations may not be structured as effective sales organizations, while Robershotte and Zalesny (2001) conclude that one high-priority recommendation for future research should be to develop an effective model for selecting and training military recruiters. Similarly, Orvis and Asch (2001) recommend that military recruiting organizations should rethink recruiting management, including the use of incentives and rewards. In considering the military recruiting sales force, it is not clear that these issues have been resolved in a manner that maximizes the probability of recruiting success.

RECRUITING PRACTICES IN PRIVATE INDUSTRY

One way to evaluate the effectiveness of military recruiting activities is to compare them to the recruiting strategies used in private industry for the same target population. Yet in many ways, this comparison is easier said than done. The difficulty of obtaining accurate, factual information in the private sector makes the foundation for the comparison weak. Moreover, the size differences in the number recruited affect what is feasible. Personnel qualifications for military service and congressional restrictions on inducements to join limit military recruiting strategies in ways that are different from the private sector.

Many civilian employers do not publicly discuss their recruiting practices. In some cases, the organization's own practices are not well defined or understood. In other situations, the recruiting strategies are well thought out but are considered a competitive advantage that should not be shared or given away. Much of the published information on recruiting practices is found in the popular press, which frequently omits clear definitions of processes, costs, and effectiveness.

The sheer size of the military's recruitment effort creates problems that are not often found in the civilian sector. For example, many processes are not scalable. An Internet application system that can handle 500,000 hits for information may not work for 5 million hits. Similarly, some recruiting techniques may simply not work. A small firm might use its executives to directly contact promising applicants. The military services would be hard pressed to use their highest ranking officers in that way.

In addition to the size problem, the military Services also experience a geography problem; they must recruit everywhere in the nation and then often move the new employees to distant locations for training and work assignments. Civilian employers often target their recruiting of hourly and nonexempt (from overtime regulations) employees to their existing employment locations, and many simply do not relocate hourly and nonexempt newly hired employees.

The Services must also restrict their recruiting to a specific population. As discussed in Chapter 5, some groups of people (disabled elderly, physically unfit, recovering drug addicts, former criminals, etc.) usually do not meet minimum qualifications for military service. The Services are looking at methods to attract applicants from sources that can be considered nontraditional. Although the military has experimented in reducing the standards for military service, most new recruits still meet stringent cognitive, physical, and moral standards. Furthermore, the Services cannot use inducements that are common in private industry during times of tight labor markets and booming economies.

Opportunities to Improve Recruiting Strategies

The popular press is full of innovative approaches to recruiting in the private sector. These approaches are not unknown to the military services. In fact, the military may actually use them in a more effective manner than private industry.

Radio Broadcasts

According to an article in *Workforce* (RAS Recruitment Systems, 2000), radio broadcasts are effective because they reach people who may not be actively looking for a job and are not reading job advertisements. Moreover, radio advertisements can be targeted to receptive listeners in a language and a medium they understand. The military Services are well aware of the power of radio broadcasts. Recruiting commands use a combination of paid radio and public service announcements to bring their message to potential applicants—ads placed on radio stations popular with the youth market.

Employee Referrals

Another way to reach good candidates is through existing employees. While some companies believe referred candidates are more successful in the application process, others are simply looking for another means of enticing candidates to their company. A number of companies (e.g., Capital One, MasterCard International) report paying the referring employee for successful hires. (see Hays, 1999; Leonard, 1999).

As noted earlier, referrals from current applicants who are already in a recruiter's delayed entry pool constitute a productive source of leads. In addition, the Services also make clear to those members already in the military that they should be on the lookout for prospective applicants. Informal referrals to military service occur frequently and appear to be effective. At the present time, however, there is no explicit incentive—especially no financial incentive—for either current applicants or Service members to provide referrals.

Newspaper Advertising Other Than Want Ads

At least one company (Capital One) does not believe want ads are effective in attracting the kinds of applicants the company needs. Their rationale is that it is unemployed people who read the want ads, and the best candidates are already employed. In contrast, ads in other parts of the newspaper attract a different kind of candidate (Hays, 1999).

Here, too, the military recruiting commands often use print media ads that are not traditional newspaper want ads. The military services have extensive advertising campaigns targeted toward recruitment. In addition to specific recruitment-oriented advertising, there is also general military awareness advertising and publicity about military events. While general military "awareness" advertising is not very prevalent, military publicity is quite common. Publicity can be both positive and negative in orientation, and consequently it can help or hurt recruitment efforts.

Voice Response Units

Voice response units, which allow the individual to call and interact with a computer, have grown in popularity because they allow the candidate to interact with the company at their convenience. They can be programmed for many tasks but often are used to provide information about the company and available jobs, conduct basic screening, and collect contact information (Hays, 1999). There is some anecdotal evidence, however, that in highly competitive labor markets, the impersonal nature of this approach has a negative effect on candidates.

Line Executives

Higher-level employees like line executives are often useful in conveying the importance of the candidate to the company and convincing the candidate to continue in the employment process. Nestlé Food Co. uses line executives for recruitment particularly with key professionals (Micco, 1997). While high-level military officers may not directly recruit new Service members, many of them are highly visible, particularly in times of war or military actions. It may be useful to ensure that these visible military leaders publicly convey the importance of military service to issues like national security and that the message and messenger are communicated to potential recruits.

Experiential Education Programs

Many employers use "experiential education programs" like co-op programs or internships and find them effective. In a cost-benefit analysis of school-to-work programs conducted by the National Employee Leadership Council and the National Alliance of Business, organizations were found to break even or achieve considerable gains with school-to work programs. Despite the costs of program development, start-up, ongoing program administration, recruitment of students, wages, training and supervision and training materials, the other costs associated with recruitment, training and supervision, and turnover were reduced, and higher retention and productivity benefited the organization (Overman, 1999). To a large extent, the military Services have led the way in combining training and employment since military service usually merges continued training in an occupational specialty and paid practice of that skill.

Internet

Like voice response units, the Internet provides an effective recruiting tool for many organizations that can provide information about jobs and the organization, collect candidate information, and do some screening. A tremendous amount has been written in the human resources literature regarding the use of the Internet and its effectiveness for recruiting. According to a survey by the Employee Relocation Council (ERC) conducted between June and August 1999, 89 percent of companies rely on the Internet for help in recruiting new employees. Most used the Internet to post job openings and provide information through the organization's home page, but many also used the Internet to search for resumes to contact (Albus, 1999). Based on the responses of 435 respondents to a survey by the National Association of Colleges and Employers in August 1999, 9 percent of new hires came from the Internet. This number increased to 15 percent of new hires for software developers and manufacturers (National Association of Colleges and Employers, 1999: 1, 5).

The primary appeal of Internet recruiting seems to be its 24-hour availability and cost savings. According to a 1999 research report published by Creative Good, which studied the electronic recruiting experiences of six large companies (Baxter, Cisco, Citibank, Granite Rock, Procter and Gamble, and Trilogy) a company using the Internet saved $2,000 in advertising costs and $6,000 in time spent looking for new hires by reducing the hiring cycle time by 60 days (Creative Good, 2000; NUA Internet Surveys, 1999).

Despite these advantages, the Internet has drawbacks for both the military and civilian businesses. There are a number of frequently mentioned concerns:

1. Limited Applicant Pools. The use of the Internet clearly limits the applicant pool to those who have access to equipment and the skills to navigate it. Moreover, Internet recruiting may be more effective in attracting candidates for some occupations (e.g., technical, computer) than others. Disparate impact may be an unintended consequence, particularly when the definition of the labor pool broadens dramatically with an Internet recruiting strategy. There is no evidence that the Internet taps into a better, more qualified applicant pool. Nor is there evidence that the Internet increases the probability that a candidate will be attracted to the organization or accept a job offer. A Society for Human Resource Management white paper (Overman, 1999) also raises the issue that local and regional candidates, who may be more reluctant to move for hourly, nonexempt positions, are not targeted through the Internet.

2. Negative User Reactions. Job seekers may become frustrated with delays in downloading materials and may not complete their investigations of an employer. Others may progress further but fail to complete online applications because of obstacles in the application process (unfamiliar jargon, unexplained abbreviations, need for browser "plug-in," etc.). Some job seekers prefer the personal attention that is associated with face-to-face recruiting, which is lacking from many Internet recruiting applications.

3. Increased Competition. Because the Internet is an inexpensive medium for communicating employer information, many can participate even when resources are limited. Thus, large employers and the military may find themselves competing through the Internet with smaller firms and search firms for the same individuals. In addition, information distributed through the Internet is usually available to anyone with an interest. Thus, recruiting and employment practices are revealed to competitors.

4. Inaccurate Information. The Internet provides many sources of information about an employer. Typically, only one source is controlled by the employer—its own web site. Thus, a candidate may be exposed to negative or inaccurate information from a wide variety of sources and lack the skills or knowledge to differentiate accurate information from inaccurate information.

Like private employers, the military Services have embraced the Internet as a recruiting tool and use Service members to staff chat rooms and field Internet-generated inquiries.

Improving Efficiencies in the Recruiting Process

While the military Services cannot alter minimal qualifications or use unusual inducements, they have embraced new approaches to recruiting and often could serve as a role model for private organizations facing large hiring demands. Another approach to improving recruiting outcomes and reducing costs in private industry is analyzing the process and identifying opportunities to make the process itself more efficient.

In tight labor markets, some companies find that neither traditional nor nontraditional approaches to attracting qualified candidates work when labor is in short supply. These organizations have tried to improve the recruiting process itself. Three possible approaches are centralization, outsourcing, and record keeping to support analysis of the process itself.

Centralization

Centralization of recruiting involves operating certain recruiting activities from one or a few locations. While certain functions (e.g., talking face to face with an applicant) require distribution of personnel, other functions (e.g., answering questions sent via the Internet) can be concentrated in one area and may increase the promptness and consistency of responses and reduce costs. The popular literature cites several descriptions of centralization. For example, Marriott centralized its recruitment in three locations: Washington DC, Arizona, and Massachusetts. In September 1998, Marriott placed a staff of 10 in downtown Washington DC, to hire 5,000 people to work in 127 hotels, senior living communities, food distribution centers, and airport and highway food service concession services. However, no data indicating the effectiveness of this change or the cost savings involved were provided.

Outsourcing

An alternative to centralizing and consolidating existing recruiting practices is outsourcing parts of the recruiting and selection process. For example, in 2000, the giant employer, Wal-Mart, hired Personic, an Internet software company, to hire employees via the Internet and track applicants through the entire hiring process (<http://www.personic.com/news_articles_05.cfm>). Presumably there are economies of scale when one firm concentrates on a narrow group of tasks, such as recruiting, reference checking, drug testing, etc. Moreover, skilled Service members might be better deployed in places where outside contractors cannot perform their duties. The potential efficiency of outsourcing has not gone unnoticed by the Service recruiting commands or Congress. Using contract recruiters in place of active-duty soldiers or using contractors for telemarketing activities for recruiters are innovations being tested in FY 2002.

Record Keeping to Aid Process Analysis

Another approach to improving recruiting processes is to analyze what methods work best; however, such analysis requires detailed, accurate records. Ideally, the literature would report the costs and the effectiveness of a strategy or recruiting improvement in increasing the number of qualified candidates at various steps in the process. Thus, organizations could choose recruiting strategies carefully and deploy the money saved toward more exposure with existing recruiting techniques, additional recruitment efforts, or research. However, few if any of these analyses have been published by individual companies. Instead, there are data on which strategies are most frequently used and which are rated most important. Analysis of recruiting processes requires complete and accurate data on each step of the recruiting process (Who was contacted? How was the person contacted? How many times was the person contacted? When did the person drop out of the recruiting process? Why? etc.). The military services may be the only organization in the United States capable of such detailed record keeping.

SUMMARY

The propensity among young people to enlist has declined since the mid-1980s, making recruiting more difficult. One way to facilitate recruiting is through advertising, by increasing overall propensity and as a result enlarging the pool of potential recruits. Our analysis suggests that in order to be most effective in increasing overall propensity, advertising messages should focus on the intrinsic and unique benefits of military service, such as patriotism and duty to country, and should be provided jointly across the Services. Service-specific advertising campaigns then could continue to be keyed to immediate recruiting goals and focus on more immediate extrinsic benefits, such as pay, bonuses, and education.

A second way to facilitate recruiting is to improve the recruiting system. The Services have well-structured selection and training programs for recruiters. The recruiter selection systems, however, are optimized for administrative convenience rather than for mission effectiveness. There exists in the recruiting force today huge variability in mission effectiveness. Personnel selection research suggests that marked recruiter performance gains are possible through the design of more rigorous recruiter selection systems. The recruiter training systems provide the fundamentals of successful sales techniques but offer limited opportunities for practice and feedback. Law and policy limit the types of rewards that are available to recruiting management to increase the incentive for effective performance at recruiting duties. There has been little innovation in developing effective recruiter reward systems within existing constraints.

9

Conclusions and Recommendations

There is a large number of hypothesized causes for the military's recruiting difficulties of the past several years, as well as a similar set of hypotheses about potential effects on recruiting effectiveness in the future. Our goal in this report is to identify and examine a wide variety of such potential causal factors. Some factors are potentially within the control of decision makers (e.g., pay and benefits, recruiting practices, advertising messages), and others are not (e.g., changes in the size of the cohort eligible for military service, changes in the skill levels of American youth).

In examining each of these factors, our perspective was forward-looking. Our focus was not on producing a definitive answer to the question of what caused the military recruiting shortfalls of recent years, but rather on identifying factors likely to influence future recruiting effectiveness. There was no expectation that we would uncover a single factor accounting for recruiting effectiveness, given the broad set of individual, situational, organizational, and societal influences on decisions about military service. We did hope, however, to identify a small set of important variables from among a broader array of possible factors affecting recruiting effectiveness.

This report is structured around a set of potential contributing factors. We explored each in turn, and in this chapter we draw a set of conclusions in each domain. First, we examined demand factors, inquiring about possible changes in overall force size and structure, in the aptitude levels needed for effective performance, in the physical demands of military work, in the moral and character requirements of military work, and in the levels of attrition and retention.

Second, we examined the demographic context for armed forces recruitment. The size and composition of the youth population are fundamental constraints on future recruitment efforts. We were able to forecast the size and some aspects of the composition of the youth population for the next 15 to 20 years with considerable accuracy because these persons are already born.

Third, we reviewed the four major domains in which military applicants are screened: aptitudes (indexed by the Armed Forces Qualification Test, AFQT), educational attainment (possession of a high school diploma), physical and medical qualification, and moral character (e.g., lack of a criminal record). In each of these domains, we reviewed the Services' current enlistment requirements and evaluated whether change in these requirements is likely. We then reviewed evidence regarding the current supply of youth possessing these characteristics and consider the likelihood of change over time in the proportion of youth with these characteristics.

Fourth, we examined the three major options available to the youth who make up the prime military recruiting market: joining the military, pursuing higher education, or entering the civilian labor market. We reviewed the changing landscape regarding (1) participation in postsecondary education and opportunities available to youth in the civilian labor market and (2) aspects of these alternatives that compete with the Service options or may be fruitfully combined with them.

Fifth, we examined changes in youth attitudes, values, perceptions, and influencers over the past two decades. We reviewed major findings from extensive long-term longitudinal and cross-sectional research on youth attitudes and on the relationship between youth attitudes and the propensity to enlist. We offered an integrated theory of behavioral choice that can productively guide future research on the determinants of propensity and of actual enlistment.

Sixth, we examined a range of issues involving military advertising, including goals, strategies, and messages. Advertising is a part of the broad recruitment process, and we examined this process more generally, including a comparison with recruiting practices in the civilian labor market.

DEMAND FACTORS

Changes in Force Size

The end of the Cold War resulted in an intentional 38 percent reduction in active-duty military enlisted strength, from 1.85 million in 1987 to 1.15 million in 1999. The drawdown is now complete, with enlisted strength now essentially stable. The question of projected changes for the future now comes to the fore. We acknowledge that this is an area in which unanticipated events can have dramatic effects; such events could radically alter any projections of needed force size. What we can do is examine articulated national security strategy and various planning documents and consider the implications of these for force size.

In light of the September 11, 2001, attacks and subsequent terrorist threats to the United States, it seems unlikely to us that force sizes will be reduced in the near term from their current levels. We also do not see clear evidence of factors that would result in a significant increase in net force size. For example, while it is clear that military technology will become increasingly sophisticated and may affect the knowledge and skill mix needed in the enlisted force, it is not clear that technological change will affect force size per se. We reviewed research on historical changes in the distribution of military personnel across occupational categories over the past 25 years and found very little evidence of change.

Conclusion: Although the events of September 11, 2001, have increased uncertainty regarding future demand, we found no compelling evidence that requirements for numbers of new personnel will change radically in the future. Therefore, for planning purposes, we assume that over the next 20 years military missions and structure will require about the same numbers of new personnel joining the military every year.

Changes in Levels of Required Aptitudes and Other Attributes

We reviewed current military entry requirements and examined the possibility of future changes in these requirements. We also examined research projecting aptitude requirements of future military jobs. The results of this work are mixed, with some studies projecting no significant changes in aptitude requirements, and others projecting a need for somewhat higher aptitude levels, at least in some jobs.

We also compared the aptitude levels (as indexed by AFQT scores) of recent accession cohorts with the civilian population ages 18–23, and with minimum aptitude requirements for military jobs. These analyses indicate that military enlistees compare very favorably to the civilian population and that the Services are accessing individuals well above the minimum requirements for successful performance.

Conclusions

- The Services are currently accessing recruits who have sufficient aptitude and can be trained to perform military tasks adequately. Recruits satisfying current qualification levels will meet future demands.

• There have been few major changes in the occupational distribution of first-term personnel in the past 10 years, but future military missions coupled with advances in technology are expected to require military personnel to make greater use of technology. Technological changes will make some jobs in the future easier and others more difficult, but overall minimum aptitude requirements in some occupational fields are likely to increase somewhat over the next 20 years. Because current qualification levels are so far above the minimum, the increase does not pose a problem. However, timely and responsive changes to training may be required. New systems are especially problematic as schedules slip and funding is used for other priorities.

Recommendation: We urge that the Services resist the notion that recruit aptitude and education targets must continue to be raised, and we recommend that they continuously review their performance requirements and the related training of new recruits to ensure that beginning knowledge gaps are filled when necessary and that unnecessary training is abandoned quickly. To the extent possible, training changes should be anticipatory, especially for new systems.

Changes in Rates of Retention and Reenlistment

The demand for new recruits is also in part a function of Service retention capabilities. Each Service has goals for retention at the level of specific military occupational specialties. Attrition rates that exceed expectations translate into recruitment needs. We note that overall end strength targets may be met while still experiencing shortfalls in key occupational areas.

We reviewed data on attrition rates across the services at 6, 12, 24, and 36 months and observed that attrition rates at each of these points in time have consistently increased over the past 15 years. The degree to which this reflects changes in individual reactions to the military experience or changes in the ease with which individuals seeking to leave are permitted to do so is unclear. Personnel who are dissatisfied with their Service experience return to their home towns spreading word about negative aspects of military service, which makes the job of the recruiter much more difficult.

We also reviewed surveys from each Service on reasons enlisted personnel leave the Service and career intention surveys, which monitor reenlistment plans. In some cases these surveys been developed quite recently and historical data are not available, making the results difficult to interpret. These efforts need to be expanded to develop a better basis for policy decisions.

Conclusion: Retention is important to and has an impact on readiness and recruiting. It therefore is important to better understand what factors influence retention decisions and to continue to monitor them. In addition to traditional Service efforts to minimize attrition and unplanned losses, the messages the military sends to its members also find their way to potential recruits, and vice versa. Also, the messages veterans pass along to friends and family members can either encourage or discourage enlistment.

Recommendation: We recommend that the Services:

- Fully integrate planning, budgeting, and resource allocation for both recruiting and retention, so they are complementary (e.g., applying assets to increase retention should result in reduced recruiting goals).
- Expand the body of knowledge concerning factors influencing retention. A process is needed to provide timely leading indicators that will help decision makers focus efforts and inform decision making in ways that can positively impact retention, especially for critical skill sets and for quality of life issues.

DEMOGRAPHIC TRENDS

Changes in Youth Population Size and Composition

Both size and composition of the youth population have implications for military recruitment. Looking first at size, we found that cohorts of persons reaching age 18 are expected to grow significantly over the next 10 years and then remain approximately at a plateau during the following decade. Approximately 4 million youth reached age 18 in 2000, a number that will increase to approximately 4.5 million by 2008 and trail off only slightly during the subsequent decade.

Turning to composition, we noted that the ethnic composition of the youth population will change significantly over the next 15 to 20 years. Even in the absence of changes in immigration patterns, the ethnic makeup of the youth population will change because of recent changes in the ethnic makeup of women of childbearing age and ethnic differences in fertility rates.

Based on recent fertility patterns, the percentage of young adults who are Hispanic, of whom the largest subgroup is of Mexican origin, will increase substantially. In 2000, approximately 14 percent of 18-year-olds were of Hispanic origin, a fraction that will gradually increase to approximately 22 percent over the next 15 years. This increase, paired with the fact that the high school graduation rate for Hispanics is lower than for other groups, is an important issue given the Services' interest in enlisting a high proportion of high school graduates.

A growing percentage of youth will be raised by parents who are immigrants to the United States, a result of high rates of recent immigration and relatively high fertility levels of foreign-born women. Approximately 11 percent of 18-year-olds in 2000 were born to foreign-born mothers, a fraction that will approximately double during the next 20 years.

Although immigration has been an important component of population growth in the United States as a whole over the past two decades, projections by the U.S. Bureau of the Census do not point to future increases in the proportion of youth who are foreign born over the next two decades. If anything, these estimates point to a slight decline in the proportion of 18-year-olds who will be immigrants between now and 2020.

The socioeconomic characteristics of parents, such as their levels of educational attainment, have a large effect on the aspirations and decisions of youths, especially concerning higher education. Average levels of maternal education for teenagers have increased markedly and will continue to do so over the next two decades, a result of increases over time in educational attainment in the population. This is important, given the positive relationship between maternal the education and the educational aspirations of youth. For young adults reaching age 18 in 2000, approximately 45 percent had mothers who received a high school diploma but went no farther in school, 35 percent had mothers with some education beyond high school, and the remainder dropped out before high school completion. By 2013, slightly less than 35 percent of youths will have mothers who received only a high school diploma and over 45 percent will have mothers who obtained postsecondary schooling. Within the next two decades, the majority of youth will be raised by mothers who have completed at least some college.

Enlistment in the armed forces is affected in part by previous exposure to military life. Traditionally, many youths have obtained this exposure through parents who, at some time in their lives, served in the armed forces. The proportion of young adults who have had one or more parents with military experience has fallen dramatically and will continue to fall in the coming years. Of youths who reached age 18 in 1990, approximately 37 percent were born to families in which one or more parents had served in the armed forces. Of those who reach 18 in 2000, this percentage had dropped to approximately 17 percent. Of those who will reach 18 in

2010, this percentage will drop further, to about 12 percent. This represents a large decline in exposure to military experience within the nuclear family.

Conclusion: Trends in numbers of births and in the composition of the child population have offsetting effects on potential enlistment trends. Although the annual number of births has increased in recent years, children are increasingly raised by highly educated parents and by parents who have no direct experience with the armed forces, factors that are negatively related to interest in military service. The net impact of these offsetting trends is a small increase in expected numbers of potential enlistees in the next decade, although a slight decline is expected in the subsequent decade. Thus, demographic trends do not emerge as factors that will contribute to increasing difficulty in meeting enlistment goals. Other factors discussed in this report, such as advertising and recruitment practices, will determine whether potential enlistees actually enlist at a rate necessary to meet goals.

Changes in Educational Attainment and Postsecondary Enrollment

During the 1990s, rates of college enrollment and levels of education completed increased dramatically as a result of three broad trends: (1) changes over time in parental characteristics, especially parents' educational attainment, which increased youths' resources and aspirations for education; (2) the greater inclusion in higher education of women and some ethnic minorities; and (3) increased economic incentives to attend and complete college, a result of changes in the labor market for college- and non-college-educated workers.

Rates of college enrollment increased in the 1990s for new high school graduates, whether they were employed shortly after high school or not. Enrollment rates increased for both two- and four-year institutions, although the increases were somewhat greater at the latter.

Two-year college enrollment rates are higher for recent Hispanic high school graduates than for their non-Hispanic counterparts. Four-year enrollment rates are much higher for non-Hispanic than for Hispanic youths. Non-Hispanic white youths have traditionally had higher four-year enrollment rates than black youths, but these rates have converged to some degree in recent years. As college enrollments increase, the pool of youth who are both eligible and interested in military service decreases.

Conclusion: The dramatic increase in college enrollment is arguably the single most significant factor affecting the environment in which military recruiting takes place.

TRENDS IN YOUTH QUALIFICATIONS AND ENLISTMENT REQUIREMENTS

Enlistment Requirements

We reviewed current enlistment standards, and the evidentiary basis for these standards. Current Department of Defense (DoD) guidance requires a minimum of 90 percent new accessions to have a high school diploma, a requirement supported by a strong positive relationship between receipt of a diploma and completion of the first term of military service and by cost-benefit analysis. Minimum aptitude requirements (by statute) are that no more than 20 percent fall into AFQT category IV (i.e., between the 10th and 30th percentile of the youth population); DoD guidance recommends a maximum of 4 percent accessions within this category. DoD guidance suggests that at least 60 percent fall into AFQT categories I–IIIA (i.e., in the top 50 percent of the youth population).

The value of aptitude requirements is well documented by research linking AFQT scores to both training and job performance. The 60 percent AFQT I–IIIA goal, however, is not specifically justified as an absolute minimum. Evidence of a relatively small difference in performance between personnel in categories IIIA and IIIB suggests that a modest reduction in the 60 percent goal to include more IIIBs would not have a large effect on performance. The physical, medical, and moral character qualifications appear to be policy based, rather than research based.

Research on projected changes in military work suggests that there may be increases in aptitude requirements for some military jobs. We see no reason to project a change in the value of the high school diploma requirement, given the strong link to attrition and the high costs of attrition. We also see no reason to project a change in the physical, medical, or moral requirements.

Success in Meeting Qualifications Goals

In the education domain, through 1998, the Services were able to meet their recruiting targets while sustaining a rate of accession of high school graduates above 90 percent. The Services have made up the gap in diploma graduates by recruiting mostly persons holding GEDs and other credentials rather than recruiting nongraduates, even though the attrition profiles of such individuals are not that different. This downward trend in success in recruiting high school diploma graduates is especially noteworthy given the population trends in high school graduation rates. High school graduation rates have continued to rise for the past 10 years, so the negative trends in military applicants and accessions appear to reflect a decrease in propensity rather than the supply of qualified youth.

In the aptitude domain, up through 1998 the Services were able to meet the goal of 60 percent highly qualified (i.e., AFQT Category I–IIIA) accessions, while maintaining a very low rate (below 2 percent) of Category IV accessions. The rate of highly qualified accessions is now only a few percentage points below the DoD guideline of 60 percent. As noted above, there is not a strong research basis for 60 percent as the standard, and we suggest that it could be relaxed slightly to substitute somewhat more IIIB accessions for IIIA accessions.

There are limited summary data on military applicants who have various physical or moral characteristics that might make them ineligible for military service. The data are more complete for accessions, for which waivers are given for a variety of conditions. The rate of waivers for moral character (crime, drug use, etc.) was quite high between 1980 and 1991; it then dropped off considerably later in the 1990s. In contrast, the rates of waivers for physical problems (mostly overweight) and other problems (dependents, etc.) have both increased during the 1990s, although their overall rates remain low.

Conclusion: The percentage of highly qualified enlistments has declined somewhat in the past 10 years. This decline, however, has been from a very high rate of highly qualified enlistments (which exceeded DoD targets) to a point for some Services just below these targets. Thus, despite recruiting difficulties and some shortfalls in recent years, the enlisted force remains highly qualified.

Projections About the Supply of Highly Qualified Youth

For the nation as a whole, high school graduation rates have risen. Rates have also risen for minority groups, and especially for black students. For them, graduation rates were around 78 percent during the early 1970s, and by 1999 they had risen to just above 87 percent. Graduation rates still remain low for Hispanic students; they have risen only from about 65 percent to just over 70 percent in the past 20 years. Thus the supply of youth with high school diplomas has improved slightly over the past few decades.

We reviewed longitudinal data on trends in reading, mathematics, and science achievement over 30 years. Scores in all three domains have been stable for the past 10 years. Projections are that there will be neither a sizable increase nor a sizable decrease in the supply of high-aptitude youth.

Conclusion: The potential supply of highly qualified youth in the U.S. population (in terms of both aptitude and education) will remain fairly stable over the next 10 years, and there is no reason to expect any decline over the next 20 years. If anything, the proportion of highly qualified youth may increase slightly, particularly if high school graduation rates continue to rise.

We note that if there were to be a small increase in aptitude levels, a higher proportion of the youth population would score in AFQT categories I–IIIA. If AFQT were renormed to ensure that categories I–IIIA contain 50 percent of the youth population, the result would be that the absolute aptitude level needed to be in the top 50 percent would increase.

> **Recommendation:** If the AFQT is renormed due to rising aptitudes, DoD should consider reducing Category I–IIIA targets to avoid an inadvertent reduction in the supply of Category I–IIIA youth.

Our review covered data on the rates of various disqualifying physical, medical, and moral factors in the youth population, documenting changes in drug use, obesity, and asthma rates. These characteristics require waivers for accession. There are two possibilities: one is that there will be an increase in the number of waivers requested and granted; the other is that a higher proportion of the youth population will be ineligible for military service.

Conclusion: Based on recent population trends, there may be further increases in certain population characteristics over the next 20 years that require waivers for accession, including both moral behavior and health conditions. Drug use, obesity, and asthma are among the most pervasive and serious of these characteristics.

> **Recommendation:** DoD should initiate cost-performance trade-off studies regarding the physical, medical, and moral standards (and waivers for such) in order to develop more specific guidance for minimum standards in this area.

TRENDS IN YOUTH OPPORTUNITIES

Joining the Military

The U.S. military is similar to civilian occupations in that it competes for the youth population based on compensation, benefits, and training and educational opportunities. We presented an overview of the conditions of military service, emphasizing similarities and differences between the military experience and the education and employment experiences that constitute the major alternative courses of action for youth following high school. We reviewed the military compensation system and contrasted it with the civilian compensation system. The compensation system, consisting of pay, allowances, special and incentive pays, and retirement, is complex. The U.S. military continues to offer many in-kind benefits, such as housing and subsistence, not typically found in civilian employers' benefit structures. Moreover, military compensation is based on needs (e.g., pay depends on marital or dependent status), which is not the case in the civilian sector.

Conclusion: The compensation and benefit structure in the military typically provides less flexibility and choice than is found in the private sector.

> **Recommendation:** We recommend that the compensation and benefits structure, including the 20-year vesting of military retirement, should be reviewed with the purpose of making the Services more attractive in today's labor market.

We reviewed research on the competitiveness of military compensation and benefits. Given the military goal of 60 percent of accessions being high school graduates in the upper half of the population aptitude distribution, the key target group for military recruitment consists of individuals with characteristics that also make them eligible for college. Research indicates that regular military compensation (an index of pay and housing and subsistence allowances) is less than what similarly qualified individuals with some college could earn in the civilian labor market, and markedly less than the earnings of college graduates. Even more crucially, research indicates a growing gap in the lifetime earnings of college graduates versus nongraduates.

The most important research we examined on the effects of recruiting resources on enlistments comes from the econometric literature on military recruiting. The effect of recruiting resources on enlistments is summarized in a measure called an elasticity, which indicates the percentage increase in recruits one can expect when a particular recruiting resource increases by 10 percent. If the elasticity of enlistments with respect to recruiters is 0.5, for example, a 10 percent increase in recruiters would result in a 5 percent increase in enlistments. As these recruiting resources differ in costs, cost data and elasticity data are combined to produce estimates of the marginal costs of one additional highly qualified recruit. The marginal cost estimates are roughly similar for recruiters, educational benefits, and enlistment bonuses. This suggests that the Services are currently using an efficient mix of these resources.

This econometric research does not suggest that the elasticity of the factors studied has changed (a possible exception is evidence of reduced effectiveness of the Army College Fund benefits). In other words, the authors of the research do not find that advertising has become less effective or that recruiters have become less effective. The increase in the rate of college attendance accounts for some of the decline in recruiting highly qualified youth, but a sizable amount of the decline in enlistments is not explainable by the traditional factors examined in this research. Note, though, that the most current research focuses on the period 1987–1997 and thus covers only a portion of the post-drawdawn period. Advertising expenditures have increased dramatically during this period, in which a number of the services have experienced difficulty meeting recruiting goals. Thus there is no clear evidence as to the elasticity of such factors as advertising in the post-drawdown era.

The effect of an individual resource or factor on recruiting cannot be evaluated without accounting for all other factors affecting recruiting. Hence, one cannot necessarily infer that if advertising expenditures increased over a particular period but recruiting or propensity declined over the period, that advertising has no, or even a negative, effect. Other factors could have dominated the change in the outcome variable of interest. Although there have been relatively few careful, multivariate studies of the effects of various factors and resources on recruiting over the 1990s, the few that have been conducted suggest that most resources, including advertising, have remained effective. Econometric researchers find that measured elasticities of advertising in the Army in the 1990s (through 1997) are consistent with earlier estimates. In some of the other Services, most notably the Marine Corps, the results have not been as clear. But this is probably because advertising expenditures in the Marine Corps have been smaller, less variable, and therefore more difficult to measure precisely; the small Marine Corps mission makes it more difficult to distinguish between demand-constrained results and the true enlistment supply curve.

Conclusion: A number of resources appear to be viable mechanisms for increasing the numbers of enlistments of highly qualified youth.

> **Recommendation:** We recommend that the Services and DoD periodically evaluate the effects of increased investment in recruiters, educational benefits, enlistment bonuses, and advertising as well as the most efficient mix of these resources. We consider increasing compensation across the board the least cost-effective mechanism for bringing in new recruits or increasing retention.

Employment as an Alternative

We reviewed opportunities available to youth in the civilian labor market. A large proportion (approximately two-thirds) of 18-year-olds choose to continue their education within 12 months of graduating from high school. Large portions of college students, probably the majority, hold part-time jobs while they are attending school. About 25 percent enter the full-time civilian labor market immediately upon high school graduation. Over time, there has been an increase in the proportion of youth who choose full-time participation in school as opposed to those who choose full-time work. Most students today are combining school and work compared with 30 years ago.

We attempted a comparison of military and civilian work with regard to recruitment methods, entry standards, compensation, benefits, and training. The dominant message is that while these factors are relatively standardized and readily describable in the military context, they vary widely across jobs and organizations in the civilian sector and more particularly in the private sector. Thus when comparing military service to private sector employment, one cannot reach an overall conclusion that one option is superior to the other in all aspects. Moreover, it is extremely difficult based on the available data to conclude that the military service or private sector employment is more attractive on such critical factors as compensation, benefits, and training. Rather, comparisons would have to be made by an individual between specific opportunities: e.g., service in the Army in a specific occupational specialty compared with work in a particular company and a particular job.

In the area of compensation, one of the central questions is the extent to which compensation deters (or aids) enlistments. However, compensation is one area in which comparing military salaries to civilian salaries is difficult because of the problems in interpreting the results. Averages across job levels and companies result in aggregations that may not represent any situation realistically, due to the wide variation in civilian pay practices. Another problem with comparing wage data is that salaries are only one component of the total compensation package. Despite the difficulty of comparisons of like information, beliefs exist that pay levels in the private sector are higher for comparable jobs.

Conclusion: There is a high degree of variability in private sector employment opportunities, in terms of entry standards, training provided, compensation, and benefits. Benefits are a significant part of the total compensation package both for workers in industry and for military enlistees. While job-by-job comparisons are difficult, average military compensation is, in fact, higher for high school graduates (without college) than average private sector pay for individuals with comparable experience. However, average pay for college graduates is substantially higher in the private sector.

Recommendation: For aspects of military service that are more attractive than private sector employment, the Services should broadly promote that information. A few examples: (a) vacation time in the military that is markedly greater than in most private sector jobs, (b) military pay that is competitive with private sector pay for high school graduates, (c) the job security provided by a contract for a tour of duty, and (d) the intrinsic rewards associated with service to country.

Education as an Alternative

About 95 percent of high school seniors expect to go to college, and 69 percent expect to earn at least a bachelor's degree. Increasing numbers of them are taking advantage of an array of available opportunities to obtain postsecondary credentials. Roughly two-thirds of American youth participate in some form of postsecondary education by their late 20s. In 1999, about two-thirds of high school graduates enrolled in a 2- or 4-year institution in the same year they graduated from high school. All of these indices—college plans, immediate enrollment after high school degree completion, and eventual participation in postsecondary education—show a long-term and consistent upward trend over the past several decades through the mid-1990s. While college aspirations and enrollment are increasing, it is also the case that a sizable proportion leave school for some period of time and subsequently return (stopout), and others leave and do not return (dropout).

While enrollment in community colleges has remained stable for new high school graduates (about 10 percent), the rate of community college participation has grown among older students. Many are pursuing specific knowledge and skills rather than a degree. For most pursuing a 2-year degree, the degree is an end in itself rather than a step toward a 4-year degree; about 23 percent transfer to a 4-year college.

If military recruitment were limited to the non-college-bound, this great reduction in the target population would be exceedingly problematic. In fact, however, in recent years the majority of high school senior males with high military propensity have also planned to complete four years of college. Nevertheless, it is also the case that average levels of military propensity are lower among the college-bound than among others, so the rise in college aspirations has added to recruiting difficulties.

We reviewed a range of programs currently available for obtaining college credit and pursuing an associate's or bachelor's degree while in the Service. Increasing numbers of two- and four-year colleges are offering Internet programs leading to a degree. Distance learning programs offered by postsecondary institutions may be an avenue for overcoming some of the barriers to obtaining a postsecondary education while in the Service (e.g., sea duty for Navy men and women).

When considering the value of educational incentives to enlistment, it should be kept in mind that the proportions of young people potentially attracted by college incentives are larger than ever before, and that they already constitute the majority of all new recruits in recent years.

Cost issues are particularly relevant to the higher education opportunities of lower-income youth. We note that there has been significant growth in the availability of merit-based financial aid that benefits middle-class students more than lower-income students. The proportion of need-based financial aid has not kept pace with the rising cost of college.

The rising expectations of students and parents regarding participation in higher education and the continued increase in postsecondary enrollment indicate that opportunities for higher education are a focal issue for the vast majority of youth.

That some of those aspiring to higher education do not pursue it immediately upon high school graduation suggests some combination of delaying due to uncertainty about career goals and barriers to access, such as financial need. The military has long been responsive in many ways to these concerns through various programs that provide support for education prior to, during, or after completion of a tour of duty. But increased parental and societal pressures for pursuing higher education suggest that increased opportunities for the simultaneous pairing of military service and higher education could enhance recruiting effectiveness.

Conclusion: If highly qualified youth can be drawn into the military, it will increasingly be through a concurrent military service-education combination. To make the military more attractive to the college-bound population, there should be realistic paths to degree completion. Particularly useful are programs such as college first (before entry to the military) or the opportunity to combine the experience of higher education and military service so that they can be accomplished simultaneously.

Recommendation: We recommend that:

- Recruiting efforts focusing on college aspirants should be continued and perhaps expanded.
- DoD investigate mechanisms for cost-effective recruiting of the college stopout/dropout market.
- DoD continue to link Service programs with existing postsecondary institutions offering distance degree programs.

YOUTH ATTITUDES, VALUES, PERCEPTIONS, AND INFLUENCERS

Youth Attitudes and Values

Youth analysts are increasingly speaking of a new phase in the life course between adolescence and adulthood, an elongated phase of semiautonomy, variously called postsecondary adolescence, youth, or emerging adulthood. During this time young people are relatively free from adult responsibilities and able to explore diverse career and life options.

Conclusion: The period during which youth make career and life decisions has been significantly prolonged, with many young people in their mid-20s still undecided about their life goals and plans.

> **Recommendation:** The military should investigate mechanisms for cost-effective recruiting of individuals who are somewhat older than the traditional target of high school seniors.

The best data on youth attitudes over an extended period of time comes from the Monitoring the Future project, which has carefully examined representative samples of youth from the mid-1970s to the present. First, with few exceptions, ratings of the importance of various life goals show a high degree of stability over two decades: the rank ordering of the goals is virtually unchanged. "Finding purpose and meaning in my life" tops the list of important life goals among young men and (even more so) young women. Second, ratings on the importance of a set of 24 job characteristics are also very similar over two decades. One area of change is that "having more than two weeks' vacation" has risen in importance. Third, views about the importance of work in young people's lives have been largely stable. There has been a slow and modest decline in the proportion considering work as a central part of life (roughly 0.5 percent per year. We also note an important change emerging from an analysis of data from the Youth Attitudes Tracking Study: namely, a considerable decrease in the value attached to duty to country.

Conclusion: In general, the past two decades have not seen dramatic changes in youth ratings of important goals in life, preferred job characteristics, preferred work settings, and views of military service. One exception is a steady decline in the importance placed on "doing something for my country." Answers to recruiting problems are not likely to be found by seeking to discover and capitalize on characteristics unique to a new generation. There is no evidence for abrupt generational shifts in values.

Recommendation: We recommend that recruiting strategies be based on long-term research rather than assumptions of generational shifts presented in the popular literature.

Civic Duty and Volunteerism

We reviewed research on issues of civic duty and volunteerism, with a focus on the links between these issues and military service. Despite concerns about declining social capital and civic society, many young Americans are involved in community service activities, both during and in the years immediately after high school. Participation in community service, volunteering, and political activities in adolescence and early adulthood have lasting consequences with regard to fostering civic concerns, community service, and various other forms of participation in society. In view of the increasingly global world, it is noteworthy that the activities that are engaging the interest and altruism of young people and inspiring the more politically active among them are often not national in scope. They are oriented to local needs, as well as to extranational or universal human rights and other objectives.

Conclusion: There is evidence of continuing involvement in community service and volunteering among American youth. Since "finding purpose and meaning in my life" tops the list of important life goals, and since military service can offer such purpose and meaning, recruiting strategies should capitalize on this aspect of military service.

Youth Influencers

Two questions framed our inquiry: Who influences youth propensity to enlist in military service, and how do young people incorporate those influences into their career plans and decisions? Across the studies we reviewed, the single most compelling observation is that parents (and particularly mothers) have a critical influence on their sons' and daughters' career aspirations and achievements. Others to whom youth turn include peers, other adults, and counselors.

Closer examination of the empirical data suggests that parents, peers, counselors, and recruiters may influence youth career decisions. Formulating direction, a defined occupational career objective or trajectory, requires cognitively processing information, but acting in the direction of a particular career also requires motivation.

Parents are uniquely positioned to provide encouragement, affirmation, and legitimization of a young person's aspirations and career decisions. Mothers have been the family voice on relationship issues. The implication for military recruitment is that efforts to enlist parental support hold the potential for yielding enlistment dividends.

Moreover, the fact that mothers are influential suggests that their ideological perspective to military service ought not be ignored in designing effective recruitment messages. Accurately ascertaining contemporary women's ideological stance toward the military and fashioning recruitment messages that appeal to rather than counter their perspectives may hold additional promise for more effective recruiting.

Conclusion: Parents, peers, school counselors, and recruiters provide information and support that can influence youth career decisions. School counselors and recruiters provide information that influences youth decisions, while parents and peers provide support in the decision-making process. Parents have a critical influence on their sons' and daughters' career aspirations and achievements; mothers are extraordinarily important influencers in the youth career decision-making process. Military recruitment effectiveness may be improved by increasing and targeting recruitment information specifically designed for parents, with particular attention to those designed for mothers.

PERSPECTIVES ON INTENTIONS AND INFLUENCE PROCESSES

According to behavioral theory, there are three immediate antecedents to any behavior—in this case, military enlistment: the intention to perform the behavior (propensity), the possession of the skills needed to perform the behavior, and environmental factors influencing its performance. This suggests two diverging lines of inquiry for examining military enlistment. First, among those with the propensity to join the military, research can focus on skill deficiency and environmental factors that may interfere with an individual's acting on the propensity. Second, for those who are qualified but have no propensity to enlist, research can focus on identifying and intervening to change the factors that determine propensity.

According to behavioral theory, only a limited number of variables need to be considered in order to understand the formation of any given intention. There is considerable empirical evidence that most of the variance in an intention to perform any given behavior can be accounted for by: (a) the attitude toward performing the behavior, (b) the norms governing performance of the behavior, or (c) a sense of self-efficacy with respect to performing the behavior in question (i.e., a subjective belief that one can perform the behavior successfully.) Moreover, each of these three psychosocial variables is, in turn, a function of underlying beliefs: about the consequences of performing the behavior; about the behavioral expectations and behaviors of specific significant others; and about the barriers and facilitators of behavioral performance. All other variables are assumed to have, at best, an indirect effect on intention by influencing one or more of these underlying beliefs.

With few exceptions, theories of behavior have not been applied to predicting and understanding the propensity to enlist. Even when analyses are conducted at the individual level, the critical theoretical determinants of intentions and behavior have rarely been assessed or have been assessed inappropriately. Instead, individual-level analyses have attempted to predict propensity from a large array of demographic, personality, and psychosocial variables (e.g., attitudes toward institutions, values) that, at best, may have indirect effects on the propensity to enlist.

Most attempts to predict propensity have occurred at the aggregate level, examining the relation between the proportion of people with a propensity to enlist (or who have actually enlisted) and a large array of demographic, economic, and psychosocial variables (e.g., percent unemployed, civilian/military pay differentials, educational benefits offered, percentage of the population holding a given belief, attitude, or value) over time. Such aggregate-level analysis can disguise important effects at the individual level. For purposes of designing interventions to increase the proportion of the population with a propensity to enlist at any given point in time or to increase the likelihood that those with a propensity will, in fact, enlist, individual-level analyses that identify the critical determinants of propensity are critical. These types of analyses have not been done. Thus, the most relevant data for guiding the development of effective messages to increase propensity are currently not available.

Conclusion: Empirically supported theories of behavioral prediction can help explain why some people do and others do not form intentions to join the military. They can also help to explain why some people do and others do not act on those intentions. Effective research to examine the determinants of enlistment decisions and of the propensity to enlist requires data conducive to individual-level as well as aggregate-level analysis.

Recommendation: We recommend that:

- Advertising campaigns and other messages to increase propensity should be based on sound empirical evidence that identifies the beliefs to be targeted.

• Ongoing surveys to assess the critical determinants of propensity should be conducted on a regular basis. These surveys should allow for individual-level analyses.

We undertook some secondary analysis of existing data from the Youth Attitude Tracking Study, casting data from that study in a behavioral theory framework. Results suggest that intrinsic incentives (e.g., duty to country, ability to stay close to one's family, equal opportunity for women and minorities) may be at least as important, if not more important, than extrinsic incentives (e.g., pay, money for education) as determinants of propensity. With the exception of advertising by the Marine Corps, most advertising has focused on extrinsic incentives, paying little attention to intrinsic incentives.

Conclusion: There has been a steady decline in the perception that many valued job attributes will be obtained from the military and a corresponding increase in the perception that they will be obtained from civilian employment or equally from both. The erosion in beliefs that valued outcomes are more likely to be obtained in the military than in civilian jobs is a factor contributing to the decline in propensity.

Recommendation: In order to increase propensity to enlist, the military should develop strategies to stop the erosion in beliefs about the values it can provide and to reclaim "ownership" of certain valued attributes.

• One way to increase the pool of youth with a propensity to enlist is to increase the importance young adults place on patriotic values, such as "doing something for my country" and "self-sacrifice," as well as on the importance they place on the "opportunity for adventure."

• A second way to increase propensity is to strengthen beliefs that certain valued attributes are more likely to be obtained in the military than in civilian jobs. In particular, attention should be focused on issues of patriotism, opportunities for adventure, and extrinsic motivations, including pay, vacations, and parental support and approval.

RECRUITING, RETENTION, AND ADVERTISING

Advertising

We framed our treatment of advertising in terms of our discussion of behavioral theory, with its focus on propensity to enlist as a key variable. An analysis of the relationship between propensity and enlistment yielded some interesting observations. While the proportion of youth with a propensity toward the military has decreased, those with a strong positive propensity are highly likely to actually enlist. One possible role for advertising is to help reinforce the current level of propensity among those already highly disposed to enlist. Another is for differentiation among the Services, as they compete for individuals with positive propensity for military service in general.

However, given current enlistment goals and the current level of the propensity to enlist, the military cannot meet its enlistment goals by directing recruiting efforts only toward youth with a positive propensity. Indeed, a very sizable proportion of military enlistments (46 percent) now come from individuals with prior negative propensity. Thus another role for advertising is to provide information concerning the role that military service plays in protecting and furthering the goals of society. If successful, this could serve the purpose of increasing the number of youth with a taste for military service.

An analysis of propensity data indicates that while the proportion of youth with a negative propensity has remained reasonably stable over the past two decades, there have been dramatic changes in the two subcategories making up the negative propensity group. The proportion indicating they will "probably not" enlist has decreased, while the proportion indicating "definitely not" has increased. This suggests an opportunity for advertising that would provide "taste-defining" information concerning such issues as duty to country, public service, and the noble virtues associated with military service.

A central insight is that if advertising affects enlistment decisions, it does so by first affecting propensity to enlist. Advertising can help maintain or increase propensity levels in the population of interest; other recruitment activities determine whether or not propensity is translated into an enlistment decision. Thus a key and underutilized role of military advertising is to support the overall propensity to enlist in the youth population and maintain propensity at a level that will enable greater productivity in military recruiting. It follows from this that an important part of evaluating the effectiveness of advertising is to monitor its effects on propensity.

A finding, for example, that a particular advertising campaign had no effect on enlistments leaves open two possibilities: the campaign had no effect on propensity, or the campaign did increase propensity but some other factors kept that change in propensity from being converted into increased enlistments. Being able to differentiate between these two possibilities should be of considerable value to the Services. Evaluations of the effects of advertising on propensity need to be done with care, avoiding such errors as measuring propensity immediately after exposure to an advertising message, when the interest is, in fact, in lasting effects on propensity.

Conclusions:

• Propensity for military service has declined, with the most dramatic change being an increase in the proportion of youth responding "definitely not." Advertising can be aimed at the potentially complementary goals of increasing yield among those with a positive propensity and changing propensity among those with a negative one. The decline in the proportion of youth with a positive propensity suggests that the military cannot rely solely on attempting to increase yield in this market but must also devote efforts to changing propensity among those with a negative one.

• When thinking about military service, intrinsic factors are foremost in the minds of only about a quarter of the youth population. Youth who place low importance on duty to country tend to have low propensity for military service.

Recommendation: We recommend that:

- A key objective of the Office of the Secretary of Defense advertising should be to increase the propensity of the youth population to enlist.
- Advertising strategies should increase the weight given to the intrinsic benefits of military service.
- The evaluation of advertising message strategies should include the monitoring of their influence on the propensity to enlist as well as trying to isolate the influence on actual enlistments. Evaluation strategies should also be based on regular surveys that track youth responses on a range of beliefs and values related to military service and the propensity to enlist.

In the final analysis, the success of an advertising effort is largely constrained by the nature and quality of the product being promoted. In the case of service in an all-volunteer military as a "product" being sold to a prospective enlistee, an essential aspect is the current and prospective military mission. Ultimately, it is the responsibility of national leadership to determine the military's mission and to articulate that mission and the value of military service in ways that make it attractive to the public—including mothers, fathers, and prospective recruits.

Recruiting in the Civilian Labor Market

Paralleling our investigation into pay, benefits, and training in the civilian labor market, our investigation into recruiting practices also revealed a high degree of variability. This variability makes it is difficult to make generalizations about recruiting practices and inducements.

One feature of the civilian market is the potential for rapid change in recruiting approaches and inducements. The need for unusual inducements appears to be closely related to the supply of qualified candidates. When the supply is tight, the effort to identify and to use perquisites that are valued by candidates increases. Often an organization's recruiting effectiveness is altered because of a decrease in the unemployment rate, and improved results may occur without any real changes or improvements in what is offered to candidates.

The civilian sector has a variety of options at its disposal, and recruiting activities can be tailored to the candidates being recruited. For example, many of the monetary benefits (e.g., thrift plans with matching funds or stock options) that the private sector can offer are simply not available to the military Services without congressional approval. Other strategies, like attracting nontraditional candidates, are not options for the military because of existing rules about who may enlist.

One area in which military appears to compete favorably is the strategies used to reach candidates. The military Services use the same strategies (e.g., the Internet) for finding qualified candidates that civilian businesses employ. In some cases, national military advertising for new recruits penetrates a larger portion of the candidate population than is reached by any employer in the private sector.

Useful recommendations based on the above observations are difficult to make. Suggesting that the military postpone extensive recruiting until the unemployment rate falls does not take care of the need to have a continuous supply of new recruits. Similarly, the military Services, like other government jobs, will not generate any form of stock options in the foreseeable future. Nor are the Services likely to offer a nonstandardized array of potential inducements, even if congressional approval were forthcoming.

Conclusions:

- Recruitment practices and inducements in the civilian sector are characterized by high across-firm variability and by the rapid tailoring of practices and inducements to meet immediate needs.
- The Services are currently using a wide range of feasible recruiting strategies. In some instances the military approach (e.g., Internet recruiting) is more extensive than similar practices in the private sector.

Military Recruitment

The Services have well-structured selection and training programs for recruiters. The Services' recruiter selection systems, however, are optimized for administrative convenience rather than for mission effectiveness. There exists in the recruiting force today huge variability in mission effectiveness. Personnel selection research suggests that marked recruiter performance gains are possible through the design of more rigorous recruiter selection systems.

The Services' recruiter training systems provide the fundamentals of successful sales techniques but offer limited opportunities for practice and feedback. Law and policy limit the types of rewards that are available to recruiting management to increase the incentive for effective performance at recruiting duties. There has been little innovation in developing effective recruiter reward systems within existing constraints.

Conclusion: Improved recruiter selection, training, and reward processes have the potential to dramatically increase recruiter productivity

Recommendation: The Services should:

- Develop and implement recruiter selection systems that are based on maximizing mission effectiveness.
- Develop and implement training systems that make maximum use of realistic practice and feedback.
- Explore innovative incentives to reward effective recruiting performance.

RECOMMENDATIONS

This volume examines a wide range of factors hypothesized as potentially contributing to recent or future military recruiting difficulties. Our hope is to identify a smaller set of important factors from among this broader array.

Several factors we examined did not emerge as likely contributors to current or future recruiting difficulties. First, aptitude and education requirements of military occupations are not likely to increase greatly in the aggregate. The type of individual the military has been recruiting will continue to be able to meet the training and job demands of military work in the foreseeable future. Second, there is no evidence that aptitude levels in the youth population have been decreasing or will decrease in the foreseeable future; thus there will be an adequate supply of youth with the aptitude to meet the requirements of military work. Third, demographic trends do not suggest an inability to meet recruiting demands. A decrease in the proportion of youth with characteristics that make them likely to have a positive propensity toward military service is offset by an increase in the overall size of the youth population; as a result, the number of youth with a positive propensity is expected to remain stable. Fourth, youth values and attitudes have remained quite stable in most domains, suggesting that, with key exceptions noted below, it is not the case that changes in these factors constitute a new impediment to recruiting effectiveness. Fifth, it is not the case that the civilian labor market has become an increasingly more attractive option. Enlisted military compensation exceeds civilian earnings for those with a high school education.

Two classes of factors appear linked to recruiting outcomes. The first class involves "doing more," meaning investing more resources in traditional recruiting activities. The second class involves "doing differently," meaning engaging in new recruiting activities or modifying the way traditional activities are carried out.

In terms of doing more, research indicates that recruiting success is responsive to additional expenditures in the domains of the number of recruiters, dollars spent on advertising, size of enlistment bonuses, dollars spent on funding subsequent education, and pay. The marginal cost of increasing recruiting effectiveness via pay is markedly higher than that for the other options.

In terms of doing differently, we make several recommendations. First, in the important domain of education, perhaps the most dramatic attitudinal and behavioral change over the past several decades is the substantial increase in educational aspirations and college attendance. We suggest that increasing mechanisms for permitting military service and pursuit of a college degree to occur simultaneously are central to recruiting success in light of the higher education aspirations of youth and their parents. Also, in view of the numbers of college dropouts and stopouts and the numbers of youth delaying the traditional activities marking the transition from adolescence to adulthood (e.g., career choice, mate choice), we suggest increasing attention to individuals who are somewhat older than the traditional target of high school seniors.

Second, in the domain of advertising, we suggest attention to three key issues. One is the balance between a focus on military service as a whole and Service-specific advertising. Advertising theory and research suggest the value of supporting overall propensity for military service in addition to Service-specific advertising. Another is a balance between a focus on the extrinsic rewards of military service (e.g., funds for college) and intrinsic rewards, including duty to country and achieving purpose and meaning in a career. While many youth are responsive to an extrinsic focus, an additional segment of the youth population sees intrinsic factors as the primary appeal of military service. A final issue is the role of parents in the enlistment decisions of their sons and daughters. Their key role suggests that attention be paid to the effects of advertising on parental perceptions of military service.

Third, in the domain of recruiting practices, we suggest that attention be paid to the selection and training of recruiters. There are substantial differences in recruiter performance, yet the process of staffing the recruiting services does not focus centrally on selecting individuals on the basis of expected productivity. We also suggest exploring options for rewarding and enhancing recruiter performance.

Recruiting is a complex process, and there is no single route to success in achieving recruiting goals. Nonetheless, we believe that progress has been made toward a better understanding of the challenge, and that useful avenues for exploration have been identified.

References

Ajzen, I. (1985). From intentions to actions: A theory of planned behavior. In J. Kuhl and J. Bechmann (Eds.), *Action control: From cognition to behavior* (pp. 11–39). New York: Springer-Verlag.

Ajzen, I. (1991). The theory of planned behavior. *Organizational Behavior and Human Decision Processes, 50,* 179–211.

Ajzen, I., and Fishbein, M. (1980). *Understanding attitudes and predicting social behavior.* Englewood Cliffs, NJ: Prentice-Hall.

Ajzen, I., and Madden, T.J. (1986). Prediction of goal-directed behavior: Attitudes, intentions, and perceived behavioral control. *Journal of Experimental Social Psychology, 22,* 453–474.

Albus, S.M. (1999, December). New hire survey. *Mobility,* 44, 48.

Alexander, K., Eckland, B.K., and Griffin, L.J. (1975). The Wisconsin model of socioeconomic achievement: A replication. *American Journal of Sociology, 81*(September), 342.

Alwin, D.F. (1998). The political impact of the Baby Boom: Are there persistent differences in political beliefs and behaviors? *Generations,* (Spring), 46–54.

American Management Association. (2001). 2001 AMA Survey on Workplace Testing: Basic skills, job skills, and psychological measurement. Available: <http://www.amanet.org/research/pdfs/bjp_2001.pdf> [October 16, 2001].

Anderson, R. (2002). Youth and information technology. In J.T. Mortimer and R. Larson (Eds.), *The future of adolescent experience: Societal trends and the transition to adulthood.* New York: Cambridge University Press.

Arbuckle, J.L. (1996). Full information estimation in the presence of incomplete data. In G.A. Marcoulides and R.E. Schumacker (Eds.), *Advanced structural equation modeling* (pp.243–277). Mahwah, NJ: Lawrence Erlbaum.

Armor, D.J., and Roll, C.R. (1994). Military manpower quality: Past, present, and future. In National Research Council, B.F. Green, and A.S. Mavor (Eds.), *Modeling cost and performance for military enlistment: Report of a workshop* (pp. 13-34). Committee on Military Enlistment Standards. Commission on Behavioral and Social Sciences and Education. Washington, DC: National Academy Press.

Arnett, J.J. (2000). Emerging adulthood. A theory of development from the late teens through the twenties. *American Psychologist, 55,* 469–480.

Asch, B.J. (2001, November 16). *Policy options for recruiting the college market.* Presentation for the Committee on the Youth Population and Military Recruitment, National Academy of Sciences/National Research Council. Washington, DC.

Asch, B.J., Fair, C., and Kilburn, M.R. (2000). *An assessment of recent proposals to improve the Montgomery G.I. Bill.* (Report No. DB-301-OSD/FRP). Santa Monica, CA: RAND.

Asch, B.J., Hosek, J.R., and Warner, J.T. (2001). *An analysis of pay for enlisted personnel.* (Report No. DB 344-OSD). Santa Monica, CA: RAND.

Asch, B.J., Kilburn, M.R., and Klerman, J.A. (1999). *Attracting college-bound youth into the military: Toward the development of new recruiting policy options.* (Report No. MR-984-OSD). Santa Monica, CA: RAND.

Astin, A. (1993). *What matters in college?* San Francisco: Jossey-Bass.

Bachman, J.G., Freedman-Doan, P., and O'Malley, P.M. (2000a). *Youth attitudes and military service: A sourcebook of findings from the Monitoring the Future Project.* Ann Arbor, MI: University of Michigan Institute for Social Research, Survey Research Center.

Bachman, J.G., Freedman-Doan, P., Segal, D.R., and O'Malley, P.M. (2000b). Distinctive military attitudes among U.S. enlistees, 1976–1997: Self-selection versus socialization. *Armed Forces and Society, 26*(4), 561–585.

Bachman, J.G., Freedman-Doan, P., and O'Malley, P.M. (2001a). Should U.S. military recruiters write off the college-bound? *Armed Forces and Society, 27*(3), 461–476.

Bachman, J.G., Johnston, L.D., and O'Malley, P.M. (2001b). *The Monitoring the Future project after 27 years: Design and procedures.* (Monitoring the Future Occasional Paper No. 54). Ann Arbor, MI: University of Michigan Institute for Social Research.

Bachman, J.G., Segal, D.R., Freedman-Doan, P., and O'Malley, P.M. (1998). Does enlistment propensity predict accession? High school seniors' plans and subsequent behavior. *Armed Forces and Society, 25*(1), 59–80.

Bandura, A. (1977). Self-efficacy: Toward a unifying theory of behavioral change. *Psychological Review, 84,* 191–215.

Bandura, A. (1986). *Social foundations of thought and action: A social cognitive theory.* Englewood Cliffs, NJ: Prentice-Hall.

Bandura, A. (1991). Self-efficacy mechanism in physiological activation and health-promoting behavior. In J. Madden (Ed.), *Neurobiology of learning, emotion and affect* (pp. 229–269). New York: Raven.

Bandura, A. (1994). Social cognitive theory and exercise of control over HIV infection. In R.J. DiClemente and J.L. Peterson (Eds.), *Preventing AIDS: Theories and methods of behavioral interventions* (pp. 25–29). New York: Plenum Press.

Bearden, R.M., and Fedak, G.E. (2000). *A literature review of measures potentially applicable for Navy recruiter selection* (NPRST Tech. Rep. Draft). Millington, TN: Navy Personnel Research, Studies, and Technology.

Becker, M.H. (1974). The health belief model and personal health behavior. *Health Education Monographs, 2,* 324–508.

Becker, M.H. (1988). AIDS and behavior change. *Public Health Reviews, 16,* 1–11.

Bellah, R.N., Madsen, R., Sullivan, W.M., Swidler, A., and Tipton, S. (1985). *Habits of the heart: Individualism and commitment in American life.* New York: Harper and Row.

Blau, P., and Duncan, O.D. (1967). *The American occupational structure.* New York: John Wiley and Sons.

Borden, N.H. (1942). *The economic effects of advertising.* Chicago, IL: Richard D. Irwin.

Borman, W.C., Horgen, K.E., and Penney, L.M. (2000). *Overview of ARI-recruiting research* (ARI Research Note 2000-07). Alexandria, VA: U.S. Army Research Institute for the Behavioral and Social Sciences.

Borman, W.C., Penny, L.M., Horgen, K.E., Birkeland, S.A., Hedge, J.W., White, L.A., and Held, J.D. (2001). *Recruiting research in the U.S. Army and Navy.* Paper presented at the International Workshop on Military Recruitment and Retention in the 21st Century. The Hague, The Netherlands.

Bowman, T. (2001, July 23). Rising AWOL trend confounds military: Services scramble to find reasons why personnel are leaving. *Baltimore Sun*, p. A1.

Browning, G. (2002). Army plans new tech for recruiting site. *Federal Computer Week.*

Burnfield, J.L., Handy, K., Sipes, D.E., and Laurence, J.H. (1999). *Moral character enlistment standards: Documentation, policy and procedure review.* Alexandria, VA: Human Resources Research Organization.

Campbell, J.R., Hombo, C.M., and Mazzeo, J. (2000). *Trends in academic progress.* (Report No. NCES 2000-469). Washington, DC: NAEP.

Center for Human Resource Research. (2000). *National Longitudinal Survey of Youth (NLSY79).* Available <http://www.chrr.ohio-state.edu/surveys.html>.

Choy, S.P. (2002). *Access and persistence: Findings from 10 years of longitudinal research on students.* Washington, DC: American Council on Education.

Coleman, J.S. (1961). *The adolescent society.* London: Routledge and Kegan Paul.

College Board. (2001). *Trends in student aid.* (Report No. 992992). New York: The College Board.

Creative Good. (2000). *E-recruiting: Online strategies in the war for talent—Summary.* Available <http://www.creativegood.com> [February 21, 2000].

Csikszentmihalyi, M., and Schneider, B. (2000). *Becoming adult: How teenagers prepare for the world of work.* New York: Basic Books.

Cyert, R., and Mowery, D. (1987). *Technology and employment: Innovation and growth in the U.S. Economy.* Washington, DC: National Academy Press.

Dao, J. (2001, June 14). Defense Department panel seeks changes to keep military personnel. *New York Times.*

Defense Manpower Data Center. (2001). *Overview of the 1999 survey of active duty personnel* (Report No. 2000-008). C. Helba, Arlington, VA: Author.

DeFleur, L. (1978). Personal correspondence and data tables.

Dreeben, R. (1968). *On what is learned in school.* Reading, MA: Addison-Wesley.

Easterlin, R.A., and Crimmins, E.M. (1991). Private materialism, personal self-fulfillment, family life, and public interest: The nature, effects and causes of recent changes in the values of American youth. *Public Opinion Quarterly, 55*, 499–533.

Eighmey, J. (2000, December 7). *Communication strategy considerations and military advertising.* Presentation for the Committee on the Youth Population and Military Recruitment, National Academy of Sciences/National Research Council. Irvine, CA.

Featherman, D.L. (1981). The life-span perspective. In *The National Science Foundation's 5-year outlook on science and technology* (Vol. 2, pp. 621–648). Washington, DC: U.S. Government Printing Office.

Fendrich, J. (1993). *Ideal citizens.* Albany, NY: State University of New York Press.

Fishbein, M. (1995). Developing effective behavior change interventions: Some lessons learned from behavioral research. In T.E. Backer, S.L. David, and G. Soucy (Eds.), *Reviewing the behavioral sciences knowledge base on technology transfer* (NIDA Research Monograph No. 155, NIH Publication No. 95-4035) (pp. 246–261). Rockville, MD: National Institute on Drug Abuse.

Fishbein, M. (2000). The role of theory in HIV prevention. *AIDS Care, 12*(3), 273–278.

Fishbein, M., and Ajzen, I. (1975). *Belief, attitude, intention, and behavior: An introduction to theory and research.* Reading, MA: Addison-Wesley.

Fishbein, M., Guenther-Grey, C., Johnson, W., Wolitski, R.J., McAlister, A., Reitmeijer, C.A., O'Reilly, K., and the AIDS Community Demonstration Projects. (1996). Using a theory-based community intervention to prevent AIDS risk behaviors: The CEC's AIDS community demonstration projects. In S. Oskamp and S.C. Thompson (Eds.), *Understanding and preventing HIV risk behavior: Safer sex and drug use* (pp. 177–206). Thousand Oaks, CA: Sage.

Fishbein, M., Middlestadt, S.E., and Hitchcock, P.J. (1991). Using information to change sexually transmitted disease-related behaviors: An analysis based on the theory of reasoned action. In J.N. Wasserheit, S.O. Aral, and K.K. Holmes (Eds.), *Research issues in human behavior and sexually transmitted diseases in the AIDS era* (pp. 243–257). Washington, DC: American Society for Microbiology.

Fishbein, M., Triandis, H.C., Kanfer, F.H., Becker, M.H., Middlestadt, S.E., and Eichler, A. (2001). Factors influencing behavior and behavior change. In A. Baum, T.R. Revenson, and J.E. Singer (Eds.), *Handbook of health psychology* (pp. 3–17). Mahwah, NJ: Lawrence Erlbaum.

Fisher, J.D., and Fisher, W.A. (1992). Changing AIDS-risk behavior. *Psychological Bulletin 111*, 455–474.

Fletcher, J.D. (2000). *Defense training and the second revolution in learning.* Briefing presented at a conference on Distance Teaching and Learning, Madison, WI.

Flynn, J.R. (1998). I.Q. gains over time: Toward finding the causes. In U. Neisser (Ed.), *The rising curve: Long-term gains in I.Q. and related measures* (pp. 25–66). Washington, DC: American Psychological Association.

Ford, L.A., Campbell, R.C., Campbell, J.P., Knapp, D.J., and Walker, C.B. (1999). *21st century soldiers and noncommissioned officers: Critical predictors of performance* (FR-EADD-99-45). Alexandria, VA: Human Resources Research Organization.

Galbraith, J.K. (1967). *The new industrial state.* Boston: Houghton Mifflin.

Gallup Organization. (1999). *National survey of working America, 1999.* Report prepared for the National Career Development Association. Princeton, NJ: Gallup Organization.

Garbarino, J., Kostelny K., and Barry, F. (1997). Value transmission in an ecological context: The high risk neighborhood. In J.E. Grusec and L. Kuczynski (Eds.), *Parenting and children's internalization of values* (pp. 307–332). New York: John Wiley and Sons.

Gerrard, M., Gibbons, F.X., and Bushman, B.J. (1996). Relation between perceived vulnerability to HIV and precautionary sexual behavior. *Psychological Bulletin, 119*, 390–409.

Goldberg, M.S., and Warner, J.T. (1987). Military experience, civilian experience, and the earnings of veterans. *Journal of Human Resources, 22*(winter), 62–81.

Golfin, P.A., and Blake, D.H. (2000). *Tech prep and the U.S. Navy* (CRM D0000399.A1). Alexandria, VA: Center for Naval Analyses.

Gribben, M. (2001). *Trends in distribution of military personnel across occupational categories.* Paper written for the Committee on the Youth Population and Military Recruitment, National Academy of Sciences/National Research Council. Washington, DC.

Grossman, E.M. (2001, May 26). Recruitment, retention shortfalls in select areas strain each service. *Indianapolis Star.*

Haller, A.O., and Miller, I.W. (1971). *The occupational aspiration scale.* Cambridge: Schenkman.

Haller, A.O., and Portes, A. (1973). Status attainment processes. *Sociology of Education, 46*(Winter), 51–91.

Handel, M.J. (2000). *Is there a skills crisis? Trends in job skill requirements, technology, and wage inequality in the United States.* The Levy Economics Institute, Bard College, New York, No. 295.

Hays, S. (1999). Capital One is renowned for innovative recruiting strategies. *Workforce,* 78(4), 92–94.

Hebert, A.J. (2001). Air Force desire for UCAV pushing program nearly as fast as possible. *Inside the Air Force.*

Hellenga, K. (2002). Social space, the final frontier: Adolescents and the Internet. In J.T. Mortimer and R. Larson (Eds.), *The future of adolescent experience: Societal trends and the transition to adulthood.* New York: Cambridge University Press.

Heller, D.E. (2000, April 12). *Merit and need-based aid: Recent changes in state and institutional policy.* Briefing presented at the Center for the Study of Higher and Postsecondary Education, Advisory Committee on Student Financial Assistance, University of Michigan. Boston.

Hey, R.P. (2001). New push for robotic aircraft: Military is arming drones for use in bombing runs and dangerous missions. *Christian Science Monitor.*

Hogan, P.F., and Dall, T. (1996). *An econometric analysis of navy television advertising effectiveness.* Paper prepared for Chief of Naval Recruiting Command. Falls Church, VA: The Lewin Group.

Hogan, P.F., and Mackin, P. (2000, November). Briefing to the Ninth Quadrennial Review of Military Compensation Working Group.

Hogan, P.F., Mehay, S., and Hughes, J. (1998, June). *Enlistment supply at the local market level.* Paper presented at the Western Economic Association Annual Meeting. Lake Tahoe, NV.

Holden, G. (1991). The relationship of self-efficacy appraisals to subsequent health related outcomes: A meta-analysis. *Social Work in Health Care,* 16, 53–93.

Hom, P.W.R., Katerberg, R., and Hulin, C.L. (1979). Comparative examination of three approaches to the prediction of turnover. *Journal of Applied Psychology, 64,* 280–290.

Hornik, R., and Woolf, K.D. (1999). Using cross-sectional surveys to plan message strategies. *Social Marketing Quarterly, 5,* 34–41.

Hotchkiss, L., and Borrow, H. (1996). Sociological perspective on work and career development. In D. Brown, L. Brooks, and Associates (Eds.), *Career choice and development, third edition* (pp. 281–334). San Francisco: Jossey-Bass.

Immerwahr, J. (2000). *Great expectations: How the public and parents—White, African American and Hispanic—view higher education.* Available <http://www.highereducation.org/reports/expectations/expectations.pdf> [May 22, 2002].

International Communications Research. (1998). *Visions of success study.* Media, PA: Author.

Jencks, C., and Phillips, M. (1998). *The black and white test score gap.* Washington, DC: Brookings.

Johnson, M.K. (2001). Change in job values during the transition to adulthood. *Work and Occupations, 28,* 315–345.

Johnson, M.K., Beebe, T., Snyder, M., and Mortimer, J.T. (1998). Volunteerism in adolescence: A process perspective. *Journal of Research on Adolescence, 8,* 309–332.

Johnston, L.D., O'Malley, P.M., and Bachman, J.G. (2001). *Monitoring the Future national survey results on drug use, 1975–2000. Volume I: Secondary school students. Volume II: College students and adults ages 19–40.* (Report No. 01-4924 and 01-2925). Bethesda, MD: National Institute on Drug Abuse.

Joint Chiefs of Staff. (2000). *Joint vision 2020.* Washington, DC: U.S. Government Printing Office.

Kerckhoff, A.C. (2002). The transition from school to work. In J.T. Mortimer and R. Larson (Eds.), *The changing experience of adolescence: Societal trends and the transition to adulthood.* New York: Cambridge University Press.

Klopp, G.A., and Hemenway, M. (2001, August 7). *Examining the continuum of recruiting, training, and initial assignment in the U.S. Navy.* Briefing presented to the Recruiting Research Consortium.

Kraus, S.J. (1995). Attitudes and the prediction of behavior: A meta-analysis of the empirical literature. *Personality and Social Psychology Bulletin, 21*, 5874.

Ladd, E.C. (1995, January). Exposing the myth of the generation gap. *Reader's Digest.*

Leonard, W. (1999, August). Employee referrals should be cornerstone of staffing efforts. *HR News, 44*(8),54.

Levy, D.G., Thie, H., Robbert, A.A., Naftel, S., Cannon, C., Ehrenberg, R.H., and Gershwin, M. (2001). *Characterizing the future defense workforce.* Santa Monica, CA: RAND.

LinemenOnline (2001). *Pay rates from across the country and around the world: A wage and benefits reference for the lineman.* Statistics from the U.S. Bureau of Labor Statistics. Available <http://www.linemenonline.com/pay_scale.html> [May 25, 2002].

Mannino, D.M., Homa, D.M., Akinbami, L.J., Moorman, J.E., Gwynn, C., and Redd, S.C. (2002, March 29). Surveillance for asthma—U.S. 1980–1999. In *Surveillance summaries,* Morbidity and Mortality Weekly Report, Centers for Disease Control and Prevention.

Mare, R.D. (1995). Changes in educational attainment and school enrollment. In R. Farley (Ed.), *State of the union: America in the 1990s. Economic trends* (Vol. 1, pp. 155–213). New York: Russell Sage Foundation.

McAdam, D. (1988). *Freedom summer.* New York: Oxford University Press.

McCloy, R.A., Hogan, P.F., Diaz, T., Medsker, G.J., Simonson, B.E., and Collins, M. (2001). *Cost effectiveness of Armed services Vocational Aptitude Battery (ASVAB) use in recruiter selection. Final report.* (Report No. FR-01-38). Alexandria, VA: Human Resources Research Organization.

Micco, L. (1997, March). HR executives optimistic about wage inflation. *HR News,* 13.

Middlestadt, S.E., Bhattacharyya, K., Rosenbaum, J., and Fishbein, M. (1996). The use of theory based semi-structured elicitation questionnaires: Formative research for CDC's Prevention Marketing Initiative. *Public Health Reports, 111*(Supplement 1):18–27.

Mitchell, S. (2000). *American attitudes.* Ithaca, NY: New Strategist.

Mortimer, J.T. (2001). Personal correspondence and data tables.

Mortimer, J.T., and Finch, M.D. (1996). *Adolescents, work, and family: An intergenerational developmental analysis.* Thousand Oaks, CA: Sage.

Mortimer, J.T., and Lorence, J. (1979). Work experience and occupational value socialization: A longitudinal study. *American Journal of Sociology, 84,* 1361–1385.

Mortimer, J.T., Lorence, J., and Kumka. (1986). *Work, family and personality: Transition to adulthood.* Norwood, NJ: Ablex.

Mortimer, J.T., Zimmer-Gembeck, M., Holmes, M., and Shanahan, M.J. (2002). The process of occupational decision-making: Patterns during the transition to adulthood. *Journal of Vocational Behavior, 61,* 1–27.

Murray, M.P., and McDonald, L.L. (1999). *Recent recruiting trends and their implications for models of enlistment supply.* (Report No. MR-847-OSD/A). Santa Monica, CA: RAND.

National Association of Colleges and Employers. (1999). *Job outlook 2000.* Bethlehem, PA: National Association of Colleges and Employers.

National Center for Health Statistics. (1999). *Prevalence of overweight among children and adolescents: U.S. 1999.* Atlanta: Centers for Disease Control and Prevention.

National Center for Health Statistics. (2002). *Birth records* (Natality Statistics microdata files). Available <http://www.cdc.gov/nchs/births.htm>.

National Commission on Excellence in Education. (1983). *A nation at risk: The imperative for educational reform.* Washington, DC: U.S. Government Printing Office.

National Research Council. (1991). Evaluating the quality of performance measures: Criterion-related validity evidence. In A.K. Wigdor and B.F. Green (Eds.), *Performance assessment for the workplace* (Vol. 1, pp. 141–183). Washington, DC: National Academy Press.

National Research Council. (1994). *Modeling cost and performance for military enlistment.* Committee on Military Enlistment Standards, B.F. Green, and A.S. Mavor (Eds.), Commission on Behavioral and Social Sciences and Education. Washington, DC: National Academy Press.

National Research Council. (1999). *The changing nature of work: Implications for occupational analysis.* Committee on Techniques for the Enhancement of Human Performance: Occupational Analysis. Washington, DC: National Academy Press.

Navy Personnel Research and Development Center. (1998). *Sailor 21: A research vision to attract, retain, and utilize the 21st century sailor.* San Diego, CA: Navy Personnel Research and Development Center.

NUA Internet Surveys. (1999, October). *Top sites make job application difficult.* Available <http://www.nua.com/surveys/index.cgi?f=VS&art_id=905355316&rel=true> [October 22, 2002].

Oesterle, S., Johnson, M.K., and Mortimer, J.T. (1998). *Volunteerism during the transition to adulthood.* Paper presented at the annual meeting of the American Sociological Association.

Office of Management and Budget. (1997). *1997 standard occupational classification revision. Federal Register: 62.* Available <http://stats.bls.gov/soc/soc_jul7.htm> [May 8, 2001].

Office of the Assistant Secretary of Defense. (1999). *Population representation in the military services.* Washington, DC: Author.

Office of the Assistant Secretary of Defense. (2000). *Population representation in the military services.* Washington, DC: Author.

Office of the Assistant Secretary of Defense. (2001). *Population representation in the military services.* Washington, DC: Author.

Orvis, B.R., and Asch, B.J. (2001). *Military recruiting: Trends, outlook, and implications* (MR-902-A/OSD). Santa Monica, CA: RAND.

Orvis, B.R., Gahart, M.T., and Ludwig, A.K. (1992). *Validity and usefulness of enlistment intention information.* (R-3775-FMP). Santa Monica, CA: RAND.

Otto, L.B. (2000). Youth perspectives on parental career influence. *Journal of Career Development, 27*(2), 111–118.

Otto, L.B., and Haller, A.O. (1975). Evidence for a social psychological view of the status attainment process: Four studies compared. *Social Forces, 57*(3), 887–914.

Overholser, C.E., and Kline, J.M. (1975). Advertising strategy from consumer research. In D.A. Aaker (Ed.), *Advertising management: Practical perspectives* (p. 82). Englewood-Cliffs, NJ: Prentice-Hall.

Overman, S. (1999). School-to-Work Partnerships (SHRM white paper). Alexandria, VA: Society for Human Resource Management. Available <http://www.shrm.org/white papers/documents/default.asp?page=61161.asp> [November 2, 2002].

Penney, L.M., Horgen, K.E., and Borman, W.C. (2000). *An annotated bibliography of recruiting research conducted by the U.S. Army Research Institute for the Behavioral and Social Sciences* (Tech. Rep. 1100). Alexandria, VA: U.S. Army Research Institute for the Behavioral and Social Sciences.

Personnel Research Associates. (1999). Unpublished data from survey of Fortune 100 companies. Rolling Meadows, IL: Personnel Research Associates.

Petraitis, J., Flay, B.R., and Miller, T.Q. (1995). Reviewing theories of adolescent substance use: Organizing pieces of the puzzle. *Psychological Bulletin, 17*(1), 67–86.

Prochaska, J.O., and DiClemente, C.C. (1983). Stages and processes of self-change in smoking: Towards an integrative model of change. *Journal of Consulting Clinical Psychology, 51*, 390–395.

Prochaska, J.O., and DiClemente, C.C. (1986a). Toward a comprehensive model of change. In W.R. Miller and N. Neather (Eds.), *Treating addictive behaviors: Processes of change* (pp. 3–27). New York: Plenum Press.

Prochaska, J.O., and DiClemente, C.C. (1986b). The transtheoretical approach. In J.C. Norcross and M.R. Goldfried (Eds.), *Handbook of psychotherapy integration* (pp. 300–334). New York: Basic Books.

Prochaska, J.O., Redding, C.A., Harlow, L.L., Rossi, J.S., and Velicer, W.F. (1994). The transtheoretical model of change and HIV prevention: A review. *Health Education Quarterly, 21*(4), 471–486.

Prochaska, J.O., DiClemente, C.C., and Norcross, J.C. (1992). In search of how people change: Applications to addictive behaviors. *American Psychologist, 47*, 1102–1114.

Putnam, R.D. (1995a). Bowling alone: America's declining social capital. *Journal of Democracy, 6*(1), 65–78.

Putnam, R.D. (1995b). Bowling alone, revisited. *The Responsive Community, 5*(2), 18–33.

RAS Recruitment Systems. (2000). On the radio. *Workforce*, 79(1), 24.

Robershotte, M., and Zalesny, M.D. (2001, August 7). *Development of a strategic recruiting research roadmap.* Briefing presented to the Recruiting Research Consortium.

Rosenstock, I.M. (1974). The health belief model and preventive health behavior. *Health Education Monographs, 2*, 354–386.

Rosenstock, I.M., Strecher, V.J., and Becker, M.H. (1994). The health belief model and HIV risk behavior change. In R.J. DiClemente and J.L. Peterson (Eds.), *Preventing AIDS: Theories and methods of behavioral interventions* (pp. 5–24). New York: Plenum Press.

Rumsey, M.G. (1995). *The best they can be: Tomorrow's soldiers.* Army 2010 Conference: Future soldiers and the quality imperative, Cantigny.

Rutter, M. (1980). *Changing youth in a changing society.* Cambridge, MA: Harvard University Press.

Ryu, S., and Mortimer, J.T. (1996). The "Occupational Linkage Hypothesis" applied to occupational value formation in adolescence. In J.T. Mortimer and M.D. Finch (Eds.), *Adolescents, work, and family: An intergenerational developmental analysis* (pp. 167–190). Thousand Oaks, CA: Sage.

Schneider, B. (2001, May 16). *Comparing youth from different generations: Scientific findings and the popular literature.* Briefing for the Committee on the Youth Population and Military Recruitment, National Academy of Sciences/National Research Council. Washington, DC.

Schneider, B., and Stevenson, D. (1999). *The ambitious generation. America's teenagers motivated but directionless.* New Haven, CT: Yale University Press.

Segal, D.R., Bachman, J.G., Freedman-Doan, P., and O'Malley, P.M. (1999). Propensity to serve in the U.S. military: Temporal trends and subgroup differences. *Armed Forces and Society, 25*(3), 407–427.

Sellman, W.S. (2001). *U.S. Military recruiting initiatives.* Paper presented at the International Workshop on Military Recruitment and Retention in the 21st Century. The Hague, The Netherlands.

Sewell, W.H., and Hauser, R.M. (1972). Causes and consequences of higher education: Models of the status attainment process. *American Journal of Agricultural Economics, 54*, 851–861.

Sewell, W.H., and Hauser, R.M. (1975). *Education, occupation, and earnings: Achievement in the early career.* New York: Academic Press.

Sewell, W.H., Haller, A.O., and Ohlendorf, G.W. (1970). The educational and early occupational status achievement process: Replication and revision. *American Sociological Review, 35*, 1014–1027.

Sewell, W.H., Haller, A.O., and Portes, A. (1969). The educational and early occupational status achievement process. *American Sociological Review, 34*, 82–92.

Shanahan, M.J. (2000). Pathways to adulthood in changing societies: Variability and mechanisms in life course perspective. *Annual Review of Sociology, 26*, 667–692.

Sheppard, B.H., Hartwick, J., and Warshaw, P.R. (1988). The theory of reasoned action: A meta-analysis of past research with recommendations for modifications and future research. *Journal of Consumer Research, 15*, 325–343.

Society for Human Resource Management. (2000). *Employees like their pay, but not how it's determined.* L. Rivenbark, *HR News*, May 24, 2000. Available <http://www.shrm.org/hrnews/articles/default.asp?page=052400a.htm> [May 22, 2000].

Society for Human Resource Management. (2001). *2001 benefits survey.* Alexandria, VA: Author.

Stafford, F. (1991). Partial careers: Civilian earnings and the optimal duration of an Army career. In C.L. Gilroy, D.K. Horne, and D.A. Smith (Eds.), *Military compensation and personnel retention: Models and evidence* (pp. 81–308). Alexandria, VA: Army Research Institute for the Behavioral and Social Sciences.

Stanford University New Service. (1980, September 25).

Strecher, V.J., DeVellis, B.M., Becker, M.H., and Rosenstock, I.M. (1986). The role of self-efficacy in achieving health behavior change. *Health Education Quarterly, 13*(1), 73–91.

Strickland, W. (2000, May 18–19). *Surveys of new recruits.* Briefing for the Committee on the Youth Population and Military Recruitment, National Academy of Sciences/National Research Council. Washington, DC.

Thirtle, M.R. (2001). *Educational benefits and officer-commissioning opportunities available to U.S. Military service members.* (MR-981-OSD). Santa Monica, CA: RAND.

Today's Military. (2001). *Common military questions.* Available <http://www.todaysmilitary.com/q_a.shtml#training> [July 2, 2001].

Tracey, Vice Admiral P.A. (2001). Correspondence, U. S. Department of Defense, Washington, DC.

Triandis H.C. (1972). *The analysis of subjective culture.* New York: John Wiley and Sons.

U.S. Bureau of Labor Statistics. (1999a, November 30). *BLS releases new 1998–2008 employment projections.* Washington, DC. Available <http://stats.bls.gov/emphome.htm>.

U.S. Bureau of Labor Statistics. (1999b). *Monthly labor review.* Available <http://www.bls.gov/emplt981.htm> [October 12, 2001].

U.S. Bureau of Labor Statistics. (2001). Data extracted from the National Employment, Hours, and Earnings Database for all nonfarm employees, not seasonally adjusted. Available <http://www.bls.gov>.

U.S. Bureau of Labor Statistics. (2002a). Household data annual averages. Table 1, Employment Status of the Civilian Non-institutional Population, 1939 to Date (Labor Force Statistics for the Current Population Survey). Available <http://www.bls.gov/cps/home.htm#empstat>.

U.S. Bureau of Labor Statistics. (2002b). Household data annual averages. Table 7, Employment Status of the Civilian Non-institutional population 25 years and over by educational attainment, sex, race, and Hispanic origin. (Labor Force Statistics for the Current Population Survey). Available <http://www.bls.gov/cps/home.htm#empstat>.

U.S. Bureau of Labor Statistics. (2002c). *National longitudinal surveys handbook 2002.* Washington, DC: U.S. Department of Labor, Bureau of Labor Statistics.

U.S. Bureau of the Census. (1999). *Statistics about business size (including small business) from the U.S. Bureau of the Census.* Available <http://www.census.gov/epcd/www/smallbus.html#EmpSize> [May 25, 1999].

U.S. Bureau of the Census. (2000a). *Annual projections of the resident population by age, sex, race, and Hispanic origin. Middle series data from National Population Projections, Detailed Files: 1999 to 2020.* Available <http://www.census.gov/population/www/projections/natdet-D1A.html>.

U.S. Bureau of the Census. (2000b). *Annual public employment survey.* Available. ,http://www.census.gov/govs/www/apesfed.html> [May 26, 2000].

U.S. Bureau of the Census. (2001). *Statistical abstract of the United States.* Washington, DC: U.S. Government Printing Office.

U.S. Bureau of the Census. (2002a). Current Population Surveys, June 1971, 1973–1977, 1979–1988, 1990–1992, 1994–1995, 1998, 2000 [machine-readable data files]/conducted by the U.S. Bureau of the Census for the Bureau of Labor Statistics. Washington, DC: U.S. Bureau of the Census [producer and distributor], 1973–2001. Santa Monica, CA: Unicon Research Corporation [producer and distributor of CPS Utilities].

U.S. Bureau of the Census. (2002b). Current Population Surveys, October 1968–2000: School Enrollment [machine-readable data files]/conducted by the U.S. Bureau of the Census for the U.S. Bureau of Labor Statistics. Washington, DC: U.S. Bureau of the Census [producer and distributor], 1968–2001. Santa Monica, CA: Unicon Research Corporation [producer and distributor of CPS Utilities].

U.S. Bureau of the Census. (2002c). *Statistical abstract of the United States, 2001.* Available <http://www.census.gov/prod/2002pubs/01statab/>.

U. S. Bureau of the Census. (2002d). Current Population Surveys, March Supplement (various years). Conducted by the U.S. Bureau of the Census for the Bureau of Labor Statistics. Washington, DC: U.S. Bureau of Labor Statistics.

U.S. Department of Defense. (2001). *Quadrennial defense review report. September 30, 2001.* Washington, DC: Author.

U.S. Department of Defense. (2002). *Population representation in the military services: Fiscal year 2001.* Washington, DC: Office of the Assistant Secretary of Defense (Force Management and Personnel).

U.S. Department of Education. (1997). National Center for Education Statistics, *Access to postsecondary education for the 1992 high school graduates,* by L. Berkner and L. Chavez (NCES 98-105), Washington, DC: U.S. Government Printing Office.

U.S. Department of Education. (1998). National Center for Education Statistics, *Stopouts or stayouts? Undergraduates who leave college in their first year,* by L.J. Horn (NCES 1999-087), Washington, DC: U.S. Government Printing Office.

U.S. Department of Education. (1999). National Center for Education Statistics Issue Brief, *Students who prepare for college and a vocation,* by L. Hudson and D. Hurst (NCES 1999-072), Washington, DC: U.S. Government Printing Office.

U.S. Department of Education. (2000). National Center for Education Statistics, *Distance education at postsecondary education institutions: 1997–98,* by Lewis, L., Farris, E., Snow, K., and Levin, D. NCES 2000-013, Washington, DC: U.S. Government Printing Office.

U.S. Department of Education. (2001). National Center for Education Statistics, *The condition of education 2001,* NCES 2001-072, Washington, DC: U.S. Government Printing Office.

U.S. Department of Justice. (1999). Sourcebook of Criminal Justice Statistics Online. In Federal Bureau of Investigation *Crime in the United States, 1972–1998* (also available <http://www.albany.edu/sourcebook>). Washington, DC: U.S. Department of Justice.

U.S. General Accounting Office (1997). *Military attrition: DoD could save millions by better screening enlisted personnel* (GAO/NSIAD-97-39). Report to the Chairman and the Ranking Member, Subcommittee on Personnel, Committee on Armed Services, U.S. Senate. Washington, DC: Author.

Uggen, C., and Janikula, J. (1999). Adolescent volunteer work and arrest. *Social Forces, 78,* 331–360.

van den Putte, B. (1991). 20 Years of the Theory of Reasoned Action of Fishbein and Ajzen: A Meta-Analysis. Unpublished manuscript, University of Amsterdam, The Netherlands.

Verba, S., Schlozman, K.L., and Brady, H.E. (1995). *Voice and equality: Civic volunteerism in American politics.* Cambridge, MA: Harvard University Press.

Warner, J.T. (2000). *Briefing of the Ninth Quadrennial Review of Military Compensation (QRMC).* Washington, DC: U.S. Department of Defense.

Warner, J.T. (2001, July 26). *Navy college fund evaluation study: Overview of findings.* Presentation for the Committee on the Youth Population and Military Recruitment, National Academy of Sciences/National Research Council. Woods Hole, MA.

Warner, J.T., Payne, D., and Simon, C.J. (1998, June). *Navy college fund study.* Paper presented at the Western Economic Association Annual Meeting. Lake Tahoe, NV.

Warner, J.T., Simon, C.J., and Payne, D. (2001, June 5). *The military recruiting productivity slowdown: The roles of resources, opportunity cost, and the tastes of youth.* Paper presented at the Western Economic Association Annual Meeting. San Francisco, CA, p. 25.

Wilson, J. (2000). Volunteering. *Annual Review of Sociology, 26,* 215–240.

Wilson, M., and Lehnus, J. (1999). *The future plans and future behaviors of YATS youth.* Arlington, VA: Defense Manpower Data Center.

Woelfel, J., and Haller, A.O. (1971). Significant others, the self-reflexive act and the attitude formation process. *American Sociological Review, 36*(February), 74–87.

Wright, J.S., and Warner, D.S. (1963). *Speaking of advertising.* New York: McGraw Hill.

Yates, M., and Youniss, J. (1996). Community service and political-moral identity in adolescents. *Journal of Research on Adolescence, 6,* 271–284.

Youniss, J., and Ruth, A. (2002). Approaching policy for adolescent development in the twenty-first century. In J.T. Mortimer and R. Larson (Eds.), *The changing adolescent experience: Societal trends and the transition to adulthood.* New York: Cambridge University Press.

Youniss, J., and Yates, M. (1997). *Community service and social responsibility in youth.* Chicago, IL: University of Chicago Press.

Youniss, J., Bales, S., Christmas-Best, V., Diversi, M., McLaughlin, M., and Silbereisen, R. (2002). Youth civic engagement in the twenty-first century. *Journal of Research on Adolescence, 12*(1), 121–148.

Youniss, J., McLellan, J.A., and Yates, M. (1997). What we know about engendering civic identity. *American Behavioral Scientist, 40,* 620–631.

Appendix A

Evaluation of the Youth Attitude Tracking Study (YATS)

Letter Report from the Committee on the Youth Population and Military Recruitment

May 2000

THE NATIONAL ACADEMIES
Advisers to the Nation on Science, Engineering, and Medicine

National Academy of Sciences
National Academy of Engineering
Institute of Medicine
National Research Council

Commission on Behavioral and Social Sciences and Education
Committee on the Youth Population and Military Recruitment

June 16, 2000

ADM Patricia A. Tracey
Deputy Assistant Secretary of Defense for Military Personnel Policy
Room 3E767, The Pentagon
Washington, DC 20310-4000

Dear Admiral Tracey:

As you know, the National Research Council is conducting a 4-year study on military advertising campaigns and long-term planning in the Department of Defense (DoD) and the several Armed Services. The recent shortfall in military recruiting has prompted DoD to examine changes in youth attitudes and consider new strategies for attracting youth to the military. In response to a request from the Office of the Assistant Secretary of Defense (OASD), the Committee on the Youth Population and Military Recruitment was formed to examine the demographic trends, cultural characteristics, attitudes, and educational attainments of American youth in order to help military planners improve recruiting for the military (the committee's task statement and committee roster are attached).

One aspect of the committee's charge is to periodically evaluate surveys and interpretive reports provided by contractors. At our first meeting, we were asked by OASD to review and comment by June 2000 on the data collection in and analyses of the Youth Attitude Tracking Study (YATS), a survey administered by the Defense Manpower Data Center (DMDC) that measures youth attitudes toward military service.

This letter is based on the committee's evaluation of materials provided by DMDC, including YATS trend and focused reports, questionnaires, data from 1998 and 1999, and discussions with individuals who have examined YATS data. We examined (1) the need for further analysis of YATS data, (2) methodological issues associated with various approaches to data collection, (3) the current administration of YATS, (4) YATS item content and areas for new questions, and (5) accessibility of

YATS data to DoD policy makers and decision makers. We begin with a brief description of the YATS survey followed by our findings and recommendations.

BACKGROUND: YATS FRAMEWORK

YATS is a computer-assisted telephone interview (CATI) of 10,000 young American men and women between 16 and 24 years of age. DMDC began administering the survey in 1975, and it has been administered annually since then. In order to limit the interview length to 30 minutes, the survey questions are grouped into sections and asked of subsets of the survey participants. That is, the questionnaire is partitioned such that different respondents are asked different subsets of items within a block of items on a given topic. The 1999 YATS questionnaire covers a wide range of issues, including propensity to enlist. The overall complexity of the questionnaire is demonstrated in Table 1, which lists and briefly describes the major questionnaire sections.

YATS data are used by the Office of the Under Secretary of Defense (Personnel and Readiness), military recruiting commands, and all of the Armed Services and their advertising agencies to track trends in youth attitudes, understand the effect of these trends on recruiting, and evaluate the effectiveness of recruiting programs. Information about YATS data and analyses is disseminated through briefings, presentations, and reports. The most recent analysis, *Youth Attitude Tracking Study 1998: Propensity and Advertising Report* (Wilson et al., 2000), provides information on the demographics of the youth population, propensity for military service, reasons for joining or not joining the military, and the effect of recruiting efforts (including advertising). In addition, Bruce Orvis and his colleagues at RAND have evaluated youth and recruiting issues using YATS data (Orvis et al., 1996).

FINDINGS AND RECOMMENDATIONS

The recommendations below represent the collective judgment of the committee after careful evaluation of YATS and its existing analytic products and discussions with people who have done extensive analyses of the YATS data. As our work continues, we may offer other recommendations.

Further Data Analyses

Further analyses of the YATS data could yield valuable insights about the causes of current recruiting shortfalls and possible remedial actions.

TABLE 1 Contents of YATS

Category	Description
Background Items	Gender, age, previous military service, education (a number of questions), employment status (a number of questions), future plans and unaided propensity (a number of questions), aided propensity (a number of questions).
Military Benefits and Incentives	One question on competitiveness of military pay and a number of questions about the attractiveness of educational incentives, prohibition of smoking, and terms of enlistment.
College Programs	A number of questions about the service academies, ROTC, and sources of money for college.
Current Events I	Questions about combat interest and Kosovo.
Current Events II	Questions about likelihood of acceptance by the military and career options in the military.
Junior ROTC	Questions about participation in high school programs.
Civilian vs. Military Perception	Questions concerning importance of 26 issues including money for education, development of self-discipline, opportunity for travel, working as part of a team, getting experience preparatory for a career, working in a high technology environment, etc.
Media Habits and Internet Use	Numerous questions on use of television, radio, newspapers, magazines, and the Internet.
Advertising Awareness	General question concerning recall of any military advertising, recall of any advertised services, etc.
Slogan Awareness	Aided recall of advertising theme lines.
Advertising Response and Information Seeking	Questions concerning readership of direct mail pieces, and actions taken to call for information or contact recruiters.
Influences I	Questions concerning persons with whom the study participant discussed the possibility of military service and the attractiveness of military service.
Influences II	Questions concerning personal sources of information, military related movies, and perceptions of military life.
Background I	Questions concerning the ASVAB test, MEPS test, and high school grades.
Background II	Questions concerning parents' education, marital status of study participant, etc.

Most of the recommended analyses involve studying potential causal links[1] between the propensity to enlist and various youth and societal factors, although other types of analyses may also yield important in-

[1]Care must be exercised in making conclusions about causality when correlational data are used.

sights regarding youth attitudes and behaviors that could aid in recruitment strategies.

1. We recommend that DoD consider comprehensive bivariate and multivariate analyses that attempt to relate trends in the propensity to enlist to possible causal factors, such as trends in youth attitudes and values, trends in demographic characteristics of youth, trends in youth influencers (such as family members with military experience), trends in military recruiting resources (e.g., number of recruiters, advertising, and enlistment incentives), trends in military operations and conflicts, and trends in such exogenous variables as civilian pay, unemployment rates, and college demand and incentives. It would also be valuable to examine differences in propensity by respondents' planned occupational field.

According to DoD, existing studies and analyses of YATS tend to focus on snapshot tabulations, single-variable trends, and relationships among, at most, two or three variables of interest. Standing alone, univariate and bivariate analyses can generate misleading and inappropriate conclusions from the YATS data and sole reliance on such analyses is not appropriate for the kinds of information needed by DoD. We are recommending more comprehensive multivariate analyses that might use several different techniques, including cross-sectional multiple regression, multivariate time-series regression, and even combined time-series cross-sectional regression insofar as the data permit these kind of analyses. Analyses should also use alternative definitions of propensity (e.g., those definitely planning to enlist) and different age groups that might generate closer relationships to actual enlistment behaviors. In addition, since there is a perception that private industry provides better training in some kinds of jobs and that the military might be the only place to obtain training in other kinds of jobs (e.g., combat), it would be useful to understand more about how propensity differs by different occupational choices of youth.

As part of this work, it might also be appropriate to conduct factor analyses of some of the youth attitude and value data, such as reasons for joining/not joining, reasons for increased/decreased interest, and importance of various career attributes or life goals. Not only might this help identify broader themes in and structures of youth attitudes and values, and whether these themes and structures change over time, but it would also serve as a useful data reduction technique that will aid in the construction of multivariate models.

We recognize that there are some limitations in carrying out further analyses of the YATS data. First, these further analyses, as is the case with all work for YATS, would focus on propensity rather than actual enlistment behavior, and although propensity is related to enlistment, it is not a one-to-one correspondence. Second, there are some serious data limitations, including changes in attitudes surveyed and in question wording over time, and significant amounts of missing data, most of which arise from the practice of asking questions only of randomly selected subsets of respondents. Under the current structure, the questionnaire is partitioned across participant subgroups such that different respondents are asked different subsets of items within a block of items on a given topic. This last data feature makes cross-sectional multivariate analyses (e.g., regression or factor analysis) very difficult due to inadequate numbers of observations.

Some of these problems might be overcome by factor analyzing pairwise correlation matrices or by aggregating data into higher level units of analysis (e.g., geographic regions). However, the two primary options for factor analysis of partitioned data are problematic. One option is analysis of a pairwise correlation matrix, where each correlation is based on the subsample responding to the particular pair of items. The other option is to impute missing values for each item for those respondents who were not asked particular items. In both cases, the observed patterns can be affected by the implicit assumptions that the approaches make about the responses of those individuals whose responses are missing. Furthermore, because the questionnaire is partitioned in this manner, it is difficult to determine how measures of values and interests relate to propensity to enlist among various subgroups within the participant sample.

2. In planning further analyses of the YATS data, we recommend that some attention be given to already completed analyses based on other data sets, such as Monitoring the Future surveys at the University of Michigan Survey Research Center.

Some of those analyses are similar to what we are recommending, and they may offer useful leads for some of the more promising approaches that might be applied to the YATS data.

In spite of some understandable limitations in the data, the committee believes that further secondary analyses of YATS can offer greater insights into some of the reasons for changes in propensity to enlist, which in turn can be tested and validated by similar analyses using other surveys (such as Monitoring the Future) or by analyzing actual enlistment trends.

3. We recommend that a procedure be established (e.g., an advisory board) to periodically review and evaluate the adequacy of analytic approaches to YATS data.

Methodology

In considering further analyses of the YATS data, we recognize that the current methodology for collecting YATS data involves procedures that necessarily limit the ability to apply multivariate techniques. Most of these limitations arise from the fact that YATS is administered only once a year and needs to gather enormous amounts of data from one large sample. As noted above, many of the questionnaire items are asked of only subsamples of the total survey sample. Despite this careful effort, the questionnaire is nevertheless daunting. The introductory boilerplate is quite formal and, for some survey participants, it may take 2 minutes or more before the first question is asked. Also, within the body of the questionnaire, many of the questions are both lengthy and conditional. For example, question Q545A consists of 66 words and asks survey participants to speculate on how military enlistment as a whole might change if a particular change were made in the terms of service.

Changes in design should be guided by the analytic goals and models established by DoD and its contractors who study and use YATS data. The questions asked and the grouping of the questions should be consistent with these analytic goals.

4. We recommend that DoD consider improving YATS methodology by implementing changes in data collection and sampling techniques.

We understand that DoD is already planning to administer the YATS surveys on a more frequent basis to improve the timeliness and responsiveness of results. This change provides opportunities for other methodological changes that could enhance the usefulness of the YATS data. For example, while every YATS survey will probably contain a common core of questions, especially those relating to propensity to enlist and various background questions, other questions do not need to be asked at every survey administration. This approach would enable gathering complete data from all respondents on the block of items in question without placing an undue time burden on respondents. It is likely that there are certain types of information, particularly with respect to attitude and value items, that do not need to be assessed on a continuous basis.

5. We recommend that DoD consider alternating blocks of in-depth questions on different administrations, so that certain attitude items (e.g., life goals) are assessed in one or two administrations during the year but not others.

There are other methodological issues that might arise from more frequent (e.g., monthly) administrations of the YATS that should be considered. For example, sample sizes will probably be smaller, thereby introducing greater sampling error, especially when studying subgroups defined by demographic characteristics, such as race, gender, and age. Some of these problems can be overcome with careful nonproportional sampling designs (i.e., oversampling some demographic groups), and some consideration might be given to somewhat longer survey intervals with larger sample sizes (e.g., every 2 months). Another possibility is to do monthly surveys of time-sensitive information (propensity, advertising-related questions) and less frequent surveys (yearly, biannual) with larger samples for those items that are less time sensitive (general attitudes and values, views of the military, etc.).

6. We recommend that whenever a survey is designed to partition the questions so that not all questions are asked of all respondents, consideration should be given to randomly assigning interrelated blocks of information to the same subgroups, such as asking one subgroup all life/career goal questions, another subgroup slogan recognition, another subgroup Service-specific questions, and so forth.[2] Consideration should also be given to maintaining sufficient sample size and content within a block of relevant questions so that multivariate analysis can be conducted without serious missing data problems.

A Portfolio of Surveys

7. We recommend that a portfolio of surveys at different time intervals replace the current annual YATS administration.

A variety of approaches might be considered. One approach, continuous tracking, offers the benefit of revealing month-to-month changes that might be related to changes in program activities by the military and related changes in economic and social conditions. It would provide more timely data that could be used analytically in an attempt to better understand the effects of various information campaigns and communication techniques. Although the week-to-week sample sizes would be smaller for continuous tracking, the total sample size would build over time and enable the useful tracking of appropriate moving averages for key variables. For some issues, specific studies could be designed to provide detailed perspectives on the interests, motives, role models, and perceptions

[2]Each respondent should be given multiple blocks of items with each block paired with every other block for some subset of respondents to allow cross-block comparisons.

of American youth as they relate to their life choices and the possibilities of choosing military service. For other issues, one-on-one in-depth interviews about the tradeoffs seen in various career and life choices could be used to reveal the specific criteria used to make choices and the value structure underlying those choices.

7a. We recommend that DoD consider using a continuous tracking survey methodology for such issues as propensity to enlist, advertising awareness, awareness of direct response campaigns, involvement in high school activities, and perceptions of the military.

7b. We recommend that DoD consider conducting specific national surveys every 2 years on such issues as values relating to careers, family life, consumption, lifestyle, leisure, education, interest in information technology, and public service.

7c. We recommend that DoD consider conducting specific, smaller-scale studies as needed, to examine issues such as trade-off analyses of specific "offers" with respect to combinations of terms of service, educational benefits, and pay, among other things. Further, small-scale studies are recommended to examine: (1) postponement of focus or commitment and a sense of direction in life, and (2) perspectives on family life and community involvement.

7d. We recommend that DoD consider using in-depth qualitative studies to offer insight about the decision-making processes of potential recruits.

Finally, communication from the military reaches today's youth in a larger context, and it would be useful to understand the other voices (e.g., universities, corporations) and to examine how communication from the military is viewed in that larger context. Although other organizations exist in other markets and decision-making realms, military communication is competing for attention and will inevitably be seen by youth as fitting or not fitting in a contemporary context.

8. We recommend that DoD consider examining communication strategies through a specific study of how a variety of organizations communicate with youth.

Item Content

The committee examined the content of the YATS survey to identify areas in which additional useful information might be collected. We understand that the overall length of the survey and the attendant administration costs are important considerations; therefore, we addressed strategies for deleting as well as adding items.

9. We recommend that linkages to the propensity to enlist established by prior analyses be used as a basis for retaining or deleting items.[3]

10. To support the development and selection of alternative communication strategies, we recommend that DoD expand the coverage of: (1) perspectives on service to the country, (2) understanding of the mission of the military, and (3) life values and motives.

11. Since college is a major competitor to the military, we recommend that DoD consider including questions comparing benefits attributed to the military and to college education.

12. We recommend that items involving attribute importance (i.e., the importance of traits associated with military or civilian jobs) be reconsidered, because an importance rating (how important a trait is to a person) provides no information on whether the attribute is viewed as positive or negative.

13. We recommend that DoD review the content of current survey items and consider adding questions that identify the characteristics of three career choices—military, work, and education—and measure youths' evaluations of these characteristics. For example, questions on total compensation (e.g., pay, health benefits, 401K plans, stock options, value of education and training), pay-for-performance plans, recognition programs, currency of technology, job security, career progression, work/family balances, etc. could be added.

Accessibility of YATS Data

Accessibility and timeliness are especially important in view of more frequent survey interval; much of the advantage of more frequent surveys will be lost if the information is not made available more quickly to researchers and policy makers. Since YATS interviewing is conducted

[3]Some items with no link to propensity might be retained if they provide information about targeted groups and methods to facilitate communication with them.

using CATI techniques, the raw data is available in computer form as soon as the interviews are completed. After a reasonable period for validity checks and creation of summary variables, DoD should make the database available on-line so that interested users could retrieve information, such as summary statistics, simple tabulations and cross-tabulations, and so forth.

In addition, it would be especially useful to let users retrieve information from the current survey and compare it with similar information from earlier surveys. Although the software for such comparisons will be somewhat more involved than that needed to provide access to a single survey, it seems especially important to have this capability in view of the more frequent periods of administration.

14. We recommend that DoD consider doing more to improve the accessibility of YATS data to policy makers and their advisers, thereby improving the timeliness of the information.

15. We recommend that DoD consider making the YATS and related databases available online, so that people with proper clearance could access the database for basic results. DoD should also consider being more proactive in making YATS and related data available to the research community. In doing so, confidentiality needs to be ensured.

We appreciate the opportunity to examine the YATS methodology and analyses and hope that our recommendations provide useful input to your efforts to improve YATS. We would be pleased to discuss these recommendations with you and your staff at your convenience.

Sincerely yours,
Paul Sackett, *Chair*
Committee on the Youth Population and Military Recruitment
cc: Wayne S. Sellman
Attachments: Statement of Task and Membership Roster

REFERENCES

Orvis, B.R., Sastry, N., and McDonald, L.L.
1996 *Military recruiting outlook: Recent trends in enlistment propensity and conversion of potential enlisted supply.* Santa Monica, CA: RAND.

Wilson, M.J., Greenlees, J.B., Hagerty, T., Hintze, D.W., and Lehnus, J.D.
2000 *Youth Attitude Tracking Study 1998: Propensity and advertising report.* Arlington, VA: Defense Manpower Data Center.

Committee on the Youth Population and Military Recruitment

STATEMENT OF TASK[4]

Both current shortfalls in recruiting and long-term demographic trends in the youth population raise important questions for military planners. A three-year ad hoc committee will study the implications of these trends for recruiting, selection, and training. The first phase of the committee's work, focused on recruiting issues, will be a study of the demographic, sociological, and psychological attributes of contemporary youth and projected attribute profiles twenty years into the future. The committee will examine a broad range of questions about the nature of the 21st century youth population, the characteristics of sub-populations that are likely to influence receptivity to recruitment efforts, the changing nature of work, and the effectiveness of various advertising approaches and incentive programs.

Committee on the Youth Population and Military Recruitment

MEMBERSHIP ROSTER

Paul R. Sackett, *Chair*, University of Minnesota, Twin Cities
David J. Armor, George Mason University, Fairfax, VA
Jerald G. Bachman, University of Michigan, Ann Arbor
John S. Butler, University of Texas at Austin
John Eighmey, Iowa State University, Ames
Martin Fishbein, University of Pennsylvania, Philadelphia
Carolyn Sue Hofstrand, T. DeWitt Taylor Middle-High School, Pierson, FL
Paul F. Hogan, Lewin-VHI Inc., Fairfax, VA
Carolyn Maddy-Bernstein, Louisiana State University, Baton Rouge
Robert D. Mare, University of California at Los Angeles
Jeylan T. Mortimer, University of Minnesota, Twin Cities
Carol A. Mutter, LTG, retired, U.S. Marine Corps
Luther B. Otto, North Carolina State University, Raleigh
William J. Strickland, HumRRO, Alexandria, VA
Nancy T. Tippins, GTE Corporation, Irving, TX

---

[4]This is an abbreviated version of the task statement described more fully in the Introduction on pages 9 and 10.

Appendix B

The Scientific Basis of the Popular Literature on Generations

Letter Report from the Committee on the Youth Population and Military Recruitment

March 2002

THE NATIONAL ACADEMIES

Advisers to the Nation on Science, Engineering, and Medicine

National Academy of Sciences
National Academy of Engineering
Institute of Medicine
National Research Council

Division of Behavioral and Social Sciences and Education
Committee on the Youth Population and Military Recruitment

March 5, 2002

Lieutenant General John A. Van Alstyne
Deputy Assistant Secretary of Defense for Military Personnel Policy
Room 3E767, The Pentagon
Washington, DC 20310-4000

Dear General Van Alstyne:

The Committee on Youth Population and Military Recruitment was established in 1999 at the request of the Office of Accession Policy in the Office of the Under Secretary of Defense to examine a broad range of questions concerning the characteristics of the 21st century youth population and subpopulations that are likely to influence recruitment efforts. The committee is now completing its work on a major report that examines such issues as the implications for recruiting and advertising strategies of trends in youth values, the appeal of options available to youth that compete with military service, and the changing nature of work in the military and civilian sectors. In addition to its major task, the committee has also been asked to write occasional, narrowly focused letter reports addressing specific issues of concern to the department that are within the committee's overall mandate.

Last fall, the Office of Accession Policy requested a letter report that assesses the scientific quality of the popular literature characterizing various generations, with a particular focus on "millennials"— a term coined and described in detail by Neil Howe and William Strauss (2000) for those born in or after 1982. The impetus for this request was the extensive use of that literature, and most particularly of the concept of millennials, for developing and implementing recruiting strategies by the Department of Defense (DoD), although its scientific quality had not been examined. The most frequently cited references are the books by Howe and Strauss; the latest in the series and the one most frequently used by military recruiters and advertisers is entitled *Millennials Rising* (Howe and Strauss, 2000). Two recent examples of reliance on the popular literature for descriptions of today's youth and their differences from earlier generations are the Army's recruiting research program for designing recruit advertising (Parlier, 2001) and a DoD-sponsored program for structuring survey research on youth attitudes for use by military recruiters and advertisers (Hoskins, 2001). In addition, the Strategic Studies Institute (Wong, 2000), has recently completed an analysis of attrition behavior of junior officers based on popular characterizations of "boomers" and "Xers" as portrayed by Howe and Strauss (1993), Legree (1997), and Zemke et al. (2000). The use of this literature is not limited to the United States: a report of the Department of Defense Canada (Wait, 2001) also draws on work on generations by Adams (1997, 2000) to contrast "elders," "boomers," and "Gen Xer's."

In this letter we examine two key ideas presented in the popular literature: there are distinct generations with sharp differences among them, and there are large and dramatic differences among youth cohorts in different generations. For example, on "the next great generation" (Howe and Strauss, 1993):

> As a group millennials are unlike any youth generation in living memory.... Over the next decade the millennial generation will entirely recast the image of youth from downbeat and alienated to upbeat and engaged—with potentially seismic consequences for America.

We also consider the quality of the research and analysis used to support these ideas. The committee reviewed eight books that are heavily cited by the DoD in analyzing and characterizing generations and in designing recruiting and advertising strategies: *Generations* (Strauss and Howe, 1991), *Generation X: Tales of an Accelerated Culture* (Copland, 1991), *13th Gen* (Howe and Strauss, 1993), *The Official Guide to Generations* (Mitchell, 1995), *American Generations* (Mitchell, 1998), *The Fourth Turning* (Strauss and Howe, 1998), *Generations at Work: Managing the Clash of Veterans, Boomers, Xers and Nexers in Your Workplace* (Zemke et al., 2000), and *Millennials Rising* (Howe and Strauss, 2000).

Concept of Generations

The popular generational literature (e.g., the "baby boomers," "generation X," "the millennials") synthesizes information from demographic projections, surveys based on selective samples, magazine articles, and newspaper reports. This information is not a part of the peer-reviewed scientific literature.

The style of this literature is engaging and entertaining. Its message for millennials is often positive, even refreshingly optimistic. It communicates a buoyant enthusiasm for youth, for the future, and for the country.

There is intuitive appeal to these ideas. Indeed, everyday language is sprinkled with associated labels. It is commonplace to speak of the "founding fathers," the "Depression-era generation," and "boomers." Although such generalizations are the stuff of casual talk and common in the popular press, they are not the focused concepts and explanatory devices of social science research, and they rarely stand up under the careful scrutiny of that research. The seminal work by Ryder (1964; 1965) and the exhaustive review by Riley et al. (1988) explore the theoretical and methodological complexity and confusions in comparing the characteristics or life-course experiences of members of one age group with those of another without specifically defining explanatory variables but rather indexing them only by age and date. That is, one cannot use only people's ages and fixed dates to compare cohorts; one must specify the events and experiences that are hypothesized to lead to cohort differences and systematically test those hypotheses.

The recent popular accounts of "generations" should not be confused with serious efforts by demographers and economists to describe and interpret patterns of cohort change. That the cohorts born during the 20th century have varied radically in size is an established demographic fact. A number of theoretical and empirical investigations have examined the implications of variation in the sizes of birth cohorts for social and economic welfare. Some of this work investigates the effects of the absolute sizes of birth cohorts; other work focuses on the sizes of birth cohorts relative to the sizes of the birth cohorts of their parents (e.g., Easterlin, 1987). Although the empirical evidence for hypotheses generated from this research is mixed (see Macunovich [1998] for a review), it constitutes a valuable effort to systematically assess the economic effects of demographic changes. Unlike the popular literature, it does not draw arbitrary and abrupt lines between generations, but rather attempts to quantify the ways that cohorts may vary on independent and dependent variables. Unlike the popular literature, the scholarly work presents testable empirical claims that researchers have can investigate using data drawn from multiple time periods and locales. These studies neither propose nor provide evidence in support of the currently popular notions of generational differences regarding values, attitudes, and beliefs.

Historical and sociological scholarship has identified distinctive characteristics of individuals in particular birth cohorts who have been subject to important social changes, such as the Depression (see Elder, 1974). They do this retrospectively. Thus, Thomas and Znaniecki (1918) used diaries and letters; Lazarsfeld et al. (1944; 1960) used panel studies; and Elder (1974) used longitudinal analysis. If one studies multiple cohorts and identifies systematic relationships between cohort characteristics and subsequent patterns of behavior, one can use those analyses to make predictions about subsequent cohorts. But such relationships cannot be inferred without advance specification of explanatory variables and collection of systematic data on multiple cohorts. The popular literature does not base its predictions on such systematic analysis.

The popular literature postulates a recurring cycle of generational characteristics. The literature is unclear about the relationship between this cycle and specific events. However, there is no scientific or historical basis for postulating or predicting that key events and social changes occur in a predictable cycle or that any given event will prove influential in the long term. An event such as the terrorist attack on the World Trade Center may prove such a historical spark. But one cannot know this in advance. Thus, there is no scientific basis for the claim in some of the popular literature that parallels between characteristics of earlier generations and today's youth is so striking that the destiny of millennials will unfold along predictable lines. Furthermore, the definition of a generation in the popular literature is inconsistent in bracketing generations, i.e., defining the beginning and end points of a generation. For example, the generational scheme defined and used by Howe and Strauss posits 1982 as the transition from generation X to the milllennial generation, while the scheme used by New Strategist Publications, another popularizer of the generational concept, posits 1977 (Mitchell, 1995).

Differences Between Youth Cohorts

The claim that one "generation" differs from another is at times taken to imply that there are sharp changes in attitudes and behaviors from one generation to the next. An example is a recent DoD sponsored survey in which 15- to 19-year-olds were labeled millennials and compared on the variable "teamwork" with 20- to 21-year-olds who were labeled generation X. The focus of the survey was youth attitudes as portrayed in the popular literature and the implications for military recruiters and advertisers (Hoskins, 2001). Yet scientific research shows that most indicators of youth attitudes and values change slowly and smoothly over time, if they change at all (Sax et al., 1999). Narrow and specific behaviors certainly do change, such as those driven by the technologies of cell phones and the Internet. These certainly are relevant to such issues as the choice of media for advertising messages, but these changes do not imply fundamental changes in attitudes and values.

Long-term data on youth have been collected by several longitudinal surveys: Monitoring the Future Project, National Longitudinal Survey of Youth, the Youth Attitude Tracking Survey, the Alfred P. Sloan Study (University of Chicago), and the Youth Development Study (University of Minnesota). For youth attitudes, the most comprehensive of these is the

Monitoring the Future Project, which has carefully examined representative samples of youth from the mid-1970s to the present (Bachman et al., 2001) and earlier volumes in the same series (e.g., Bachman et al., 2000). Monitoring the Future is a nationwide survey of youth attitudes and behaviors conducted by the Survey Research Center at the University of Michigan. A national stratified random sample of high school seniors, ranging in size from 14,000 to 19,000 annually, have been asked to answer the same questions over the past 25 years.

Several findings from this project are important for military recruiting. First, ratings of the importance of various life goals (e.g., finding purpose and meaning, making a lot of money) show a high degree of stability over two decades: the rank ordering of the goals is virtually unchanged. Second, ratings on the importance of a set of 24 job characteristics (e.g., difficult and challenging, high status and prestige, amount of vacation) are also very similar over two decades. Third, views about the importance of work in young people's lives have been largely stable: there has been a slow and modest decline in the portion considering work as a central part of life, but only of roughly one-half of 1 percent per year. When changes occur gradually, an analysis that focuses on comparing generations can mistakenly create the impression of a sharp change. Collapsing the data into two artificial averages creates the illusion of a sharp change between generations when, in fact, no sharp change occurred.

There has also been a slight decline in the proportion of respondents who consider that work is important (1) as a chance to make friends and (2) as an opportunity to make contact with a lot of people. These latter two job characteristics were rated as very important by only 39 percent and 26 percent of the respondents, respectively, in the most recent surveys (a finding that may run contrary to the notion that millennials are more team oriented).

Fourth, one example of a gradual change of great importance is youth interest in postsecondary education. Monitoring the Future and other studies that track the value youth place on postsecondary education (e.g., Current Population Survey [U.S. Department of Commerce, 1999]; Bachman et al., 2001; Youth Attitude Tracking Study [Wilson et al., 2000]) uniformly report a continuing upward trend in how contemporary youth value postsecondary education. Indeed, longitudinal studies of U.S. youth indicate that value for postsecondary education is the single most compelling differentiating factor for contemporary youth. The policy implications of this finding are important for both military recruiting strategy and the design of programs that combine military service and higher education opportunities.

Quality of Research and Analysis

The kind of data and analysis used in the popular literature on youth generations also raises serious questions about the quality and usefulness of that work. Scientific approaches to the study of youth involve systematic collection and evaluation of data; the popular approaches tend to use selective data. The representativeness of a sample is crucial when collecting survey or interview data. Social scientists devote careful attention to ensuring that samples represent the population of interest. In contrast, *Millennials Rising* draws heavily on a 1999 survey of 655 students in the class of 2000 in four public high schools in Fairfax County, Virginia, one of the wealthiest counties in America. For the three high schools named in the survey (the fourth was not listed), the SAT math scores were well above the national average of 514. For one school the verbal score was slightly higher than the national average of 505, whereas the scores for the other two schools were 51 and 76 points higher than the national average (see www.fairfax.gov). In short, generalizing the picture of youth attitudes and behaviors emerging from four high schools in Fairfax County to the rest of the country is not scientifically sound.

There are numerous other interpretational problems with the data in the popular literature. One is the use of retrospective comparisons, in which people are asked to compare current young people with their recollection of earlier cohorts: to compare a 15-year-old today with a 15-year-old of 10–15 years ago. Scientific research has clearly demonstrated that such recall is prone to bias (see, e.g., Dawes, 1988). Another frequent problem in the popular literature is the presentation of single point-in-time data, with the implicit suggestion that the data reflect some type of change, although there are no data reported for previous cohorts.

Conclusions

In sum, the committee concludes that two critical features of the popular writing on generations run counter to scientific findings. First, the notion of distinct generations with clear differences between them is not supported by social science research. Second, contrary to claims of large and dramatic differences among youth cohorts in different generations, high-quality longitudinal research documents a high degree of stability in youth attitudes and values. Change is limited, and when it does occur, it occurs gradually. In addition, the popular literature is often based on selective, not systematic, data and analysis and on nonrepresentative samples.

The committee does not believe that it is appropriate to give credence to popular portrayals of "generations" as a key explanatory concept for understanding youth attitudes and behaviors. More particularly, we question specific claims regarding characteristics of the current youth cohort, such as claims that this cohort differs dramatically on dimensions such as teamwork, achievement, modesty, and good conduct. We do not assert that all of the conclusions drawn in the popular literature are incorrect; some of the insights and impressions may prove to be accurate. We do advise against uncritical acceptance of claims for generational characteristics, and we encourage careful examination of the scientific bases for any such claims.

We hope this brief letter is helpful to you in evaluating DoD's recruitment strategies.

Sincerely yours,
Paul Sackett, *Chair*
Committee on the Youth Population and Military Recruitment
Attachments: Membership Roster

REFERENCES

Adams, M.
1997 *Sex in the snow, Canadian social values at the end of the millennium.* Toronto: Penguin Group.
2000 *Better rich than happy?* Toronto: Penguin Group.

Bachman, J.G., Freedman-Doan, P., and O'Malley, P.M.
2000 *Youth, work, and military service: Findings from two decades of monitoring the future national samples of American youth. Technical Report.* Ann Arbor, MI: Institute for Social Research.

Bachman, J.G., Johnston, L.D., and O'Malley, P.M.
2001 *Monitoring the future: Questionnaire responses from the nation's high school seniors.* Ann Arbor, MI: Institute for Social Research.

Copland, D.
1991 *Generation X: Tales for an accelerated culture.* New York: St. Martins Press.

Dawes, R.M.
1988 *Rational choice in an uncertain world.* Chicago: Harcourt Brace Jovanovich, Inc.

Easterlin, R.A.
1987 *Birth and fortune.* Second Edition. Chicago: University of Chicago Press.

Elder, G.H., Jr.
1974 *Children of the great depression.* Chicago: University of Chicago Press.

Hoskins, J.
2001 Department of Defense Youth Poll: Wave One. Unpublished paper.

Howe, N., and Strauss, W.
1993 *13th Gen: Abort, retry, ignore, fail?* New York: Vintage Books.
2000 *Millennials rising: The next great generation.* New York: Vintage Books.

Lazarsfeld, P.F., Berelson, B., and Gaudet, H.
1960 *The people's choice.* New York: Columbia University Press. (Original published 1944).

Legree, P.J.
1997 *Generation X: Motivation, morals, and values.* (ARI Special Report, June 1997). Alexandria, VA: U.S. Army Research Institute for the Behavioral and Social Sciences.

Macunovich, D.J.
1998 Fertility and the Easterlin hypothesis: An assessment of the literature. *Journal of Population Economics, 11*:53–111.

Mitchell, S.
1995 *The official guide to generations.* New York: New Strategist.
1998 *American generations.* New York: New Strategist.

Parlier, G.
2001 Manning the Army of the future. Recruiting update on the Youth Population and Military Recruitment. Unpublished paper. Program Analysis and Evaluation Directorate, U.S. Army Recruiting Command, Alexandria, VA.

Riley, M.W., Foner, A., and Waring, J.
1988 Sociology of age. In *Handbook of sociology,* Neil J. Smelser, ed. Newbury Park, CA: Sage.

Ryder, N.B.
1964 Notes on the concept of a population. *American Journal of Sociology, 69*:447–463.
1965 The cohort as a concept in the study of social change. *American Sociological Review, 30*:843–861.

Sax, L.J., Astin, A.W., Korn, W.S., and Mahoney, K.
1999 *The American freshman: National norms for fall 1999.* Los Angeles: Higher Education Research Institute, UCLA.

Strauss, W., and Howe, N.
1991 *Generations: The history of America's future, 1584–2069.* New York: William Morrow.
1998 *The fourth turning: An American prophecy.* New York: Broadway Books.

Thomas, W.J., and Znaniecki, F.
1918 *The Polish peasant in Europe and America.* Chicago: University of Chicago Press.

U.S. Department of Commerce
1999 *Current population survey.* Washington, DC: U.S. Department of Commerce, U.S. Bureau of the Census.

Wait, T.
2001 *Youth in Canada, population projection to 2026.* Ottawa, Canada: Department of National Defence Canada.

Wilson, M.J., Greenlees, J.B., Hagerty, T., Hintz, D.W., and Lehnus, J.D.
2000 Youth Attitude Tracking Study 1998: Propensity and Advertising Report. Submitted to the Defense Manpower Data Center. CEDS/YATS, dsw01-96-c-0041.

Wong, L.
2000 *Generations apart: Xers and Boomers in the Officer Corps.* Carlisle, PA: Strategic Studies Institute. Available <http://carlisle-www.army.mil/usassi/ssipubs/pubs2000/apart/apart.pdf>.

Zemke, R., Raines, C., and Filipczak, R.
1999 *Generations at work: Managing the clash of veterans, Boomers, Xers, and Nexters in your workplace.* New York: American Management Association.

Committee on the Youth Population and Military Recruitment

MEMBERSHIP ROSTER

Paul R. Sackett, *Chair*, University of Minnesota, Twin Cities
David J. Armor, George Mason University, Fairfax, VA
Jerald G. Bachman, University of Michigan, Ann Arbor
John S. Butler, University of Texas at Austin
John Eighmey, Iowa State University, Ames
Martin Fishbein, University of Pennsylvania, Philadelphia
Carolyn Sue Hofstrand, Taylor High School, Volusia County, FL
Paul F. Hogan, Lewin-VHI Inc., Fairfax, VA
Carolyn Maddy-Bernstein, Louisiana State University, Baton Rouge
Robert D. Mare, University of California at Los Angeles
Jeylan T. Mortimer, University of Minnesota, Twin Cities
Carol A. Mutter, LTG, retired, U.S. Marine Corps
Luther B. Otto, North Carolina State University, Raleigh
William J. Strickland, HumRRO, Alexandria, VA
Nancy T. Tippins, GTE Corporation, Irving, TX

Appendix C

Biographical Sketches

Paul R. Sackett (*Chair*) is professor in the Department of Psychology at the University of Minnesota, Twin Cities. His research interests revolve around legal, psychometric, and policy aspects of psychological testing, assessment, and personnel decision making in workplace settings. He has served as the editor of *Personnel Psychology*, as president of the Society for Industrial and Organizational Psychology, as co-chair of the Joint Committee on the Standards for Educational and Psychological Testing, as a member of the National Research Council's Board on Testing and Assessment, and as chair of the American Psychological Association's Board of Scientific Affairs. He has a Ph.D. in industrial and organizational psychology from Ohio State University.

David J. Armor is professor of public policy in the School of Public Policy at George Mason University, where he teaches statistics and social policy and conducts research in education, military manpower, and family policy. He began his research in military manpower while at the Rand Corporation. Between 1986 and 1989 he served as principle deputy and acting assistant secretary for force management and personnel in the Department of Defense. He was a member of the National Research Council's Committee on Military Enlistment Standards. He has a Ph.D. in sociology from Harvard University.

Jerald G. Bachman is program director and distinguished senior research scientist in the Survey Research Center of the Institute for Social Research at the University of Michigan, Ann Arbor. His scientific publications focus on youth and social issues. His current research interests include drug use and attitudes about drugs; youth views about military service; other values, attitudes, and behaviors of youth; and public opinion as related to a number of other social issues. He is a principal investigator on the Monitoring the Future project and the principal investigator on the Youth Attitudes about Military Service project. He has a Ph.D. in psychology from the University of Pennsylvania.

John Sibley Butler is a professor of management and sociology at the University of Texas at Austin. He is chair of the Department of Management in the Graduate School of Business. He has written extensively on military organizations and race relations and the military. He has also explored many entrepreneurial issues, including the development of business incubators by immigrant groups and the general area of entrepreneurship. His primary research areas include formal organizations/organizational behavior and sociology of economics, with an emphasis on organizations and entrepreneurships. He has a Ph.D. in sociology from Northwestern University.

Marilyn Dabady (*Senior Research Associate*) is study director for the National Research Council's Panel on Methods for Assessing Discrimination at the Committee on National Statistics. Her background is in social psychology, organizational behavior, and human resource management. Currently, her main areas of interest are interpersonal and intergroup relations; prejudice, stereotyping, and discrimination; and diversity management. She has a bachelor's degree in psychology from the University of Albany, State University of New York, and M.S. and Ph.D. degrees in psychology from Yale University.

John Eighmey is professor of journalism and mass communication in the Greenlee School of Journalism and Communication at Iowa State University. He is an authority on advertising and marketing communication, consumer research, and communication management. He has held senior management positions at the Federal Trade Commission in Washington, DC, and at Young & Rubican, a worldwide advertising agency based in New York City. He has a Ph.D. in marketing from the University of Iowa.

Martin Fishbein is the Harold C. Coles distinguished professor of communications in the Annenberg School at the University of Pennsylvania. His areas of expertise include attitude theory and measurement, communication and persuasion, behavioral prediction and change, and intervention development, implementation, and evaluation. He also has carried out studies of the relations among beliefs, attitudes, intentions, and behaviors in field and laboratory settings. He has a Ph.D. in psychology from the University of California at Los Angeles.

Carolyn "Sue" Hofstrand is the director of counseling and guidance at Taylor High School in Volusia County, Florida. She is a nationally certified school counselor with experience at elementary, middle, high, and postsecondary schools. She serves as director of a comprehensive counseling program and is involved in leadership positions in counseling organizations at the state, local, and national levels. She has a master's degree in education, counseling, and guidance from North Dakota State University, where she serves on the board of visitors for the College of Education and Human Resources.

Paul F. Hogan is vice president and senior economist at The Lewin Group in Fairfax, Virginia. He has more than 20 years of experience in applying microenonomics, statistics, and operations research methods to problems in labor economics, including labor supply and demand, efficient staffing methods, and performance and cost measurement. He served as the senior analyst on the President's Military Manpower Task Force and as director of Manpower Planning and Analysis in the Office of the Secretary of Defense, the office charged with staffing methods and criteria used by military departments to determine demands for personnel. His doctoral studies include economics, econometrics, and finance at the University of Rochester.

Carolyn Maddy-Bernstein is an education consultant whose work includes research and service in career and technical education, guidance and counseling, and educating students who are members of special populations. From 1988 to 1999, she served as director of the Office of Student Services for the National Center for Research in Vocational Education at the University of Illinois, Urbana-Champaign. As a Louisiana State University faculty member, she taught and worked on research and service projects funded by the Governor's Office of Workforce Development and the Department of Education. She has been a public school teacher, counselor, and administrator and has worked at the secondary and postsecondary levels. She has a Ph.D. in education from Virginia Polytechnic Institute and State University.

Robert D. Mare is professor in the Department of Sociology at the University of California at Los Angeles. He serves as director of the California Center for Population Research and a co-principal investigator at the California Census Research Data Center. He is an expert in the areas of demography, quantitative methodology, stratification/mobility, and education. He has dealt with such issues as educational attainment, school enrollment, and intergenerational educational mobility; changes in marriage patterns and their implications for inequality; the effects of differential fertility and social mobility on inequality; and the determinants and consequences of trends in the youth labor force. He is a former editor of *Demography*, the official journal of the Population Association of America. He has a Ph.D. in sociology from the University of Michigan.

Anne S. Mavor (*Study Director*) is the staff director for the Committee on Human Factors and the Committee on the Youth Population and Military Recruitment. Her previous National Research Council work has included studies on occupational analysis and the enhancement of human performance, modeling human behavior and command decision making, human factors in air traffic control automation, human factors considerations in tactical display for soldiers, scientific and technological challenges of virtual reality, emerging needs and opportunities for human factors research, and modeling cost and performance for purposes of military enlistment. For the past 25 years her work has concentrated on human factors, cognitive psychology, and information system design. She has an M.S. in experimental psychology from Purdue University.

Jeylan T. Mortimer is professor of sociology and director of the Life Course Center at the University of Minnesota, Twin Cities. Her interests focus on adolescent development and the social psychology of work. She directs the Youth Development Study, a longitudinal assessment of the implications of early work experience for adolescent development, the transition to adulthood, socioeconomic attainment, and early adult mental health. She has a Ph.D. in sociology from the University of Michigan.

Carol A. Mutter is a retired lieutenant general of the United States Marine Corps. Her experience has been in research, development, and acquisition, as well as financial management, logistics, personnel administration, and equal opportunity. In her most recent Marine Corps assignment, she was the senior Marine Corps personnel management executive, making policy for and managing the careers and quality of life of all Marines and civilians working for the Marine Corps. She currently serves on the National Advisory Council of the Alliance for National Defense, the Advisory Board for the Indiana Council on World Affairs, and is the National President of the Women Marines Association, as well as a senior fellow at the Joint Forces Staff College. She has an M.A. in national security and strategic studies from the Naval War College in Newport, RI.

Luther B. Otto is William Neal Reynolds distinguished professor of sociology emeritus at North Carolina State University, Raleigh. His research focuses on youth and careers. He directed the Career Development Study, a detailed study of the early career histories of 7,000 young men and women from the time they were juniors in high school through age 30. He has published numerous articles, chapters, and books on youth and careers. He served two four-year terms on the Basic Social-Cultural Research Review Committee of the National Institutes of Health. He has been active in professional associations, has served in a number of editorial capacities, and frequently consults with federal and state agencies and private foundations on issues related to youth, education and work. He has a Ph.D. in sociology from the University of Wisconsin-Madison.

William J. Strickland is vice president of the Human Resources Research Organization (HumRRO) in Alexandria, Virginia. He also directs its Workforce Analysis and Training Systems Division. He is a retired Air Force colonel who was director of human resources research at the U.S. Air Force Armstrong Laboratory. In that position, he was responsible for all Air Force research in the areas of manpower and personnel, education and training, simulation and training devices, and logistics. Earlier in his career, he commanded an Air Force recruiting squadron, was the chief of market research for Air Force recruiting, and was the deputy director for operations for Air Force recruiting. A fellow of the American Psychological Association, he is a past president of its Division of Military Psychology. He has a Ph.D. in industrial and organizational psychology from the Ohio State University.

Nancy T. Tippins is president of the Selection Practice Group of Personnel Research Associates, Inc. in Arlington Heights, IL. She is responsible for the development and execution of the firm's strategies related to employee selection, assessment, and development. Prior to joining the firm, she spent over 20 years managing personnel research functions involved in selection methods, staffing policies and procedures, equal employment opportunity and affirmative action, outplacement and downsizing, human resource services, and surveys for GTE, Bell Atlantic, and Exxon Company, USA. She has a Ph.D. in industrial and organizational psychology from the Georgia Institute of Technology.

Index

F

G

H

Q

R

V

W

Y